# Foundations of Mathematics

## Symposium Papers Commemorating the Sixtieth Birthday of Kurt Gödel

Edited by



Jack J. Bulloff

Thomas C. Holyoke · S.W. Hahn

Springer-Verlag New York Inc. 1969

Managing Editor:

Jack J. Bulloff
State University of New York at Albany · Albany, N.Y.

Technical Editors:

Thomas C. Holyoke
Antioch College · Yellow Springs, Ohio

Samuel W. Hahn
Wittenberg University · Springfield, Ohio

The Photograph of Kurt Gödel is by Orren J. Turner



ISBN 978-3-642-86747-7 ISBN 978-3-642-86745-3 (eBook)
DOI 10.1007/978-3-642-86745-3

© by Springer-Verlag, Berlin · Heidelberg 1969

Library of Congress Catalog Card Number 68-28757

Softcover reprint of the hardcover 1st edition 1969

Title No. 1518

*Kurt Gödel*

## Biographical Data

KURT GÖDEL was born on April 28, 1906 in Brünn (Brno), Czechoslovakia (at that time part of the Austro-Hungarian Monarchy). He studied mathematics and physics at the University of Vienna and took the Ph. D. degree there in 1930. He was Dozent in mathematics at the University of Vienna 1933—1938. He has been a member of the Institute for Advanced Study from 1940—1953, and several times before 1940. He joined its faculty in 1953. He became an American citizen in 1948 and received honorary degrees from Yale and Harvard in 1951 and 1952. He is a member of the National Academy of Sciences and of the American Philosophical Society and corecipient of the 1951 Einstein Award.

# Preface

Dr. KURT GÖDEL's sixtieth birthday (April 28, 1966) and the thirty-fifth anniversary of the publication of his theorems on undecidability were celebrated during the 75th Anniversary Meeting of the Ohio Academy of Science at The Ohio State University, Columbus, on April 22, 1966. The celebration took the form of a Festschrift Symposium on a theme supported by the late Director of The Institute for Advanced Study at Princeton, New Jersey, Dr. J. ROBERT OPPENHEIMER: "Logic, and Its Relations to Mathematics, Natural Science, and Philosophy." The symposium also celebrated the founding of Section L (Mathematical Sciences) of the Ohio Academy of Science.

Salutations to Dr. GÖDEL were followed by the reading of papers by S. F. BARKER, H. B. CURRY, H. RUBIN, G. E. SACKS, and G. TAKEUTI, and by the announcement of in-absentia papers contributed in honor of Dr. GÖDEL by A. LÉVY, B. MELTZER, R. M. SOLOVAY, and E. WETTE. A short discussion of "The II Beyond Gödel's I" concluded the session.

This Anniversary Volume has been prepared under the sponsorship of Section L. Thanks are due to Mr. John H. MELVIN, Executive Secretary of the Ohio Academy of Science, and to Dr. ROBERT C. STEPHENSON, Chairman of the 75th Anniversary Committee, The Ohio State University Research Foundation, for the many things they did to make the meeting possible. Dr. CARL KAYSEN, Director of the Institute for Advanced Study, provided items from his files which have made for a better and more attractive volume. Valuable suggestions were made to authors and editors by Dr. KLAUS PETERS of Springer-Verlag. Finally, it is a pleasure to acknowledge the helpful advice offered by the referees, by colleagues of Dr. GÖDEL, and by members of the Association for Symbolic Logic.

The Editors

## List of Contributors

STEPHEN F. BARKER, The Johns Hopkins University, Baltimore, Maryland / USA.

HASKELL B. CURRY, University of Amsterdam, Amsterdam / The Netherlands.

AZRIEL LÉVY, Hebrew University, Jerusalem / Israel, and Stanford University, Stanford, California / USA.

BERNARD MELTZER, University of Edinburgh / Scotland.

HERMAN RUBIN, Purdue University, Lafayette, Indiana / USA.

GERALD E. SACKS, Massachusetts Institute of Technology, Cambridge, Massachusetts / USA.

ROBERT M. SOLOVAY, The Rockefeller University, New York, New York, and University of California, Berkeley, California / USA.

GAISI TAKEUTI, University of Illinois, Urbana, Illinois, and The Institute for Advanced Study, Princeton, New Jersey / USA.

EDUARD WETTE, Uckerath, and Radevormwald / West Germany.

## Contents

# Greetings

## Greetings to Gödel Symposium from Dr. J. Robert Oppenheimer

"It is an honor and a pleasure for me to help to celebrate KURT GÖDEL, his anniversaries, and his great work, which has not only immeasurably deepened and enriched the understanding of the logical structure of so much abstract and mathematical argument, but illuminated the role of limitation in human understanding in general. I salute the scholars, many of high distinction, who have assembled to honor GÖDEL by presenting their discoveries and their views, and I hope that the contributions to this occasion will bring great pleasure to the man who has inspired it."

## Greetings from Dr. Gödel

"I wish to convey my greetings to the symposium, and in particular to the speakers. I am sorry I cannot attend, but I am looking forward with interest and pleasure to the volume in which the lectures will be presented."

# Tribute to Dr. Gödel

JOHN VON NEUMANN

The true significance of Dr. GÖDEL's work was perhaps most clearly and succinctly summarized by the late Professor JOHN VON NEUMANN in his remarks in March 1951, on the occasion of the presentation of the ALBERT EINSTEIN Award to Dr. GÖDEL. The textual reproduction of those remarks follows:

"KURT GÖDEL's achievement in modern logic is singular and monumental — indeed it is more than a monument, it is a landmark which will remain visible far in space and time. Whether anything comparable to it has occurred in the logic of modern times may be debated. In any case, the conceivable proxima are very, very few. The subject of logic has certainly completely changed its nature and possibilities with GÖDEL's achievement.

GÖDEL's name is associated with many important achievements in detail, and with two absolutely decisive ones. The occasion is such that I think I should only talk about the two latter.

The nature of the first one is easy to indicate, although its exact technical character and execution escape an adequate characterization without the specialized and rather intricate techniques of formal logic.

GÖDEL was the first man to demonstrate that certain mathematical theorems can neither be proved nor disproved with the accepted, rigorous methods of mathematics. In other words, he demonstrated the existence of *undecidable* mathematical propositions. He proved furthermore that a very important specific proposition belonged to this class of undecidable problems: The question, as to whether mathematics is free of inner contradictions. The result is remarkable in its quasi-paradoxical 'self-denial': It will never be possible to acquire *with mathematical means* the certainty that mathematics does not contain contradictions. It must be emphasized that the important point is, that this is not a philosophical principle or a plausible intellectual attitude, but the result of a rigorous mathematical proof of an extremely sophisticated kind.

The formulation that I gave above has coarsened the result and obliterated some of the fine points of its rigorous formulation, but if one is to state the theorem without having recourse to the difficult technical language of formal logic this is, I think, the best approximation that one can achieve.

Gödel actually proved this theorem, not with respect to mathematics only, but for all systems which permit a formalization, that is a rigorous and exhaustive description, in terms of modern logic: For no such system can its freedom from inner contradiction be demonstrated with the means of the system itself.

Gödel's second decisive result can only be stated in the terminology of formal logic and of an important but rather abstruse modern mathematical discipline: Set theory. Two surmised theorems of set theory, or rather two principles, the so-called 'Principle of Choice' and the so-called 'Continuum Hypothesis' resisted for about 50 years all attempts of demonstration. Gödel proved that neither of the two can be disproved with mathematical means. For one of them we know that it can not be proved either, for the other the same seems likely, although it does not seem likely that a lesser man than Gödel will be able to prove this.

I will not attempt a detailed evaluation of these achievements, I will limit myself to repeat: In the history of logic, they are entirely singular. No indemonstrability within mathematics proper had ever been rigorously established before Gödel. The subject of logic will never again be the same."

# Bibliography of Kurt Gödel


1. Die Vollständigkeit der Axiome des logischen Funktionenkalküls. Monatshefte für Mathematik und Physik **37**, 349—360 (1930). See item 28.
2. Einige metamathematische Resultate über Entscheidungsdefinitheit und Widerspruchsfreiheit. Anzeiger der Akademie der Wissenschaften in Wien **67**, 214—215 (1930). See item 28.
3. Diskussion zur Grundlegung der Mathematik. Erkenntnis **2**, 147—151 (1931/32).
4. Über formal unentscheidbare Sätze der Principia Mathematica und verwandter Systeme I. Monatshefte für Mathematik und Physik **38**, 173—198 (1931). Italian translation by EVANDRO AGASSIZ in: Introduzione ai problemi dell'assiomatica, pp. 203—228. Milano: Società editrice Vita e pensiero 1961. (Pubblicazioni dell'Università cattolica del Sacro Cuore. Ser. III. Scienze filosofiche, 4). See also items 27, 28.
5. Zum intuitionistischen Aussagenkalkül. Anzeiger der Akademie der Wissenschaften in Wien **69**, 65—66 (1932).
6. Ein Spezialfall des Entscheidungsproblems der theoretischen Logik. Ergebnisse eines mathematischen Kolloquiums, ed. by KARL MENGER. **2**, 27—28 (1929/30).
7. Über Vollständigkeit und Widerspruchsfreiheit, ibid. **3**, 12—13 (1930/31). See item 28.
8. Eine Eigenschaft der Realisierungen des Aussagenkalküls, ibid. 20—21.
9. Eine Interpretation des intuitionistischen Aussagenkalküls, ibid. **4**, 39—40 (1931/32).
10. Über Unabhängigkeitsbeweise im Aussagenkalkül, ibid. 9—10.
11. Zur intuitionistischen Arithmetik und Zahlentheorie, ibid. 34—38. See item 27.
12. Bemerkung über projektive Abbildungen, ibid. **5**, 1 (1932/33).
13. Über die Länge von Beweisen, ibid. **7**, 23—24 (1934/35). See item 27.
14. Zum Entscheidungsproblem des logischen Funktionenkalküls. Monatshefte für Mathematik und Physik **40**, 433—443 (1933).
15. On Undecidable Propositions of Formal Mathematical Systems (Mimeographed notes of lectures given in 1934). 30 p. See item 27.
16. The Consistency of the Axiom of Choice and of the Generalized Continuum-Hypothesis. Proc. Nat. Acad. Sci. USA **24**, 556—557 (1938).
17. Consistency-Proof for the Generalized Continuum-Hypothesis. Proc. Nat. Acad. Sci. USA **25**, 220—224 (1939).
18. The Consistency of the Continuum Hypothesis. Annals of Mathematics Studies **3**, 66 p. Princeton, New Jersey: Princeton University Press 1940. Second printing, revised and with some notes added, 1951, 69 p. Seventh printing, with some notes added, 1966, 72 p.
19. Russell's Mathematical Logic. In: The Philosophy of BERTRAND RUSSELL, ed. by P. A. SCHILPP, pp. 123—153. Evanston and Chicago: Northwestern University Press 1944. See item 26.
20. What is Cantor's Continuum Problem? Amer. Math. Monthly **54**, 515—525 (1947). See item 26.

21. An Example of a New Type of Cosmological Solutions of EINSTEIN's Field Equations of Gravitation. Rev. Modern Physics **21**, 447–450 (1949).
22. A Remark about the Relationship between Relativity Theory and Idealistic Philosophy. In: ALBERT EINSTEIN, Philospher-Scientist, ed. by P. A. SCHILPP, pp. 555–562. Evanston: Library of Living Philosophers 1949. German translation with some additions to the footnotes in: ALBERT EINSTEIN als Philosoph und Naturforscher, pp. 406–412. Stuttgart: Kohlhammer 1955.
23. Rotating Universes in General Relativity Theory. Proceedings of the International Congress of Mathematicians in Cambridge, Mass. **1950, I**, 175–181.
24. Über eine bisher noch nicht benützte Erweiterung des finiten Standpunktes. Dialectica **12**, 280–287 (1958). Revised English edition to appear in Dialectica.
25. Remarks before the Princeton Bicentennial Conference on Problems of Mathematics. In: The Undecidable, ed. by MARTIN DAVIS, pp. 84–88. Hewlett, N. Y.: Raven Press 1965.
26. A reprint of item 19 and a revised and enlarged edition of item 20 were published in: Philosophy of Mathematics, ed. by P. BENACERRAF and H. PUTNAM, pp. 211–232, 258–273. Englewood Cliffs, N. J.: Prentice Hall 1964.
27. English translations of items 4, 11, 13 and a revised and enlarged edition of item 15 were published in: The Undecidable, ed. by MARTIN DAVIS, pp. 4 to 38, 75–81, 82–83, 39–75. Hewlett, N. Y.: Raven Press 1965.
28. English translations of items 1, 2, 4, 7, with some notes by the author, were published in: From Frege to Gödel, ed. by JEAN VAN HEIJENOORT, pp. 583 to 591, 595–596, 596–616, 616–617. Cambrigde, Massachusetts: Harvard University Press 1967.

# Realism as a Philosophy of Mathematics

STEPHEN F. BARKER

KURT GÖDEL's great contributions to mathematical logic are highly technical in nature, and it is controversial what their philosophical implications are. It is therefore of some interest to consider a viewpoint concerning the philosophy of mathematics that GÖDEL himself adopted. This viewpoint must to some degree embody his own estimate of the philosophical significance of his work. Naturally, the opinion of any profound thinker regarding the philosophy of his subject merits attention; but this is especially true when, as in GÖDEL's case, his own work may have philosophical implications of great importance yet which are difficult to assess.

Although GÖDEL has not written very much about the philosophy of mathematics, there are some striking philosophical remarks in his essay, "RUSSELL's Mathematical Logic". There he writes that "Classes and concepts may... be conceived as real objects... existing independently of our definitions and constructions". And he goes on to say, "It seems to me that the assumption of such objects is quite as legitimate as the assumption of physical bodies and there is quite as much reason to believe in their existence. They are in the same sense necessary to obtain a satisfactory theory of mathematics as physical bodies are necessary to obtain a satisfactory theory of our sense perceptions..." [1]. Thus GÖDEL affirms the view that there are certain kinds of real objects — sets (classes) and concepts — which are to be regarded as constituting the subject-matter of mathematics. Moreover, in saying that they are real, GÖDEL feels that he is expressing an assumption: something that might be false but that it is very reasonable to affirm in the light of what we do know. He compares this philosophical 'assumption' concerning mathematics to the proposition that physical bodies exist. Our belief that physical bodies exist seems to him also to be an assumption, but a highly legitimate one — presumably because he regards it as much the best explanation of why our sensations form patterns as they do. GÖDEL regards the 'assumption' that there are mathematical objects as equally

[1] GÖDEL, KURT: Russell's mathematical logic. In: The philosophy of BERTRAND RUSSELL. Edited by PAUL SCHILLP. The library of living philosophers, p. 137. New York: The Tudor Publishing Company 1944.

legitimate — presumably because he thinks it much the best explanation of the intellectual significance that mathematics has for us.

These remarks of GÖDEL's express a strongly realistic philosophy of mathematics. By realism as a philosophy of mathematics we may understand the doctrine that laws of mathematics are fundamentally to be regarded as literal descriptions of objects of some kind. Moreover, GÖDEL goes further, and passages elsewhere in his essay confirm that this his viewpoint is realism of a particular form. He evidently holds that these sets and concepts are not located in space or time, that they exist quite independently of our minds, yet that they can be apprehended and described by us. Thus GÖDEL's viewpoint stands in opposition to nominalism, which may be defined as the view that there are no abstract, non-spatio-temporal entities whatsoever. GÖDEL's realism also stands opposed to conceptualism, which may be defined as the view that there are such abstract entities but that they are brought into being by our mental activity.

GÖDEL's particular form of realistic viewpoint concerning mathematics has a long history in philosophical thought. Because the view can be traced back to PLATO, BERNAYS has suggested calling it "platonism" — but with a small "p" so as to indicate that although the view is akin to PLATO's, those who have subsequently held it need not be regarded as strictly disciples of PLATO[2]. Thus, platonism is a form of realism according to which mathematics has for its subject-matter a realm of real, non-spatial, non-mental, timeless objects. It claims that these objects can be grasped by human reason; that we can be acquainted with them through the 'eye of reason', as it were, even though our five senses provide us no avenue of contact with them. According to this view, the intellectual significance of the formulas of mathematics lies in the fact that they describe these important objects and that what they say about them is true. The platonist believes that just as a geographer has the task of describing the interrelations in space of physical bodies such as continents, oceans, rivers, and mountains, so the task of the mathematician is to chart the logical interrelations of certain non-spatial and timeless objects. FREGE, late in the 19th century, and RUSSELL, early in the present century, were leading advocates of this platonistic form of realism in the philosophy of mathematics[3]. However, it seems fair to say that platonistic realism, and indeed realism of any form, has been

[2] BERNAYS, PAUL: Sur le platonisme dans les mathématiques. L'Enseignement mathématique **34**, 52–69 (1935).

[3] A statement of GOTTLOB FREGE's view is to be found in his: The foundations of arithmetic. Translated by J. L. AUSTIN. Oxford: Blackwell 1950; originally published in 1883. BERTRAND RUSSELL's position is expressed for example in his: Introduction to mathematical philosophy. London: G. Allen 1919.

less widely accepted as a philosophy of mathematics during recent years than it was during the early decades of the present century. The trend of thought has been away from this viewpoint.

But whether or not it is now widely accepted, how tenable does such a form of realism remain as a philosophy of mathematics? An advocate of such realism may wonder how anyone else could wish to deny that there are such things as sets, that these are non-physical and non-mental, and that it is the aim of mathematics to describe them truly. To the advocate of this form of realism, these propositions probably seem as eminently plausible as anything in the philosophy of mathematics could possibly be. Nevertheless, many writers on the philosophy of mathematics certainly do disagree with this realistic viewpoint, and want to deny some or all of these claims. Let us go on, then, to attempt a partial evaluation of this form of realism, by considering some of the direct objections that have been made to it. By direct objections are meant objections that aim to show that this form of realistic philosophy of mathematics is unsatisfactory in and of itself. If we were to obtain an over-all perspective, we would also need to consider indirect objections which consist of attempts to construct other philosophies of mathematics. By developing some other quite different philosophical viewpoint that was more plausible, one of course would indirectly undermine all forms of realism; but it is beyond the scope this paper to survey such attempts. Instead, let us concentrate merely upon direct objections to platonistic realism, objections that seek to show that it cannot be right and that do not do this by offering any alternative philosophy.

First, let us consider an objection that most people would think of first. It is the objection that this form of realistic philosophy of mathematics is too metaphysical. This realistic view postulates a realm of abstract, non-empirical entities that are supposed to be independent of our minds, yet that are supposed to be accessible to human reason. But our sense-experience does not provide us with any evidence of these entities, for we do not see, feel, taste, touch, or smell them, as we do physical bodies. An objector may urge that we therefore have no right to affirm that there are such entities, for to do so is unwarrentedly to go beyond the bounds of experience; and to do that is to indulge in extravagant metaphysics.

This first objection does not carry any decisive weight, however. Of course unnecessarily extravagant metaphysical views ought to be avoided. But the claim of someone who advocates this realistic philosophy of mathematics is that this philosophy is not unnecessarily extravagant when it affirms the reality of these mathematical objects. Someone who embraces this realistic viewpoint will do so exactly because he feels that the assumption of the reality of such a realm of objects provides the simplest adequate explanation of why mathematics is intellectually significant. To argue that we can have a right to believe in a kind of objects only if our senses provide direct evidence of them, is improperly to beg the question against the realistic philosophy of mathematics. Of course one who like PLATO or FREGE regards mathematics as treating of a realm of timeless objects can scarcely be an empiricist — he cannot believe that all knowledge or reality is directly based on sense-experience. This type of realism in the philosophy of mathematics is indeed incompatible with empiricism as a philosophy of knowledge. But that would provide a decisive argument against realism only if we had some overwhelming antecedent reason for believing in the universal correctness of empiricism — and it is not evident that we have that. The platonistic realist will hold that his own viewpoint does offer a plausible explanation of the nature of mathematical knowledge, whereas empiricism cannot; and so he will remain unmoved by this first objection.

A second and very different sort of objection against all forms of realism as a philosophy of mathematics is suggested by GÖDEL's own profound theorem about incompletability. Someone might argue that GÖDEL's incompletability theorem constitutes a serious difficulty for realism of any sort. Here the argument would be that we cannot any longer suppose that we are studying an objectively real, mind-independent subject matter, once we realize that no formal theory about that subject matter can be both complete and consistent. Here the underlying supposition would be that any realm of real objects must in principle admit of description in terms that are both consistent and complete. As a consequence of GÖDEL's work, we now know that neither set theory nor number theory admits of a formulation that is both consistent and complete; so the argument is that neither sets nor numbers can be regarded as real objects.

It is instructive to note the position that GÖDEL himself took. His essay "RUSSELL's Mathematical Logic" was published in 1944, and so comes more than a decade after his incompletability theorem. GÖDEL's own declaration in that essay in favor of a realistic philosophy of mathematics shows that he himself did not regard the incompletability of systems such as *Principia Mathematica* as constituting a decisive argument against the realistic philosophy of mathematics. Surely this attitude is indeed the correct one. From the viewpoint of a realistic philosophy of mathematics, the incompletability theorem can be regarded not as calling into question the independent reality of mathematical entities such as sets or numbers, but rather as indicating an essential limitation in the expressive power of symbolism; the limitation being that no symbolism can fully succeed in characterizing a system of objects as rich as the natural numbers. The realist may continue to believe that sets and numbers are real entities, even though he acknowledges that no consistent list of axioms can be formulated that will describe them completely. The supposition underlying this second objection need not be accepted.

A third and still different sort of objection aimed against the realistic philosophy of mathematics can be derived from a criticism which BENACERRAF has made of the logistic thesis[4]. According to the logistic thesis of WHITEHEAD and RUSSELL, the natural numbers are to be defined as certain particular sets of sets. WHITEHEAD and RUSSELL identified the natural number one with the set of all unit sets, the natural number two with the set of all couples, and so on; and they were able to show that the natural numbers defined in this fashion can possess all (or at least many) of their usual mathematical properties. However, other approaches are possible; BENACERRAF points out that there are different ways in which the various natural numbers could be identified with various sets. For example, in a set theory not based on the theory of types, the number one can be defined as the set whose sole member is the empty set; two as the set whose sole member is one; and so on. On this different basis it is also possible to develop a theory of the natural numbers according to which they will possess their usual mathematical properties. Here then are quite different ways in which the natural numbers could be identified with particular sets. And there are many other ways besides. However, if this is so, which set is the number one? Is it the set of all unit sets; is it the unit set of the empty set; or is it some other set? If we think that the number one is some definite object, then it cannot be more than one of these sets. BENACERRAF has urged that this shows the attempt to define numbers as sets to be basically misguided. There are no good reasons for saying that any particular number is any particular set, so numbers are not sets at all, he holds.

BENACERRAF then carries his reasoning further, maintaining that the essential thing about the natural numbers is that they form a recursive progression: all their other properties are derivative from this. For the purposes of counting — the basic use of the natural numbers — any system of objects, sets or not, that forms a recursive progression is as adequate as any other. What is important is not the individuality of each particular number, but the interrelated structure which the numbers jointly possess. Objects do not the job of numbers singly; the whole system performs the job or nothing does. In formulating the essential properties of the number system, you merely characterize an abstract structure. BENACERRAF treats this line of thought as in itself

[4] BENACERRAF, PAUL: What numbers could not be. The Philosophical Review **74**, 47–73 (1965).

constituting a strong reason for denying that numbers are objects of any kind.

BENACERRAF's objection is an objection against the view that numbers are objects. As it stands, it is not directly an objection against the view that sets or concepts (the entities GÖDEL mentioned) are objects. However, the realistic position as a philosophy of mathematics would be very much undermined if BENACERRAF's objection were accepted at face value; for numbers are the most central, even if perhaps not logically the most fundamental, of all the supposed entities treated by mathematics. The realistic position to be plausible needs to be able to offer an over-all view of the entities of mathematics, including numbers. Those who, like GÖDEL, have held that mathematics treats of sets or concepts surely have done so because they believed that numbers and all other mathematical entities can be defined as certain sorts of sets or concepts. So to this extent BENACERRAF's objection must be regarded as a serious challenge to the whole realistic philosophy of mathematics.

BENACERRAF's objection is a valuable one, and if the realist means to claim that we can simply discover that numbers are identical with certain more basic objects such as particular sets, then BENACERRAF's objection is decisive against him. BENACERRAF makes clear how wrong it would be to say that WHITEHEAD and RUSSELL *discovered* that the natural numbers are certain sets of sets. But does the realist need to say that sort of thing? It can legitimately be held that WHITEHEAD and RUSSELL *proposed* to identify the natural numbers with certain sets of sets, and that they *showed* that this identification can preserve the mathematical properties of the natural numbers, so that we can do with these sets of sets much of what we wanted to do with natural numbers[5]. A cautious realist might be satisfied with saying this about numbers, for numbers are not to him the fundamental objects of mathematics. The realist will want to hold that set theory is the philosophicalli basic part of mathematics, and that set theory is a system of true statements about certain real objects, sets. In order to have a simple, encompassing philosophical account of mathematics, realists usually want to treat number theory as reducible to set theory — and for their purposes it will suffice if they can propose at least one way of effecting this reduction; that there are also available other equally good ways of identifying numbers with particular sets is immaterial to the realist, for his aim is merely to establish that number theory can one way or another be fitted into his conception of what mathematics is about.

[5] W. V. QUINE has especially emphasized the view that such identifications are to be regarded as proposals for eliminating talk about one sort of entity in favor of talk only about some simpler sort of entity. See his: Word and object. The Technology Press 1960, sections 53 and 54.

Thus BENACERRAF's considerations do not wholly refute realism. However, they do tend to undermine the realist's position fairly seriously, for by emphasizing how it is the structure of the number system that is important rather than what objects compose it, BENACERRAF makes us feel the pointlessness of proposals for identifying natural numbers with objects such as sets. Such identifications can be proposed, but they hardly seem worth taking very seriously, as they divert our attention from that structural aspect of the number system which is its essential character. A realistic account of the nature of number thus loses much of its attractiveness; and realism thereby comes to seem less attractive as an over-all philosophy of mathematics.

Finally, a fourth direct objection to realism will be considered, an objection which arises out of the current situation in set theory and which calls into question the legitimacy of regarding sets themselves as real objects. Now, earlier in this century people certainly expected and hoped that there would be developed one single form of set theory, whose axioms would be intuitively plausible or even self-evident, and whose theorems would be complete and perfectly acceptable to everyone. This hope has not been supported by developments. At present, instead of finding that there is one single clearly correct system of set theory, we find that various rather different but apparently equally consistent forms of set theory have been developed by different mathematicians. One system was that of WHITEHEAD's and RUSSELL's *Principia Mathematica*; another was that of ZERMELO and FRÄNKEL; still another is the system of GÖDEL, VON NEUMANN, and BERNAYS. These are not just complementary systematizations of slightly different portions of what can be regarded as a single subject matter. Instead, their theses about sets do conflict to some extent. To be sure, a conflict between theories is not by itself anything extraordinary. Two different geographers might have different theories concerning the arrangement of mountains and rivers in some half-explored continent. The fact that these geographical theories conflicted would not by itself give us any reason for doubting that there was a real continent whose structure these geographers were studying. In such a case perhaps we would not know whose theory was right, but we would know what sorts of further inquiry were required in order to find out who was right (for example, further explorers might be sent out to inspect the terrain more carefully); and we would know what it would be like to get the answer (for example, we could envisage certain further reports which if received would decisively settle the matter). However, in the case of the conflict between alternative forms of set theory, the situation is essentially different. Here we cannot send explorers into the timeless realm of platonic reality, charging them to report back to us concerning the true configurations of the terrain.

Moreover, the alternative systems appear to be equally consistent; there is no ground for hope that all but one of them will be eliminated as inconsistent. Consequently, we have no basis and cannot have any basis for asserting that some one form of set theory is the true one. Are we then condemned to perpetual ignorance concerning the truth about sets? Instead of taking that agnostic line, it seems better to hold that no adequate meaning attaches to the notion of finding out what the true theory of sets is; and that therefore it makes little sense to claim that there is a true theory. This in turn implies that it makes little sense to suppose there to be a real realm of objects, sets, constituting the subject matter of set theory; and that undermines the position of realism.

The objection here need not presuppose any crudely sweeping positivistic principle to the effect that propositions which we cannot verify are always meaningless. In order to make this objection, all that one need claim is that talk about objects as real cannot make good sense unless methods can be envisaged by means of which alternative theories about those objects can be tested. To be sure, the objection just considered pertains only to the reality of sets, and does not relate directly to the question whether concepts or numbers are real objects. However, here again it is appropriate to observe that whatever tends to undermine some substantial portion of the realistic standpoint concerning mathematics, tends to undermine it all. The kinship between sets and numbers is close enough so that if the realistic view cannot be maintained concerning sets, then anyone's inclination to maintain it concerning concepts or numbers ought to be markedly weakened.

Several direct objections have now been considered against the realistic philosophy of mathematics, especially against its platonistic form. It has been suggested that two of them (BENACERRAF's objection and the objection based on the current situation in set theory) do have considerable force, but that the other direct objections mentioned are quite indecisive. The existence of some forceful objections goes far toward explaining and justifying the fact that realism has fewer supporters now than formerly. However, the indecisiveness of the other objections serves to indicate that realism is not as far-fetched a view nor as wide open to attack as its opponents often have supposed. Realism merits respect even if we do not embrace it, for it is a serious attempt to find a straightforward explanation of what makes mathematics significant. The principal merit of realism as a philosophy of mathematics would seem to be that it emphasizes the objectivity of mathematical inquiry, and thereby opposes the temptation to speak as though mathematical results could be arbitrarily created by the free activity of the mind. This is worth emphasizing; but in doing so, realism, especially in its platonistic form, exaggerates the point and misleadingly neglects the profound dissimilarities between propositions about physical objects and propositions about sets and numbers.

To be sure, the over-all plausibility of the viewpoint of realism cannot be fully evaluated merely on the basis of direct arguments pro and con. A full discussion of the soundness of realism as a philosophy of mathematics would also have to consider whether some other philosophical view of mathematics can be developed as a positive alternative to realism, an alternative that will be more adequate and less open to objection[6]. The situation involves competition among alternative philosophical views, and in the end that view will most deserve acceptance which outdoes competing theories by most illuminatingly and least objectionably accounting for the pattern of results and activities that constitute mathematics.

[6] The alternatives to realism used to be regarded as being intuitionism and formalism. For an elementary discussion of suggestions leading toward newer alternatives, see the last chapter of my: Philosophy of mathematics. Englewood Cliffs, N.J.: Prentice-Hall 1965. For more original proposals, see BENACERRAF, op. cit., D. S. SHWAYDER: Modes of referring and the problem of universals. Berkeley: University of California Press 1963, chapter IV; and HILARY PUTNAM: Mathematics without foundations. J. Philosophy **64**, 5–22 (1967).

# The Undecidability of $\lambda K$-Conversion

HASKELL B. CURRY

## 1. Introduction

In 1936 CHURCH[1] proved that two fundamental questions of the theory of $\lambda$-conversion[2] were recursively undecidable. The first question was whether two obs (well-formed formulas) of the system are equal (convertible); the second was whether or not an ob $X$ has a normal form. The purpose of this note is to give a new proof of an analogous result for $\lambda K$-conversion. This proof was obtained in the course of preparing the second volume of [5].

In the case of $\lambda K$-conversion an undecidability theorem holds which is stronger than that of CHURCH. In fact let a class $\mathfrak{A}$ of obs be called *expansion-invariant* just when any ob which reduces to a member of $\mathfrak{A}$ is also a member of $\mathfrak{A}$. Then the undecidability theorem may be stated, subject to clarification in the following discussion, as follows:

**Theorem.** *If $\mathfrak{A}$ and $\mathfrak{B}$ are nonvoid*[3] *and nonoverlapping classes of obs which are expansion-invariant, then $\mathfrak{A}$ and $\mathfrak{B}$ are recursively inseparable.*

In this form the theorem is due, to the best of my knowledge, to DANA SCOTT. SCOTT presented a similar theorem, with proof, in lectures at Berkeley, Calif. in 1963. Neither his theorem nor his proof, however, have been published; and I did not know of their existence until after the present proof was well along. The new proof is quite different. Whether it is simpler, in any absolute sense, I cannot say; but if one takes as known the usual techniques of combinatory logic and recursive functions, it is decidedly simpler. It is therefore better adapted for the proposed second volume of [5].

The theorem specializes to the two cases of CHURCH's theorem as applied to $\lambda K$-conversion. In both these cases one takes $\mathfrak{B}$ to be the complement of $\mathfrak{A}$. In the first case one takes $\mathfrak{A}$ to be the class of all $X$

---

[1] In [3] (for explanation of the numbers in brackets see the Bibliography).

[2] The reader is assumed to be familiar with the meaning of this term and of some other terms which are defined in [5]. In certain cases where the usage differs from CHURCH's own usage the latter usage is given in parentheses.

[3] That is we can find effectively obs $P$ and $Q$ such $P$ belongs to $\mathfrak{A}$ and $Q$ to $\mathfrak{B}$.

such that

(1) $$X = Y$$

where $Y$ is fixed; in the second case one takes $\mathfrak{A}$ to be the class of all $X$ which have a normal form.

Some preliminary explanations which are necessary to clarify the theorem will appear in § 2. Then the new proof will be presented in § 3.

## 2. Preliminary Explanations

As explained in [5] there are two forms of combinatory logic, the theory of $\lambda$-conversion (and various generalizations and extensions of it) and the synthetic theory of combinators[4]. In the former case the $\lambda$-operation, i.e. the formation of

$$\lambda x \cdot M$$

from a variable $x$ and an ob $M$ (which, in $\lambda K$-conversion, does not have to contain $x$), is primitive; in the latter this notion is defined in terms of application and certain atomic combinators. The equivalence of these two approaches has been shown[5]. Accordingly here I shall use the second approach; this means simply that the $\lambda$-operation does not have to be regarded as primitive. With certain modifications the proof can be carried through on the basis of the other approach.

We shall need to consider special combinators $\mathsf{I}$, $\mathsf{K}$, $\mathsf{S}$ such that for arbitrary obs $X$, $Y$, $Z$:

(I) $$\mathsf{I}X = X,$$

(K) $$\mathsf{K}XY = X,$$

(S) $$\mathsf{S}XYZ = XZ(YZ);$$

and also a combinator $\mathsf{B}$ (which can be defined as $\mathsf{S}(\mathsf{KS})\mathsf{K}$), such that

(B) $$\mathsf{B}XYZ = X(YZ).$$

These schemes can be understood as replacement rules. We understand a *reduction* as any series of replacements of an instance of the left side of one of these schemes by the corresponding instance of the right

[4] There is a tendency among many writers to apply the term "combinatory logic" only to the latter. But this was not the intention of [5] (cf. pp. 5–6); and the equivalence about to be mentioned is a good reason for extending the term in the way there intended.

[5] See [5] § 6E.

side[6]. We shall indicate by

(2) $$X \geqq Y$$

that $X$ reduces to $Y$ in that sense.

It has been known since 1936[7] that one can embed all of recursive arithmetic within combinatory logic. In fact, let $\mathsf{Z}_n$ be defined for all natural numbers $n$ thus:

$$\mathsf{Z}_0 \equiv \mathsf{KI}(= \lambda x y \cdot y),$$
$$\mathsf{Z}_{n+1} \equiv \mathsf{SBZ}_n(= \lambda x y \cdot x(\mathsf{Z}_n x y)).$$

Then we can take $\mathsf{Z}_n$ as the representative of the natural number $n$[8]. If $f$ is any one-place partial recursive function, then there is a combinator $F$, called the *formalization of* $f$, such that whenever

$$f(m) = n$$

then

(3) $$F\mathsf{Z}_m \geqq \mathsf{Z}_n,$$

and, if $f(m)$ is undefined, then the weak reduction of $F\mathsf{Z}_m$ continues indefinitely. Simular results hold for partial recursive functions of any number of arguments[9].

Finally we shall suppose that there is a Gödel numbering of the obs which has the usual properties; and in terms of this numbering the notions of recursiveness, recursive separability, etc. can be extended to classes, relations, etc. of obs in the usual fashion. Let the number assigned in this numbering to an ob $X$ be $\mathrm{gd}(X)$. Then there will be recursive functions $\delta$ and $\gamma$ such that

(4) $$\mathrm{gd}(XY) = \delta(\mathrm{gd}(X), \mathrm{gd}(Y));$$

and

(5) $$\mathrm{gd}(\mathsf{Z}_x) = \gamma(x).$$

The functions $\delta$ and $\gamma$ can, in fact, be taken as primitive recursive.

---

[6] This is known in [5] as weak reduction. There is also a strong reduction which includes it and is more closely allied to reduction in the theory of $\lambda$-conversion. Note that if one defines $\mathsf{I}$ as $\lambda x \cdot x$, $\mathsf{K}$ as $\lambda xy \cdot x$, etc., weak reduction is a specialization of reduction in $\lambda$-conversion. Thus expansion-invariance with respect to weak reduction is weaker than such invariance with respect to one of the other forms of reduction.

[7] KLEENE [6].

[8] This idea is basically due to CHURCH [2], p. 813; cf. [5], pp. 175—185. There are other possible representations, for instance one used by SCOTT; and one can postulate the natural numbers in terms of zero and a successor function as new primitives. Such modifications do not affect the conclusion of this paper in any essential way.

[9] These facts will be established in the second volume of [5], but they are already implicit in KLEENE (l. c.) and in [4].

## 3. Proof of the Theorem

The explanations in § 2 clarify the meaning of the theorem. With these explanations in mind we proceed to the proof.

Suppose that there is a recursive class $\mathfrak{C}$ which separates $\mathfrak{A}$ and $\mathfrak{B}$. From this assumption we shall derive a contradiction.

Let $\alpha(x)$ be the characteristic function of the Gödel numbers of $\mathfrak{C}$. We can suppose, without loss of generality, that if

$$x = \mathrm{gd}(X),$$

then

(6) $$\alpha(x) = \begin{cases} 0 \text{ if } X \text{ is in } \mathfrak{A} \\ 1 \text{ if } X \text{ is in } \mathfrak{B}. \end{cases}$$

Since neither $\mathfrak{A}$ nor $\mathfrak{B}$ is void, there will be obs $P$ and $Q$ such that $P$ is in $\mathfrak{A}$ and $Q$ is in $\mathfrak{B}$.

Let $G$ be an ob to be specified later. Then by means of the paradoxical combinator $\mathsf{Y}$ of [5; § 5G] we can find an ob $H$ such that

(7) $$H \geqq GH$$ [10].

Let $h = \mathrm{gd}(H)$, and let

$$J \equiv H\mathsf{Z}_h, \qquad j \equiv \mathrm{gd}(J).$$

Then, by (4) and (5)

$$j = \delta(h, \gamma(h)).$$

Hence if we define the numerical function $\theta(x)$ by

$$\theta(x) = \delta(x, \gamma(x)),$$

and let $L$ be the formalization of $\theta$, then

(8) $$L\mathsf{Z}_h \geqq \mathsf{Z}_j.$$

We now specify $G$. Let $\mathsf{D}$ be the Bernays ordered pair combinator, viz.

$$\mathsf{D} \equiv \lambda xyz \cdot z(\mathsf{K}y)x.$$

Then, for arbitrary obs $X$, $Y$,

(9) $$\mathsf{D}XY\mathsf{Z}_0 \geqq X, \qquad \mathsf{D}XY\mathsf{Z}_1 \geqq Y.$$

Let $A$ be the formalization of $\alpha$. We define $G$ by

(10) $$G \equiv \lambda xy \cdot \mathsf{D}QP(A(Ly)).$$

---

[10] In fact, in § 5G we form such an $H$ by weak reduction of $\mathsf{Y}G$. If we used for $\mathsf{Y}$ the paradoxical combinator of TURING [7] we could define $H$ to be $\mathsf{Y}G$.

Then we have

$$(11)\qquad \begin{aligned} J \equiv H\mathsf{Z}_h &\geqq GH\mathsf{Z}_h && \text{by (7)},\\ &\geqq \mathsf{D}QP(A(L\mathsf{Z}_h)) && \text{by (10)},\\ &\geqq \mathsf{D}QP(A\mathsf{Z}_j) && \text{by (8)}. \end{aligned}$$

Now, since $j$ is a Gödel number, we have either

$$\alpha(j) = 0 \quad \text{or} \quad \alpha(j) = 1\,.$$

If we have the former, i.e.

$$(12)\qquad \alpha(j) = 0\,,$$

then by (3)

$$A\mathsf{Z}_j = \mathsf{Z}_0\,.$$

Therefore,

$$\begin{aligned} J &\geqq \mathsf{D}QP\mathsf{Z}_0\\ &\geqq Q && \text{by (9)}, \end{aligned}$$

and hence, by the expansion invariance property, $J$ is in $\mathfrak{B}$; from this it follows by (6) that

$$(13)\qquad \alpha(j) = 1\,,$$

contrary to (12). On the other hand, if we assume (13), then in the same way we have

$$J \geqq \mathsf{D}QP\mathsf{Z}_1 \geqq P\,,$$

and hence we have (12). Thus we have a contradiction, q.e.d.

## References

Works are cited by numbers in brackets according to the list below. Journals are abbreviated according to the practice of Mathematical Reviews.


1. CHURCH, ALONZO: The calculi of lambda-conversion. Ann. of Math. Studies 6. Princeton, N.J.: Princeton University Press 1941, 2nd ed. 1951.
2. — A set of postulates for the foundation of logic. (Second paper.) Ann. of Math. (2) **34**, 839–864 (1933).
3. — An unsolvable problem of elementary number theory. Amer. J. Math. **58**, 345–363 (1936).
4. CURRY, H. B.: The paradox of KLEENE and ROSSER. Trans. Amer. Math. Soc. **50**, 454–516 (1941).
5. —, and R. FEYS: Combinatory logic. Amsterdam: North Holland Publishing Co. 1958.
6. KLEENE, S. C.: $\lambda$-definability and recursiveness. Duke Math. J. **2**, 340–353 (1936).
7. TURING, A. M.: The $\mathfrak{p}$-function in $\lambda K$-conversion. J. Symbolic Logic **2**, 164 (1937).

# The Definability of Cardinal Numbers

AZRIEL LÉVY [1]

## § 1. Introduction and Statement of the Results

One says that the sets $x$ and $y$ are *equinumerous* (in symbols, $x \approx y$) if there is a $1-1$ function mapping $x$ on $y$. The notion of the cardinal $|x|$ of $x$ is obtained from equinumerosity by abstraction. The use of $|x|$ does usually not require any special apparatus. E.g., when we say $|x| = \aleph_0$ we mean to say that there is a $1-1$ function mapping $x$ on the set of all natural numbers; when we say $|x| < |y|$ we mean that there are $1-1$ functions mapping $x$ into $y$, but none of them is onto $y$; etc. As is common in mathematics, one tends to pass from the abstract notion of cardinal numbers to real cardinal numbers, i.e., one wants to regard the cardinal numbers as objects of the mathematical system. This is where one encounters the problem of how to define the cardinal $|x|$ of $x$ as an object of set theory.

In naive set theory, as well as in type theory and in QUINE's "New Foundations", the definition of the cardinal $|x|$ of $x$ poses no problem; we use the Frege-Russell notion of cardinal and define $|x|$ as the set of all sets $y$ which are equinumerous with $x$. In axiomatic set theory, and we have here in mind mostly the Zermelo-Fraenkel version, if $x \neq 0$ then there is no set which contains all sets $y$ equinumerous with $x$. If we have the axiom of choice among our axioms, then we can develop the theory of ordinal numbers in the von Neumann way and define $|x|$ to be the least ordinal $\alpha$ equinumerous with $x$; the existence of such an $\alpha$ is guaranteed by the well-ordering theorem. If we have the axiom of foundation among our axioms, even if the axiom of choice is absent, we can, as SCOTT pointed out in [7], use an amended Frege-Russell cardinal and define $|x|$ as the set of all sets $y$ of least rank among those equinumerous with $x$. The aim of the present paper is to prove that in the absence of the axioms of choice and foundation the situation is quite different; in this case the operation $|x|$ on $x$ is undefinable in a very strong sense.

---

[1] This research has been sponsored by the Information Systems Branch, Office of Naval Research, Washington, D.C. under Contract F-61052 67 C 0055, and by the U.S. National Science Foundation, Contract GP-3926.

To formulate the results rigorously we shall consider a set theory T of the Zermelo-Fraenkel type. We shall use in T, at our discretion, also the definite article $\iota$, knowing that these uses can be eliminated. The primitive notions of T, in addition to equality, are the membership relation $x \in y$, and possibly also the void set 0. Among the axioms of T we have the axioms of extensionality, pairing, union, power-set, subsets, infinity and replacement (Axioms Ia, II—V, VII, VIII of [2]).

Let us denote with $T_1$ the theory obtained from T by adding to it the operation $|x|$ as a *new primitive notion* and

$$C_1 \qquad |x| = |y| \leftrightarrow x \approx y$$

as an *additional axiom*. In $T_1$ the cardinal $|x|$ of $x$ is already taken to be an object; however, if we want also to consider the relation $|x| = z$ between $x$ and $z$ as a legitimate relation of set theory, then it seems very natural to strengthen the axiom schemas of subsets and replacement in T so as to allow also the symbol $|x|$ in these schemas. E.g., in the axiom schema of replacement

$$\forall u \, \forall v \, \forall w (\varphi(u, v) \wedge \varphi(u, w) \rightarrow v = w) \\ \rightarrow \forall x \, \exists y \, \forall z (z \in y \leftrightarrow (\exists t \in x) \; \varphi(t, z))$$

one would allow $\varphi(u, v)$ to be any formula of the language of $T_1$ rather than only formulas of the language of T.

Let us denote the axiom schema of replacement in which $\varphi$ varies over all the formulas of $T_1$, including those in which the symbol $|x|$ occurs, with $C_2$. Let $T_2$ be the theory obtained from $T_1$ by adding to is $C_2$ as a new axiom schema. In $T_2$ we can use all formulas $\varphi$ of the language also in the axiom schema of subsets, since that axiom schema follows from the axiom schema of replacement. $T_2$ shares with T the good feature that we can apply the axiom schemas of subsets and replacement to every formula $\varphi$ of the respective language. This is not the case with $T_1$ (since there we can apply the axiom schemas only to $\varphi$'s in which the symbol $|x|$ does not occur). In this respect $T_1$ suffers from the same disadvantage as the set theories of VON NEUMANN-BERNAYS and QUINE's "New Foundations" where in the main set — or class — existence schema one is not allowed to substitute all formulas for the metamathematical variable $\varphi$.

When one considers the question of whether one can define in T the cardinality operation $|x|$, the following possibilities turn up.

I. $|x|$ *is definable in T*: There is a term $\tau(x)$ of T with the only free variable $x$ such that

$$T \vdash \tau(x) = \tau(y) \leftrightarrow x \approx y \, .$$

II. $|x|$ *is relatively definable in* $\mathsf{T}$: There is a term $\tau(z, x)$ of $\mathsf{T}$ with the only free variables $z$ and $y$ such that

$$\mathsf{T} \vdash \exists z \, \forall x \, \forall y (\tau(z, x) = \tau(z, y) \leftrightarrow x \approx y) .$$

III. $\mathsf{C}_1$ *and* $\mathsf{C}_2$ *are not creative with respect to* $\mathsf{T}$: If $\varphi$ is a sentence of $\mathsf{T}$ (i.e., $\varphi$ does not contain the symbol $|x|$) and $\mathsf{T}_2 \vdash \varphi$ then already $\mathsf{T} \vdash \varphi$.

IV. $\mathsf{C}_1$ *is not creative with respect to* $\mathsf{T}$: If $\varphi$ is a sentence of $\mathsf{T}$ and $\mathsf{T}_1 \vdash \varphi$ then already $\mathsf{T} \vdash \varphi$.

It is obvious that I entails II and that III entails IV. We shall now show that II entails III. If $\mathsf{T}$ satisfies II and $\mathsf{T}^c$ is the theory obtained from $\mathsf{T}$ by adding to the language of $\mathsf{T}$ the individual constant $c$, and by adding to $\mathsf{T}$ the axiom $\tau(c, x) = \tau(c, y) \leftrightarrow x \approx y$, then it follows from simple logical considerations that if $\varphi$ is a sentence which does not contain $c$ and $\mathsf{T}^c \vdash \varphi$ then already $\mathsf{T} \vdash \varphi$. If we define $|x|$ in $\mathsf{T}^c$ by $|x| = \tau(c, x)$ then $\mathsf{T}^c$ includes $\mathsf{T}_2$ (the individual constant $c$ is allowed in the axiom schema of replacement since it can be substituted for a parameter) and thus, if $\varphi$ is a sentence of $\mathsf{T}$ and $\mathsf{T}_2 \vdash \varphi$ then $\mathsf{T}^c \vdash \varphi$ and $\mathsf{T} \vdash \varphi$.

If III holds then $\mathsf{C}_1$ and $\mathsf{C}_2$ can be regarded as an implicit definition of $|x|$ in a very weak sense, since one of the two Leśniewski criteria, namely that of non-creativity, is satisfied (see [9], § 8.2). On the other hand, if III fails, i.e., there is a sentence $\varphi$ of $\mathsf{T}$ such that $\mathsf{T}_2 \vdash \varphi$ but not $\mathsf{T} \vdash \varphi$, then, by the completeness of the first order predicate calculus, there is a model of $\mathsf{T}$ in which $\varphi$ is false. No operation $|x|$ on this model can satisfy both $\mathsf{C}_1$ and $\mathsf{C}_2$.

As we mentioned above, if we have in $\mathsf{T}$ the axiom of choice or the axiom of foundation then I holds. IV holds for every theory $\mathsf{T}$, as follows from a strong version of HILBERT's $\varepsilon$-theorem by taking $|x| = \varepsilon_y(y \approx x)$ (see in ASSER [1] the version of the $\varepsilon$-theorem which deals with eliminating $\varepsilon$ in an $\varepsilon$-theory which assumes also (A2) of p. 42 of [1]).

A simple, non-elementary, proof of IV is as follows (it follows, essentially, ASSER [1] and MENDELSON [5]). Suppose $\varphi$ is a sentence of $\mathsf{T}$ such that $\mathsf{T}_1 \vdash \varphi$. Let $\mathfrak{A}$ be an arbitrary denumerable model of $\mathsf{T}$ and let its members be $a_0, a_1, \ldots$. We define $|x|$ on $\mathfrak{A}$ as the $a_i$ with the least subscript $i$ such that $x \approx a_i$ holds in $\mathfrak{A}$. $\mathfrak{A}$ together with the operation $|x|$ is now a model of $\mathsf{T}_1$ and hence $\varphi$ holds in $\mathfrak{A}$. Since we have shown that $\varphi$ holds in every denumerable model of $\mathsf{T}$, and since $\mathsf{T}$ has no finite models, we know, by the completeness of the first order predicate calculus, that $\mathsf{T} \vdash \varphi$.

Let $\mathsf{ZF}$ be the theory which consists exactly of the axioms mentioned above, i.e., the usual axioms of the Zeremlo-Fraenkel set theory without the axioms of choice and foundation. We shall prove the strong undefinability result that even III does not hold for $\mathsf{ZF}$ (assuming that $\mathsf{ZF}$

is consistent). This proof is given in § 2, but instead of proving the result for ZF we shall prove it for a theory ZF′ which is exactly like ZF except that it contains the axiom of foundation but, on the other hand, it admits *atoms*, i.e., objects which are not sets (otherwise known as *urelements* or *individuals*). We shall actually present in § 2 a sentence $\varphi$ of ZF′ such that $ZF_2' \vdash \varphi$ but not $ZF' \vdash \varphi$ (assuming that ZF is consistent). The latter part will be proved by the Fraenkel-Mostowski method. It will be clear from the contents of § 2 that we can use the same sentence $\varphi$, and essentially the same proofs, also for $ZF_2$ and ZF once we interpret in ZF the word "atom" as referring to sets $x$ such that $x = \{x\}$. For the general method of transferring independence proofs from ZF′ to ZF see Specker [9] and Mendelson [4].

Since III fails for ZF we know that II fails too, i.e., $|x|$ is not even relatively definable in ZF. It is now natural to ask what happens when we remove the immediate reason for the relative undefinability of $|x|$ by adding to ZF, as new axioms, all the theorems of $ZF_2$ which do not involve the undefined operation $|x|$; is $|x|$ relatively definable in the theory thus obtained? An obviously equivalent question is the question whether II holds for $ZF_2$ for a term $\tau(z, x)$ which does not involve the operation symbol $|u|$, i.e., whether there is such a term $\tau(z, x)$ for which $ZF_2 \vdash \exists z\, \forall x\, \forall y(\tau(z, x) = \tau(z, y) \leftrightarrow x \approx y)$. (The similar question of whether I holds for $ZF_2$ for a term $\tau(x)$ which does not involve the operation symbol $|u|$ is just the question whether $C_1$ and $C_2$, which introduce the operation symbol $|u|$ to ZF, satisfy Leśniewski's criterion of eliminability — [9], § 8.2). We shall show in § 3 that the answer to our question is negative by proving that if ZF is consistent so is $ZF_2$ together with the axiom schema $\neg \exists z\, \forall x\, \forall y(\tau(z, x) = \tau(z, y) \leftrightarrow x \approx y)$ for all terms $\tau(z, x)$ which do not involve the symbol $|u|$. As in § 2 we shall carry out the proof for a corresponding set theory ZF′ with atoms, rather than for ZF itself, using, again, the Fraenkel-Mostowski method.

The last question which we shall deal with is whether the relative definability of $|x|$ in T entails the definability of $|x|$ in T. This question is answered negatively in § 4, by proving that there is a particular term $\tau'(z, x)$ for ZF with the free variables $z$ and $x$ such that in the theory

$$ZF \cup \{\exists z\, \forall x\, \forall y(\tau'(z, x) = \tau'(z, y) \leftrightarrow x \approx y)\}$$

we can prove the formula $\tau(x) = \tau(y) \leftrightarrow x \approx y$ for no term $\tau(x)$ with the only free variable $x$. There, again, we shall prove it for the theory ZF′ with atoms, by means of the Fraenkel-Mostowski method.

## § 2. The Creativity of $C_1$ and $C_2$ with respect to ZF′

As mentioned in § 1 we shall deal here with the theory ZF′ which is like ZF except that it contains the axiom of foundation but it admits atoms. In ZF′ the notion $|x|$ is undefined, yet we shall use it for convenience's sake in the abstract sense mentioned in the beginning, i.e., formulas in which the symbol $|x|$ is used are understood to be abbreviations of formulas which do not use this symbol.

1. A set $A$ is said to be of the *n-th kind* if whenever $A = C \cup D$ and $C \cap D = 0$ then one of $|C|$ and $|D|$ is $\leqq \aleph_n$ and the other one is $> \aleph_n$.

2. If $A$ is a set of the $n$-th kind and $E$ is a set such that

$$|(A - E) \cup (E - A)| \leqq \aleph_n$$

then $E$ is also a set of the $n$-th kind, and $E$ is equinumerous to $A$, as easily seen.

3. A set $A$ of the $n$-th kind cannot be well-ordered.

*Proof*. On one hand $|A| > \aleph_n$; on the other hand $|A| \ngeqq \aleph_{n+1}$ since if $|A| \geqq \aleph_{n+1}$ then $A$ is the union of two sets both of which are of cardinality $\geqq \aleph_{n+1}$, contradicting 1.

4. A set $u$ is said to be *transitive* if every member of a member of $u$ is a member of $u$. The *transitive closure* of $x$ — $\mathrm{Tc}(x)$ — is the intersection of all transitive sets which contain $x$. $\mathrm{Tc}(x)$ consists exactly of $x$, the members of $x$, the members of the members of $x$, and so on. The *support* of $x$ — $\mathrm{Sp}(x)$ — is the set of all the atoms in $\mathrm{Tc}(x)$.

5. For a set $A$ of sets, we call a function $f$ on $A$ a *cardinality function* if we have for all $x, y \in A$

$$f(x) = f(y) \leftrightarrow x \approx y.$$

6. Let us consider the following sentences.

$\chi_1$: For each finite number $n$ there is a set $A$ of sets of atoms such that for every set $y$ of atoms which is of the $n$-th kind there is a $z \in A$ such that $y \approx z$.

$\chi_2$: For each finite $n$ and each cardinality function $f$ on a set $A$ as in $\chi_1$ the support of $\mathfrak{R}(f)$ (= the range of $f$) includes a set of the $n$-th kind.

$\chi_3$: There is a set $C$ which contains for every finite $n$ a set of atoms of the $n$-th kind.

The sentence $\varphi$, which we will show to be provable in $\mathsf{ZF}_2'$ but not in ZF′, is taken to be $\chi_1 \wedge \chi_2 \to \chi_3$.

*Proof of* $\chi_1 \wedge \chi_2 \to \chi_3$ *in* $\mathsf{ZF}_2'$. Let $n$ be finite and let $D_n$ be the class of all cardinalities of sets of atoms of the $n$-th kind. $D_n$ is a set, as follows from applying $\mathsf{C}_2$ to a set $A$ as in $\chi_1$ by taking $|x|$ to be the function in $\mathsf{C}_2$ (i.e., we take $v = |u|$ for $\varphi(u, v)$). Applying $\mathsf{C}_2$ again we get the set $D = \{D_n | n < \omega\}$. If $A$ is as in $\chi_1$ and $f$ is the function on $A$ defined by $f(x) = |x|$ then $\Re(f) = D_n$, hence, by $\chi_2$, $\mathrm{Sp}(D_n)$ includes a set of the $n$-th kind. Since $D_n \in D$ we have $\mathrm{Sp}(D_n) \subseteqq \mathrm{Sp}(D)$ and therefore $\mathrm{Sp}(D)$ includes a set of the $n$-th kind for every finite $n$. The power-set of $\mathrm{Sp}(D)$ obviously satisfies the requirement of $\chi_3$.

It is well known that if ZF, or ZF′, is consistent so is the theory ZF* obtained from ZF′ by adding to it the axiom

7. There are exactly $\aleph_\omega$ atoms (see, e.g. [3, p. 146]). To show that $\chi_1 \wedge \chi_2 \to \chi_3$ cannot be proved in ZF′ we shall construct an interpretation of ZF′ in ZF* under which $\chi_1$ and $\chi_2$ go over to provable sentences while $\chi_3$ goes over to a refutable sentence.

8. We denote the set of all atoms with $At$. Since there are $\aleph_\omega$ atoms we can write them as $(n, k, l, \alpha)$ where $n, k, l < \omega$ and $\alpha < \omega_{n+1}$. We write $A_{n,k,l} = \{(n, k, l, \alpha) \mid \alpha < \omega_{n+1}\}$. Let $G$ be the group of all permutations $\sigma$ of $At$ of the form

$$\sigma(n, k, l, \alpha) = (n, \sigma_1(n, k), \sigma_2(n, l), \sigma_3(n, k, l, \alpha))$$

where
$$\begin{aligned} \sigma_1(n, k) &= k \text{ for all pairs } \langle n, k\rangle \text{ except finitely many,} \\ \sigma_2(n, l) &= l \text{ for all pairs } \langle n, l\rangle \text{ except finitely many,} \\ \sigma_3(n, k, l, \alpha) &= \alpha \text{ for all } (n, k, l, \alpha) \text{ except finitely many.} \end{aligned}$$

9. Let $B$ be the set of all sets $b$ of atoms which satisfy the following condition: $b \cap A_{n,k,l} = 0$ for all triples $\langle n, k, l\rangle$ except finitely many and for these finitely many triples $\langle n, k, l\rangle$ $|b \cap A_{n,k,l}| \leqq \aleph_n$. $b, b_1, b_2$ will always stand for members of $B$. $B$ is obviously an ideal closed under the permutations of $G$, i.e., subsets of members of $B$, finite unions of members of $B$ and sets $\sigma(b)$, where $\sigma \in G$, are also members of $B$. $B$ contains also all finite subsets of $At$.

10. We denote with $G_b$ the subgroup of $G$ which consists of all permutations $\sigma$ such that if $(n, k, l, \alpha), (n, k, l', \alpha') \in b$ then $\sigma(n, k, l, \alpha) = (n, k, l, \alpha)$ (i.e., $\sigma_1(n, k) = k$, $\sigma_2(n, l) = l$, $\sigma_3(n, k, l, \alpha) = \alpha$) and for $\beta < \omega_{n+1}$ $\sigma_3(n, k, l, \beta) = \sigma_3(n, k, l', \beta)$.

11. By the axiom of foundation every permutation $\sigma$ of $At$ extends in a unique way to an automorphism of the universe such that $x \in y \leftrightarrow \sigma(x) \in \sigma(y)$ (for the proof of this see [6], a similar definition and proof are given in 66 and 68 below). This extension preserves the group action of $G$, i.e., if 1 is the identity permutation of $At$ and $\sigma\tau$ is the product of $\sigma$ and $\tau$ as permutations of $At$ then for every $x$ in the universe

12. $1(x) = x$, $(\sigma\tau)(x) = \sigma(\tau(x))$, and hence $\sigma^{-1}(\sigma(x)) = \sigma(\sigma^{-1}(x)) = 1(x) = x$.

13. $x \subseteqq y \to \sigma(x) \subseteqq \sigma(y)$, as easily seen.

14. $\sigma(x) \subseteqq x \wedge \sigma^{-1}(x) \subseteqq x \to \sigma(x) = x$.

*Proof.* By 12 and 13 we have $x = \sigma(\sigma^{-1}(x)) \subseteqq \sigma(x) \subseteqq x$, hence $\sigma(x) = x$.

15. For every $b \in B$ and $\sigma \in G$ there is a $b' \in B$ such that $G_{b'} \subseteqq \sigma G_b \sigma^{-1}$.

*Proof.* We choose

$$b' = \sigma(b) \cup \{\sigma(n, k, l, \gamma) \mid \exists l' \, \exists \delta [(n, k, l, \delta) \in b \wedge \sigma_3(n, k, l', \gamma) \neq \gamma]\}.$$

By 9, $\sigma(b) \in B$; $b' \in B$ will follow, again by 9, if we show that the set $b' - \sigma(b)$ is finite. To do this we notice that the condition

$$\exists l' \, \exists \delta [(n, k, l, \delta) \in b \wedge \sigma_3(n, k, l', \gamma) \neq \gamma]$$

on $(n, k, l, \gamma)$ is satisfied by only finitely many $(n, k, l, \gamma)$'s since, by $\sigma \in G$, there are only finitely many $\gamma$'s for which there are $n''$, $l''$, $k''$ such that $\sigma_3(n'', k'', l'', \gamma) \neq \gamma$ and, by $b \in B$, there are only finitely many triples $\langle n, k, l \rangle$ for which there are $\delta$'s such that $(n, k, l, \delta) \in b$. Let $\tau \in G_{b'}$, we have to show $\tau \in \sigma G_b \sigma^{-1}$, i.e., $\sigma^{-1} \tau \sigma \in G_b$. If $x \in b$ then $\sigma(x) \in \sigma(b) \subseteqq b'$; since $\tau \in G_{b'}$ we have $\tau\sigma(x) = \sigma(x)$, hence $\sigma^{-1}\tau\sigma(x) = x$. Also, if $(n, k, l, \alpha), (n, k, l', \alpha') \in b$ and $\beta < \omega_{n+1}$ then we consider two cases.

Case (a): $\sigma_3(n, k, l, \beta) \neq \beta$ or $\sigma_3(n, k, l', \beta) \neq \beta$. Then $\sigma(n, k, l, \beta)$, $\sigma(n, k, l', \beta) \in b'$, by definition of $b'$. Put $x = (n, k, l, \beta)$, $y = (n, k, l', \beta)$ then $\sigma(x), \sigma(y) \in b'$ hence $\tau\sigma(x) = \sigma(x)$, $\tau\sigma(y) = \sigma(y)$, $\sigma^{-1}\tau\sigma(x) = x$, $\sigma^{-1}\tau\sigma(y) = y$, $(\sigma^{-1}\tau\sigma)_3(x) = \beta = (\sigma^{-1}\tau\sigma)_3(y)$.

Case (b): $\sigma_3(n, k, l, \beta) = \sigma_3(n, k, l', \beta) = \beta$. We denote with $x$ and $y$ the same atoms as above and get $\sigma(x) = (n, \sigma_1(n, k), \sigma_2(n, l), \beta)$, $\sigma(y) = (n, \sigma_1(n, k), \sigma_2(n, l'), \beta)$. By $(n, k, l, \alpha), (n, k, l', \alpha') \in b$ we have

$$\left.\begin{matrix}(n, \sigma_1(n, k), \sigma_2(n, l), \sigma_3(n, k, l, \alpha)) \\ (n, \sigma_1(n, k), \sigma_2(n, l'), \sigma_3(n, k, l', \alpha'))\end{matrix}\right\} \in \sigma(b) \subseteqq b'.$$

Since $\tau \in G_{b'}$ we get, by applying $\tau$ to $\sigma(x)$ and $\sigma(y)$

$$\tau_3(n, \sigma_1(n, k), \sigma_2(n, l), \beta) = \tau_3(n, \sigma_1(n, k), \sigma_2(n, l'), \beta);$$

let us denote this ordinal with $\lambda$. We have

$$\tau\sigma(x) = (n, \tau_1(n, \sigma_1(n, k)), \tau_2(n, \sigma_2(n, l)), \lambda),$$
$$\tau\sigma(y) = (n, \tau_1(n, \sigma_1(n, k)), \tau_2(n, \sigma_2(n, l')), \lambda).$$

Since $(n, k, l, \alpha) \in b$ we saw above that $\sigma^{-1}\tau\sigma(n, k, l, \alpha) = (n, k, l, \alpha)$, $x$ differs from $(n, k, l, \alpha)$ only in the last component, hence also the images of $x$ and $(n, k, l, \alpha)$ under the member $\sigma^{-1}\tau\sigma$ of $G$ differ only in their last component, by the definition of $G$. Thus we have

$$(*) \qquad \sigma^{-1}\tau\sigma(x) = (n, k, l, \mu) \quad \text{for some} \quad \mu < \omega_{n+1};$$

similarly $\sigma^{-1}\tau\sigma(y) = (n, k, l', \nu)$, for some $\nu < \omega_{n+1}$; our aim is to prove $\mu = \nu$. Multiplying (*) by $\sigma$ we get $\sigma(n, k, l, \mu) = \tau\sigma(x)$, similarly $\sigma(n, k, l', \nu) = \tau\sigma(y)$. If $\sigma_3(n, k, l, \mu) = \mu$ and $\sigma_3(n, k, l', \nu) = \nu$ we have $\lambda = \mu = \nu$; otherwise, without loss of generality, we assume $\sigma_3(n, k, l, \mu) \neq \mu$, and then, by definition of $b'$, $\sigma(n, k, l, \mu), \sigma(n, k, l', \mu) \in b'$. Hence, since $\tau \in G_{b'}$, $\sigma^{-1}\tau\sigma(n, k, l, \mu) = (n, k, l, \mu) = \sigma^{-1}\tau\sigma(x)$, and therefore $x = (n, k, l, \mu)$; but $x = (n, k, l, \beta)$, hence $\mu = \beta$. Now by $\sigma(n, k, l', \mu) \in b'$ and $\mu = \beta$ we have $\sigma(n, k, l', \beta) \in b'$, hence, by $\tau \in G_{b'}$, $(n, k, l', \nu) = \sigma^{-1}\tau\sigma(n, k, l', \beta) = (n, k, l', \beta)$, and thus we have $\nu = \beta = \mu$.

16. For every $y$, we say that $y$ is *b-symmetric* if for all $\sigma \in G_b$ $\sigma(y) = y$; $y$ is *symmetric* if $y$ is b-symmetric for some $b$. We say that the element $x$ is *hereditarily symmetric* if every member of its transitive closure is symmetric. Let $M$ denote the class of all hereditarily symmetric elements whose supports contain $(n, k, l, \alpha)$'s for only finitely many $n$'s and $l$'s. $M$ is, obviously, a transitive class.

17. If $x \subseteq M$, $x$ is symmetric and $\mathrm{Sp}(x)$ contains atoms $(n, k, l, \alpha)$ for only finitely many $n$'s and $l$'s then $x \in M$.

*Proof.* This follows directly from 16 and $\mathrm{Sp}(x) = \bigcup_{y \in x} \mathrm{Sp}(y)$.

18. $At \subseteq M$.

*Proof.* Follows from 16 since the atom $x$ is $\{x\}$-symmetric and $\mathrm{Sp}(x) = \{x\}$.

19. $A_{n,k,l} \in M$.

*Proof.* Follows from 17 and 18 since $A_{n,k,l}$ is $\{x\}$-symmetric for every $x \in A_{n,k,l}$ and $\mathrm{Sp}(A_{n,k,l}) = A_{n,k,l}$.

20. $x, y \in M \to \{x, y\}, \{x\}, \langle x, y\rangle \in M$.

*Proof.* If $x$ is $b_1$-symmetric and $y$ is $b_2$-symmetric then, since $\mathrm{Sp}(\{x, y\}) = \mathrm{Sp}(x) \cup \mathrm{Sp}(y)$, it is enough, by 17, to show that $\{x, y\}$ is $b_1 \cup b_2$-symmetric to get $\{x, y\} \in M$. If $\sigma \in G_{b_1 \cup b_2} \subseteq G_{b_1}, G_{b_2}$ then $\sigma(\{x, y\}) = \{\sigma(x), \sigma(y)\} = \{x, y\}$, hence $\{x, y\}$ is $b_1 \cup b_2$-symmetric. Since $\{x\} = \{x, x\}$ and $\langle x, y\rangle = \{\{x\}, \{x, y\}\}$; the rest of 20 follows.

21. $\{\langle (n, k, l, \alpha), (n, k, l', \alpha)\rangle \mid \alpha < \omega_{n+1}\} \in M$.

*Proof.* Follows from 18, 20, 17 and 10, since this set is $\{x, y\}$-symmetric, where $x \in A_{n,k,l}$, $y \in A_{n,k,l'}$, and its support is

$$A_{n,k,l} \cup A_{n,k,l'}.$$

22. $\{A_{n,k,l} \mid k < \omega\} \in M$.

*Proof.* Follows from 19 and 17 since the set is $\{x\}$-symmetric for every $x \in A_{n,k,l}$ and its support is $\bigcup_{k<\omega} A_{n,k,l}$.

23. For every permutation $\sigma$ of $At$ $\mathrm{Sp}(\sigma(x)) = \sigma(\mathrm{Sp}(x))$.

*Proof.* As a consequence of the axiom of foundation we can use induction on the membership relation, i.e., a statement is proved for all elements $x$ by assuming it for all $y \in x$ and proving it for $x$ (see TARSKI [10]). We shall now prove 23 in this way. If $x \in At$ then $\sigma(x) \in At$, $\mathrm{Sp}(\sigma(x)) = \{\sigma(x)\} = \sigma(\{x\}) = \sigma(\mathrm{Sp}(x))$. If $x$ is a set then

$$\sigma(x) = \{\sigma(y) \mid y \in x\}$$

and hence

$$\begin{aligned}\mathrm{Sp}(\sigma(x)) &= \bigcup_{y \in x} \mathrm{Sp}(\sigma(y)) = \bigcup_{y \in x} \sigma(\mathrm{Sp}(y)) = \sigma\Big(\bigcup_{y \in x} \mathrm{Sp}(y)\Big) \\ &= \sigma(\mathrm{Sp}(x)),\end{aligned}$$

where we use the induction hypothesis $y \in x \to \mathrm{Sp}(\sigma(y)) = \sigma(\mathrm{Sp}(y))$ and the simple fact that $\sigma$ distributes over union.

24. If $x \in M$ and $\sigma \in G$ then $\sigma(x) \in M$.

*Proof.* By induction on the membership relation. If $x \in At$ then $\sigma(x) \in At \subseteqq M$. If $x$ is a set then since $x \in M$ $\mathrm{Sp}(x)$ contains $(n, k, l, \alpha)$'s for only finitely many $n$'s and $l$'s. By 23

$$\mathrm{Sp}(\sigma(x)) = \sigma(\mathrm{Sp}(x)) = \{\sigma(z) \mid z \in \mathrm{Sp}(x)\},$$

hence, since $\sigma \in G$, also $\mathrm{Sp}(\sigma(x))$ contains atoms for only finitely many $n$'s and $l$'s. By the induction hypothesis $\sigma(x) = \{\sigma(y) \mid y \in x\} \subseteqq M$ hence, by 17, it is enough to prove that $\sigma(x)$ is symmetric. Since $x \in M$ is $b$-symmetric for some $b \in B$ hence $\tau\sigma(x) = \sigma(x)$ for every $\tau \in \sigma G_b \sigma^{-1}$. By 15 there is a $b' \in B$ such that $G_{b'} \subseteqq \sigma G_b \sigma^{-1}$ hence $\tau\sigma(x) = \sigma(x)$ for every $\tau \in G_{b'}$ and $\sigma(x)$ is $b'$-symmetric.

25. Let $a$ be a finite set of atoms and let $\xi$ be a permutation of $At$ such that $\xi(x) = x$ for $x \in At - a$. If $x$ is $b$-symmetric then $\xi(x)$ is $b \cup a$-symmetric.

*Proof.* Let $x$ be $b$-symmetric and let $\sigma \in G_{b \cup a}$. Since $\sigma$ moves no member of $a$ and $\xi$ moves no member of $At - a$ we have $\xi\sigma = \sigma\xi$. Hence, by 12, $\sigma(\xi(x)) = \sigma\xi(x) = \xi\sigma(x) = \xi(x)$.

26. Let $a$ and $\xi$ be as in 25. If $x \in M$ then also $\xi(x) \in M$.

*Proof.* By induction on the membership relation. If $x \in At$ this is obvious. If $x$ is a set then, by the induction hypothesis, 17 and 23, it is enough to prove that $\xi(x)$ is symmetric, but this follows from 25.

27. If $\sigma$ is a permutation of $At$ such that $\sigma(u) = u$ for each $u \in \mathrm{Sp}(x)$ then $\sigma(x) = x$.

*Proof.* Easy, by induction on the membership relation.

We shall interpret ZF′ in ZF* by interpreting "element" (i.e., set or atom) as "member of $M$", "atom" as "atom" and "$x \in y$" as "$x \in y$". Let us first verify that the interpretation of the axioms of ZF′ are theorems of **ZF***. We shall now write next to the name of each axiom its interpretation. In fact, instead of writing the exact interpretation we shall usually write a statement equivalent to it in view of the transitivity of the class $M$.

28. The axiom of extentionality: If $x, y \in M$, $x, y$ are not atoms and $\forall z(z \in x \leftrightarrow z \in y)$ then $x = y$. (Obvious).

29. The axioms of union: If $x \in M$ then $\bigcup x \in M$.

*Proof.* $\mathrm{Sp}(\bigcup x) \subseteqq \mathrm{Sp}(x)$; therefore, by 17, it is enough if we prove that if $x$ is $b$-symmetric so is $\bigcup x$. Let $z \in y \in x$ and let $\sigma \in G_b$ then $\sigma(z) \in \sigma(y) \in \sigma(x) = x$, therefore

$$\sigma(\bigcup x) = \sigma\{z \mid \exists y(z \in y \in x)\} = \{\sigma(z) \mid \exists y(z \in y \in x)\} \subseteqq \bigcup x,$$

for every $\sigma \in G_b$, and by 14 we have $\sigma(\bigcup x) = \bigcup x$.

30. The power-set axiom: If $x \in M$ then $P(x) \cap M \in M$, where $P(x) = \{y \mid y \text{ is a set} \wedge y \subseteqq x\}$.

*Proof.* If $y \subseteqq x$ then $\mathrm{Sp}(y) \subseteqq \mathrm{Sp}(x)$; therefore, by 17, it is enough to prove that if $x$ is $b$-symmetric so is $P(x) \cap M$. Let $y \subseteqq x$, $y \in M$ and $\sigma \in G_b$. By 13 $\sigma(y) \subseteqq \sigma(x) = x$. By 24 $\sigma(y) \in M$; therefore $\sigma(P(x) \cap M) \subseteqq P(x) \cap M$ for every $\sigma \in G_b$ and hence, by 14, $\sigma(P(x) \cap M) = P(x) \cap M$.

31. The axiom schema of subsets: If $x_1, \ldots, x_n, y \in M$ then there is a set $z \in M$ such that $\forall x(x \in z \leftrightarrow x \in y \wedge \varphi^{(M)}(x; x_1, \ldots, x_n))$, where $\varphi$ is any formula of set theory with no free variables other than $x, x_1, \ldots, x_n$ and $\varphi^{(M)}$ is the formula obtained from $\varphi$ by relativising the quantifiers in $\varphi$ to $M$, i.e., by replacing all quantifiers of the form $\exists u$ and $\forall u$ by $(\exists u \in M)$ and $(\forall u \in M)$, respectively.

As a consequence of 11 and 24 every $\sigma \in G$ is an automorphism of $M$, i.e., it is a $1-1$ map of $M$ on itself such that for all $x, y \in M$ $x \in y \leftrightarrow \sigma(x) \in \sigma(y)$ and $x$ is an atom if and only if $\sigma(x)$ is an atom. As a consequence we have, for every formula $\psi(y_1, \ldots, y_k)$ and $\sigma \in G$, that $y_1, \ldots, y_k \in M \to (\psi^{(M)}(y_1, \ldots, y_k) \leftrightarrow \psi^{(M)}(\sigma(y_1), \ldots, \sigma(y_k))$, as is easily shown by induction on the length of $\psi$. In our present case, since $x_1, \ldots, x_n, y \in M$ we can assume that all of $x_1, \ldots, x_n, y$ are $b$-symmetric for some $b \in B$ (since $B$ is closed under finite unions). Consider the set $z = \{x \mid x \in y \wedge \varphi^{(M)}(x; x_1, \ldots, x_n)\}$. In order to prove that $z$ is in $M$ it is enough, by 17, and $y \in M$, to prove that $z$ is $b$-symmetric. Let $x \in z$ and $\sigma \in G_b$, then $x \in y$; hence $\sigma(x) \in \sigma(y) = y$. Also, since $x \in z$ we have $\varphi^{(M)}(x; x_1, \ldots, x_n)$ hence $\varphi^{(M)}(\sigma(x); \sigma(x_1), \ldots, \sigma(x_n))$ but since $\sigma \in G_b$ $\sigma(x_i) = x_i$ for $1 \leqq i \leqq n$, hence $\varphi^{(M)}(\sigma(x); x_1, \ldots, x_n)$ and therefore $\sigma(x) \in z$. Thus we have for every $\sigma \in G_b$ $\sigma(z) \subseteqq z$ and hence, by 14 $\sigma(z) = z$ and $z$ is $b$-symmetric.

32. The axiom of infinity: There is a set $y \in M$ such that $0 \in y$ and for every $x \in y$ also $x \cup \{x\} \in y$.

*Proof.* It is easy to prove by transfinite induction that every ordinal is 0-symmetric. Since for every ordinal $\alpha$ $\mathrm{Tc}(\alpha) = \alpha + 1 = \{\beta \mid \beta \leqq \alpha\}$ we have that every ordinal is in $M$. In particular we have $\omega \in M$ and $\omega$ obviously satisfies the requirements for $y$ in 32.

33. The axiom schema of replacement: In the presence of the axiom schema of subsets the axiom schema of replacement is equivalent to the schema

$$\forall u \, \forall v \, \forall w (\varphi(u, v) \wedge \varphi(u, w) \rightarrow v = w) \\ \rightarrow \forall y \, \exists z \, \forall u \, \forall v (u \in y \wedge \varphi(u, v) \rightarrow v \in z).$$

Therefore we have to prove now

$$x_1, \ldots, x_n \in M \wedge (\forall u, v, w \in M)(\varphi^{(M)}(u, v) \wedge \varphi^{(M)}(u, w) \rightarrow v = w) \\ \rightarrow (\forall y \in M)(\exists z \in M) \, \forall u \, \forall v (u \in y \wedge \varphi^{(M)}(u, v) \rightarrow v \in z),$$

where $\varphi(u, v)$ is a formula with no free variables other than $u, v, x_1, \ldots, x_n$. First we prove:

34. Let $\psi(x_1, \ldots, x_n, y)$ be a formula with no free variables other than indicated. If $x_1, \ldots, x_n \in M$ and for these $x_1, \ldots, x_n$ there is exactly one $y \in M$ such that $\psi^{(M)}(x_1, \ldots, x_n, y)$ then $\mathrm{Sp}(y) \subseteqq \bigcup_{i=1}^{n} \mathrm{Sp}(x_i)$.

*Proof.* Assume $u \in \mathrm{Sp}(y)$, $u \notin \bigcup_{i=1}^{n} \mathrm{Sp}(x_i)$. Since $x_1, \ldots, x_n, y \in M$ $\bigcup_{i=1}^{n} \mathrm{Sp}(x_i) \cup \mathrm{Sp}(y)$ contains atoms $(n, k, l, \alpha)$ for only finitely many $n$'s and $l$'s therefore there is an atom $v$ such that $v \notin \bigcup_{i=1}^{n} \mathrm{Sp}(x_i) \cup \mathrm{Sp}(y)$. Let $\xi$ be the permutation of $At$ which just interchanges $u$ and $v$. By 26 $\xi$ is an automorphism of $M$, hence by $\psi^{(M)}(x_1, \ldots, x_n, y)$ we have $\psi^{(M)}(\xi(x_1), \ldots, \xi(x_n), \xi(y))$. Since, for $1 \leqq i \leqq n$, $u, v \notin \mathrm{Sp}(x_i)$ we get by 27 $\xi(x_i) = x_i$, hence we have $\psi^{(M)}(x_1, \ldots, x_n, \xi(y))$ and thus, by our hypothesis $\xi(y) = y$. On the other hand, since $u \in \mathrm{Sp}(y)$ we get, by 23, $v = \xi(u) \in \xi(\mathrm{Sp}(y)) = \mathrm{Sp}(\xi(y))$ but $v \notin \mathrm{Sp}(y)$ hence $\xi(y) \neq y$, contradiction.

*Proof of 33.* Let $x_1, \ldots, x_n, y \in M$ then the supports of $x_1, \ldots, x_n, y$ contain atoms $(n, k, l, \alpha)$ for only finitely many $n$'s and $l$'s; let $n_0$ and $l_0$ be strict upper bounds of these $n$'s and $l$'s. Put

$$b = \{(n, 0, l, 0) \mid n < n_0, l < l_0\} \in B.$$

Let $t = \{v \mid v \in M \wedge (\exists u \in y) \, \varphi^{(M)}(u, v)\}$. Since for $u \in y$ $\mathrm{Sp}(u) \subseteqq \mathrm{Sp}(y)$ we have, by 34, that $\mathrm{Sp}(t) = \bigcup_{v \in t} \mathrm{Sp}(v)$ contains only atoms $(n, k, l, \alpha)$

with $n < n_0, l < l_0$. Let $z = \bigcup_{\sigma \in G_b} \sigma(t)$. By the definition of $t$ $t \subseteqq M$, hence, by 24 also, $\sigma(t) \subseteqq M$ for $\sigma \in G$ and therefore $z \subseteqq M$. $z$ is obviously $b$-symmetric. $\mathrm{Sp}(z) \subseteqq \bigcup_{\sigma \in G_b} \mathrm{Sp}(\sigma(t)) = \bigcup_{\sigma \in G_b} \sigma(\mathrm{Sp}(t))$, by 23. If $(n, k, l, \alpha) \in \mathrm{Sp}(t)$ then, as we showed, $n < n_0, l < l_0$ hence $(n, 0, l, 0) \in b$ and therefore, for $\sigma \in G_b$, $\sigma_2(n, l) = l$ and $\sigma(n, k, l, \alpha) = (n, \sigma_1(n, k), l, \sigma_3(n, k, l, \alpha))$. Thus, by $\mathrm{Sp}(z) \subseteqq \bigcup_{\sigma \in G_b} \sigma(\mathrm{Sp}(t))$ we have that $\mathrm{Sp}(z)$ contains only atoms $(n, k, l, \alpha)$ with $n < n_0, l < l_0$ and therefore, by 17, $z \in M$. $z \supseteqq t$ since $1 \in G_b$ hence we have $\forall u \forall v (u \in y \wedge \varphi^{(M)}(u, v) \rightarrow v \in z)$.

35. The axiom of foundation: Every non-void set $y \in M$ has a member $z$ which is an atom or else $z \cap y = 0$. (Obvious, by the axiom of foundation of $\mathsf{ZF}'$).

Having proved the interpretations of all the axioms of $\mathsf{ZF}'$, we know that also the interpretations of all the theorems of $\mathsf{ZF}'$ are provable. Let us show now that the interpretations of $\chi_1$ and $\chi_2$ of 6 hold while the interpretation of $\chi_3$ fails.

36. The interpretation of $x \approx y$ is "there is a $1 - 1$ function $f \in M$ mapping $x$ on $y$"; we shall denote it by $x \approx^{(M)} y$. Since, by the proof of 32, all the ordinals are 0-symmetric members of $M$, it follows by 17 and the proof of 20 that also all the functions $f$ which map ordinals on ordinals are 0-symmetric members of $M$. Hence, for sets $x, y$ of ordinals we have $x \approx y \leftrightarrow x \approx^{(M)} y$ and therefore the function $\omega_\alpha$ of $\alpha$ is absolute with respect to $M$ (i.e., when we relativize the definition of $\omega_\alpha$ to $M$ we get again a definition of $\omega_\alpha$). The interpretation of "$A$ is a set of the $n$-th kind" is therefore "whenever $A = C \cup D$ with $C, D \in M$, $C \cap D = 0$ then $|C| \leqq \aleph_n$ in $M$ and $|D| > \aleph_n$ in $M$ (i.e., $C \approx^{(M)} E \subseteqq \omega_n$, $D \supseteqq F \approx^{(M)} \omega_n$ but not $D \approx^{(M)} \omega_n$) or $|D| \leqq \aleph_n$ in $M$ and $|C| > \aleph_n$ in $M$"; we denote this interpretation with "$A$ is a set of the $n$-th kind in $M$".

37. $A_{n',k',l'} \approx^{(M)} A_{n',k'',l''} \rightarrow k' = k''$.

*Proof.* Suppose $k \neq k'$ and $f \in M$ is a $1 - 1$ mapping of $A_{n',k',l'}$ on $A_{n',k'',l''}$. Since $f \in M$ $f$ is $b$-symmetric for some $b \in B$. Since $b \in B$ there are $\alpha, \beta < \omega_{n+1}$ such that for every $l < \omega$ $(n', k', l, \alpha), (n', k', l, \beta) \notin b$. Let $u = (n', k', l', \alpha)$. Let $\sigma \in G$ be given by $\sigma(v) = v$ for every $v \in At$ except that $\sigma(n', k', l, \alpha) = (n', k', l, \beta)$ and $\sigma(n', k', l, \beta) = n', k', l, \alpha)$ for $l = l'$ and for every $l$ such that $A_{n',k',l} \cap b \neq 0$. Obviously $\sigma \in G_b$ and $\sigma(u) \neq u$. Also, since $k'' \neq k'$, $\sigma(v) = v$ for every $v \in A_{n',k'',l''}$. Since $\langle u, f(u) \rangle \in f$ and $f$ is $b$-symmetric we have $\langle \sigma(u), \sigma(f(u)) \rangle \in f$, but $f(u) \in A_{n',k'',l''}$, hence $\sigma(f(u)) = f(u)$ and $\langle \sigma(u), f(u) \rangle \in f$, i.e., $f(\sigma(u) = f(u)$, contradicting the $1 - 1$-ness of $f$.

38. A set $A$ of atoms is a set of the $n$-th kind in $M$ if and only if for some $k$ and $l$ the symmetric difference between $A$ and $A_{n,k,l}$ is of cardinality $\leqq \aleph_n$ in $M$, and hence, by 2, if and only if $A \approx^{(M)} A_{n,k,l}$ for some $k$ and $l$.

*Proof.* Let us call $b \in B$ *simple* if for all $n, k, l, l', \alpha, \beta$ if $(n, k, l, \alpha)$, $(n, k, l', \beta) \in b$ then also $(n, k, l', \alpha) \in b$. For every $b \in B$ the set

$$b' = \{(n, k, l', \alpha) \mid \exists l\, \exists \beta [(n, k, l, \alpha), (n, k, l', \beta) \in b]\}$$

is a simple member of $B$ which includes $b$, as easily seen. Therefore, if an element $x$ is $b$-symmetric for some $b \in B$, it is also $b'$-symmetric for some simple $b' \in B$.

By 2, one direction of 38 will be proved once we show that $A_{n,k,l}$ is of the $n$-th kind in $M$. $A_{n,k,l} \in M$ by 19. Suppose $A_{n,k,l} = C \cup D$, $C, D \in M$, $C \cap D = 0$; then $C$ is $b$-symmetric for some simple $b \in B$. If $C \subseteqq b$ then let $f = \{\langle(n, k, l, \alpha), \alpha\rangle \mid (n, k, l, \alpha) \in C\}$. $f$ is a $1-1$ function mapping $C \subseteqq A_{n,k,l}$ on a subset $E$ of $\omega_{n+1}$. $f \in M$ by 36, 18, 20 and 17, since $f$ is clearly $b$-symmetric. Since $C \subseteqq b \cap A_{n,k,l}$ and $b \in B$ we have $|C| \leqq \aleph_n$, but $E \approx C$ hence also $|E| \leqq \aleph_n$ and there is a subset $F \subseteqq \omega_n$ such that $E \approx F$. By 36 $F \in M$ and $E \approx^{(M)} F$ hence $C \approx^{(M)} E \approx^{(M)} F \subseteqq \omega_n$. If $C \nsubseteqq b$ then let $(n, k, l, \beta) \in C - b$. If $(n, k, l, \gamma)$ is an atom which is not in $b$ then let $\sigma \in G$ be given by $\sigma(v) = v$ for every $v \in At$ except that $\sigma(n, k, l', \beta) = (n, k, l', \gamma)$ and $\sigma(n, k, l', \gamma) = (n, k, l', \beta)$ for $l' = l$ and for every $l'$ such that $A_{n,k,l'} \cap b \neq 0$. Since $b$ is simple $\sigma \in G_b$ and $\sigma(n, k, l, \beta) = (n, k, l, \gamma)$. $C$ is $b$-symmetric and $(n, k, l, \beta) \in C$, hence also $(n, k, l, \gamma) \in C$; but $\gamma$ was an arbitrary ordinal $< \omega_{n+1}$ such that $(n, k, l, \gamma) \notin b$ therefore we have $A_{n,k,l} - b \subseteqq C$, i.e., $D \subseteqq b$. Without loss of generality let us refer to that one among $C$ and $D$ which is a subset of $b$ as $C$. We have, as we have seen, $C \approx^{(M)} F \subseteqq \omega_n$; thus $|C| \leqq \aleph_n$ in $M$ and we still have to show that $|D| > \aleph_n$ in $M$. Since $D \supseteqq A_{n,k,l} - b$, and $|A_{n,k,l} \cap b| \leqq \aleph_n$ and $|A_{n,k,l}| = \aleph_{n+1}$ there is a subset $a$ of $A_{n,k,l} - b$ of cardinality $\aleph_n$ and a function $g$ mapping it on $\omega_n$. Clearly $a \in B$ and, by 18, 36, 20 and 17 $g$ is an $a$-symmetric member of $M$ which maps the subset $a$ of $D$ on $\omega_n$, and therefore $|D| \geqq \aleph_n$ in $M$. We cannot have $|D| = \aleph_n$ in $M$ since $|D| = \aleph_{n+1}$ (in set theory) and therefore there is no $1-1$ map of $D$ on $\omega_n$.

To prove the other direction of 38 let $E \in M$ be a set of atoms of the $n$-th kind in $M$. $F$ is $b$-symmetric for some simple $b \in B$. For any triple $\langle m, k, l\rangle$ we have, as we saw above, $E \cap A_{m,k,l} \subseteqq b$ and $|E \cap A_{m,k,l}| \leqq \aleph_m$ in $M$, or else $A_{m,k,l} - E \subseteqq b$ and $|E \cap A_{m,k,l}| > \aleph_m$ in $M$. Let

$$W = \{\langle m, k, l\rangle \mid A_{m,k,l} - E \subseteqq b \wedge |E \cap A_{m,k,l}| > \aleph_m \mathit{\ in\ } M\}.$$

Let $\langle m, k, l\rangle \in W$ then $|E| \geqq |E \cap A_{m,k,l}| > \aleph_m$ in $M$, but since $E$

is of the $n$-th kind in $M$ we cannot have $|E| > \aleph_{n+1}$ in $M$, hence $m \leqq n$. If $m < n$ then since $A_{m,k,l}$ is of the $m$-th kind in $M$ we cannot have $|E \cap A_{m,k,l}| > \aleph_n$ in $M$ and therefore, since $E$ is of the $n$-th kind in $M$, we have $|E \cap A_{m,k,l}| \leqq \aleph_n$ in $M$. Combining this with $|E \cap A_{m,k,l}| > \aleph_m$ in $M$ we get $|E \cap A_{m,k,l}| > \aleph_{m+1}$ in $M$ which contradicts the fact $A_{m,k,l}$ is of the $m$-th kind in $M$. Thus we proved $m = n$. $|E \cap A_{n,k,l}| > \aleph_n$ in $M$, hence, since $E$ is of the $n$-th kind in $M$, $|E - A_{n,k,l}| \leqq \aleph_n$ in $M$. If $\langle n, k', l'\rangle \neq \langle n, k, l\rangle$ then $E \cap A_{n,k',l'} \subseteqq E - A_{n,k,l}$ therefore also $|E \cap A_{n,k',l'}| \leqq \aleph_n$ in $M$ hence $\langle n, k', l'\rangle \notin W$ and $|W| \leqq 1$. Let $F = \bigcup_{\langle m,k,l,\rangle \notin W} E \cap A_{m,k,l}$. We saw above that if $\langle m, k, l\rangle \notin W$ then $E \cap A_{m,k,l} \subseteqq b$ and $|E \cap A_{m,k,l}| \leqq \aleph_m$ in $M$; hence $F \subseteqq b$. For only finitely many triples $\langle m, k, l\rangle$ is $b \cap A_{m,k,l} \neq 0$ and since for $\langle m,k,l\rangle \notin W$, $E \cap A_{m,k,l} \subseteqq b \cap A_{m,k,l}$ and $|E \cap A_{m,k,l}| \leqq \aleph_m$ in $M$ we have that $F$ can be well-ordered in $M$. Since $F \subseteqq E$ and $E$ is of the $n$-th kind in $M$ we have, by 3, $|F| \leqq \aleph_n$ in $M$ and $|E - F| > \aleph_n$ in $M$. Thus $W \neq 0$, hence $W = \{\langle n, k, l\rangle\}$ for some $n, k, l$, $E - F = E \cap A_{n,k,l}$ and $|A_{n,k,l} - E| \leqq \aleph_n$ in $M$. Therefore $|(E - A_{n,k,l}) \cup (A_{n,k,l} - E)| = |F| + |A_{n,k,l} - E| \leqq \aleph_n$ in $M$.

39. $\chi_1$: For each finite number $n$ there is a set $A \in M$ of sets of atoms such that for every set $y \in M$ of atoms which is of the $n$-th kind in $M$ there is a $z \in A$ such that $y \approx^{(M)} z$.

*Proof.* Take $A = \{A_{n,k,0} | k < \omega\}$. By 22, $A \in M$. If $y \in M$ is a set of atoms of the $n$-th kind in $M$ then, by 38, $y \approx^{(M)} A_{n,k,l}$ for some $k$ and $l$. By 21 $A_{n,k,l} \approx^{(M)} A_{n,k,0}$ hence $y \approx^{(M)} A_{n,k,0} \in A$.

40. $\chi_2$: For each finite $n$ and for each cardinality function (in the sense of $M$) $f \in M$ on a set $A \in M$ as in 39 the support of $\mathfrak{R}(f)$ includes some set of atoms of the $n$-th kind.

*Proof.* Let $A, f, n$ be as in 40. $A$ and $f$ are $b$-symmetric for some $b \in B$. Let $k \neq k'$ be such that $b \cap A_{n,k,l} = 0$, $B \cap A_{n,k',l} = 0$ for every $l$ (there are such $k, k'$ since $b \cap A_{n,k,l} \neq 0$ for only finitely many $n, k, l$). Since $A_{n,k,0}$ is a set of atoms of the $n$-th kind in $M$ there is a member $C$ of $A$ such that $C \approx^{(M)} A_{n,k,0}$ and, by 38, for some $l'$, $|(C - A_{n,k,l'}) \cup (A_{n,k,l'} - C)| \leqq \aleph_n$ in $M$. Let $\sigma \in G$ be such that $\sigma(v) = v$ for every $\sigma \in At$ except that $\sigma(n, k, l, \alpha) = (n, k', l, \alpha)$ and $\sigma(n, k', l, \alpha) = (n, k, l, \alpha)$ for all $l, \alpha$. By our choice of $k$ and $k'$, $\sigma \in G_b$. $C \in A$ and $\langle C, f(C)\rangle \in f$; since $A$ and $f$ are $b$-symmetric we have $\sigma(C) \in A$, $\langle \sigma(C), \sigma(f(C))\rangle \in f$, i.e. $f(\sigma(C)) = \sigma(f(C))$. We shall see now that it is not the case that $\sigma(C) \approx^{(M)} C$. Since $\sigma$ is an automorphism we have

$$\begin{aligned} E &= \sigma((C - A_{n,k,l'}) \cup (A_{n,k,l'} - C)) \\ &= (\sigma(C) - \sigma(A_{n,k,l'})) \cup (\sigma(A_{n,k,l'}) - \sigma(C)) \\ &= ((\sigma(C) - A_{n,k',l'}) \cup (A_{n,k',l'} - \sigma(C)). \end{aligned}$$

Since $|(C - A_{n,k,l'}) \cup (A_{n,k,l'} - C)| \leqq \aleph_n$ in $M$ there is a $1-1$ function $g \in M$ mapping $(C - A_{n,k,l'}) \cup (A_{n,k,l'} - C)$ into $\omega_n$. Since $\sigma$ is an automorphism of $M$, and by 36, $\sigma(g)$ is a $1-1$ function in $M$ mapping $E$ into $\omega_n$. From $|E| \leqq \aleph_n$ in $M$ we conclude, by 2 and 38, that $\sigma(C) \approx^{(M)} A_{n,k',l'}$; since $C \approx^{(M)} A_{n,k,0}$ and $k \neq k'$ we get, by 37, $\sigma(C) \not\approx^{(M)} C$. $f$ is a cardinality function, therefore $C \not\approx^{(M)} \sigma(C)$ entails $f(C) \neq f(\sigma(C)) = \sigma(f(C))$. Since $\sigma(v) = v$ for every

$$v \in At - \bigcup_{l<\omega} (A_{n,k,l} \cup A_{n,k',l})$$

and $\sigma(f(C)) \neq f(C)$ we get, by 27, that $\mathrm{Sp}(f(C))$ contains some atom $(n, k, l, \alpha)$ or $(n, k', l, \alpha)$ for some $l$ and $\alpha$; that atom is also a member of $\mathrm{Sp}(\mathfrak{R}(f)) \supseteqq \mathrm{Sp}(f(C))$, without loss of generality let us denote it with $(n, k, l, \alpha)$. For a given $\beta < \omega_{n+1}$ let $\tau \in G$ be given by $\tau(v) = v$ for every $v \in At$ except that $\tau(n, k, l, \alpha) = (n, k, l, \beta)$ and $\tau(n, k, l, \beta) = (n, k, l, \alpha)$. Since $A_{n,k,l} \cap b = 0$ for every $l$, $\tau \in G_b$. $f$ is $b$-symmetric, hence $\mathfrak{R}(f)$ is $b$-symmetric too. Since $(n, k, l, \alpha) \in \mathrm{Sp}(\mathfrak{R}(f))$ we have $(n, k, l, \beta) = \tau(n, k, l, \alpha) \in \tau(\mathrm{Sp}(\mathfrak{R}(f)) = \mathrm{Sp}(\tau(\mathfrak{R}(f))) = \mathrm{Sp}(\mathfrak{R}(f))$, and this holds for every $\beta < \omega_{n+1}$, i.e. $A_{n,k,l} \subseteqq \mathrm{Sp}(\mathfrak{R}(f))$. 40 follows now from 38.

41. $\neg \chi_3$: No set $C \in M$ contains for every finite $n$ a set of atoms of the $n$-th kind.

*Proof.* If $C \in M$ then, by the definition of $M$, $\mathrm{Sp}(C)$ contains atoms $(n, k, l, \alpha)$ for only finitely many $n$'s and $l$'s. Therefore there are $n'$ and $l'$ such that $(n', k, l', \alpha) \notin \mathrm{Sp}(C)$ for every $k$ and $\alpha$. As a consequence of 38 and 3 C does not contain any set of atoms of the $n'$-th kind.

## § 3. The Relative Indefinability of $|x|$

As we promised in § 1 we shall prove here that if $\mathsf{ZF}'$ is consistent so is $\mathsf{ZF}_2'$ with the axiom schema

42. $\neg \exists z \forall x \forall y (\tau(z, x) = \tau(z, y) \leftrightarrow x \approx y)$, where $\tau(z, x)$ ranges over all terms of $\mathsf{ZF}'$ (i.e., terms which do not involve primitive notions other than membership or notions defined in terms of such primitive notions).

Our procedure in the present section is to consider a theory $\mathsf{ZF}_+'$ which is like $\mathsf{ZF}'$ except that it contains, as an additional primitive notion, a unary operation symbol $\mathfrak{a}$. The axioms of $\mathsf{ZF}_+'$ are the axioms of $\mathsf{ZF}'$, where also the symbol $\mathfrak{a}$ is allowed in the axiom schemas of subsets and replacement, with the addition of the axiom $\mathsf{C}_1$ (of § 1) for a *defined* term $|x|$ of $\mathsf{ZF}_+'$ (the definition of $|x|$ is given in 55 below). All the instances of the schema $\mathsf{C}_2$ are theorems of $\mathsf{ZF}_+'$ since $|x|$ is defined in terms of the primitive notions of $\mathsf{ZF}_+'$, all of which are allowed in the axiom schema of replacement. Thus all the axioms of $\mathsf{ZF}_2'$ are theorems of $\mathsf{ZF}'_+$. We shall show that if $\mathsf{ZF}'$ is consistent then $\mathsf{ZF}'_+$ is consistent with schema 42. This entails that schema 42 is also consistent with $\mathsf{ZF}'_2$.

It is well known that if $\mathsf{ZF}$, or $\mathsf{ZF}'$, is consistent so is the theory $\mathsf{ZF}^{**}$ obtained from $\mathsf{ZF}'$ by adding to it the axiom of choice and the axiom

43. There are exactly $\aleph_0$ atoms,

(See [6]). We shall construct an interpretation of $\mathsf{ZF}'_+$ in $\mathsf{ZF}^{**}$, similar to that of MOSTOWSKI [6], under which each instance of schema 42 goes over to a theorem of $\mathsf{ZF}^{**}$.

44. Let $<$ be an ordering of the set $At$ of all atoms in the order type of the rational numbers. Let $G$ be the group of all order-preserving permutations of $At$. Let $B$ be the set of all finite subsets of $At$. For $b \in B$ let $G_b$ be the subgroup of $G$ of all $\sigma \in G$ such that $\sigma(u) = u$ for all $u \in b$. Let us say that an element $x$ is *b-symmetric* if $\sigma(x) = x$ for all $\sigma \in G_b$; $x$ is *symmetric* if it is $b$-symmetric for some $b \in B$; $x$ is *hereditarily symmetric* if all the members of $\mathrm{Tc}(x)$ are symmetric. For $u \in At$ let $At_u = \{v \mid v \in At \wedge v \leqq u\}$. Let $M$ be the class of all hereditarily symmetric elements $x$ such that $\mathrm{Sp}(x) \subseteqq At_u$ for some $u \in At$. $M$ is obviously a transitive class. It is easily seen that $M$ is closed under $G$ and hence $G$ is a group of automorphisms of $M$. Several of the lemmas of § 2 apply also here, either literally or with trivial changes; whenever we shall mention in the present section a lemma of § 2, we shall mean its obvious analogue here.

45. If $x \subseteqq M$, $x$ is symmetric and $\mathrm{Sp}(x) \subseteqq At_u$ for some $u \in At$, then $x \in M$. (This is the analogue of 17, the proof is trivial.)

46. $B \subseteqq M$ (trivial).

47. For every $x \in M$ there is a $b \in B$ such that $x$ is $b$-symmetric, and for every $c \in B$ if $x$ is also $c$-symmetric then $b \subseteqq c$. This $b$ is called the *asymmetry set* of $x$; we denote it with $\mathfrak{a}(x)$. (For the proof see MOSTOWSKI [6], 100—101 on p. 243.)

48. If $x \in M$ and $\sigma \in G$, $\mathfrak{a}(\sigma(x)) = \sigma(\mathfrak{a}(x))$. Proved in [6], (102 on p. 243).

49. If $x \in M$, $\sigma \in G$ and $\sigma(\mathfrak{a}(x)) \neq \mathfrak{a}(x)$ then $\sigma(x) \neq x$ (An easy consequence of 48.)

50. For $b \in B$ we define $At_b = \{v \mid v \in At \wedge (\exists u \in b)\, v \leqq u\}$.

51. If $x \in M$ then $\mathrm{Sp}(x) \subseteqq At_{\mathfrak{a}(x)}$.

*Proof.* Let $w \in \mathrm{Sp}(x)$, then there is a finite sequence $y_1, \ldots, y_n$, $n \geqq 1$, such that $w = y_1 \in y_2 \in \cdots \in y_n = x$. If $w \notin At_{\mathfrak{a}(x)}$ and $u$ is any member of $At - At_{\mathfrak{a}(x)}$ then there is an order preserving $\sigma \in G$ such that $\sigma(v) = v$ for every $v \in \mathfrak{a}(x)$ and $\sigma(w) = u$. Since $\sigma \in G_{\mathfrak{a}(x)}$ is an automorphism we have $u = \sigma(w) = \sigma(y_1) \in \sigma(y_2) \in \cdots \in \sigma(y_n) \in \sigma(x) = x$, therefore $u \in \mathrm{Sp}(x)$; since $u$ is an arbitrary member of $At - At_{\mathfrak{a}(x)}$ we have $\mathrm{Sp}(x) \supseteqq At - At_{\mathfrak{a}(x)}$, contradicting $x \in M$.

52. If $x \in M$, $u \in At$ and $\mathrm{Sp}(x) \subseteqq At_u$ then $\mathfrak{a}(x) \subseteqq At_u$.

*Proof.* Let $\sigma \in G$ be such that $\sigma(v) = v$ for $v \in At$, $v \leqq u$ and $\sigma(v) > v$ for $v > u$. Then, by 27, $\sigma(x) = x$, hence, by 49, $\sigma(\mathfrak{a}(x)) = \mathfrak{a}(x)$ and therefore $\mathfrak{a}(x) \subseteqq At_u$.

We interpret $\mathsf{ZF}'_+$ in $\mathsf{ZF}^{**}$ by interpreting 'element' 'atom' '$x \in y$' and '$\mathfrak{a}(x)$', as 'member of $M$', 'atom' '$x \in y$' and '$\mathfrak{a}(x)$', respectively. As in Mostowski [6] and in § 2 all the interpretations of the single axioms of $\mathsf{ZF}'$ are theorems (of $\mathsf{ZF}^{**}$). As in 31 it is also easy to show that the interpretations of the instances of the axiom schema of subsets, including those which involve $\mathfrak{a}$, are theorems of $\mathsf{ZF}^{**}$. (In the proof one must notice that every $\sigma \in G$, which is automorphism of $M$, commutes with the operation $\mathfrak{a}$, by 48.) For the axiom schema of replacement we use the same version as in 33. First we prove

53. Let $\psi(x_1, \ldots, x_n, y)$ be a formula of $\mathsf{ZF}'_+$ with no free variables other than indicated. If $x_1, \ldots, x_n \in M$ and for these $x_1, \ldots, x_n$ there is exactly one $y$ such that $\psi^{(M)}(x_1, \ldots, x_n, y)$ then $\mathfrak{a}(y) \subseteqq \bigcup_{i=1}^{n} \mathrm{Sp}(x_i)$.

53 is analogous to 34, and is proved similarly, using 27 and 49. By means of 52 we prove the relativization to $M$ of the axiom schema of replacement as in the proof of 33 (we take literally the same $t$ and $z$, and we use 51).

54. The rank $\varrho(x)$ of an element $x$ is defined by recursion on the $\varepsilon$-relation (Tarski [10]) as follows: $\varrho(x) = \sup_{y \in x}(\varrho(y) + 1)$. It is easy to verify that this is a valid definition since $\varrho(x)$ depends only on the behavior of $\varrho$ on $\mathrm{Tc}(x) - \{x\}$. It is also easy to prove that for every $x \in M$ $\varrho^{(M)}(x) = \varrho(x)$, and that for every permutation $\sigma$ of $At$ $\varrho(\sigma(x)) = \varrho(x)$ ([6], 35 on p. 215).

55. We define $|x|$ in $\mathsf{ZF}'_+$ by a modified Frege-Russell-Scott definition as follows:

$$|x| = \{y \mid y \approx x \wedge \forall z (z \approx x \rightarrow |\mathfrak{a}(z)| > |\mathfrak{a}(y)| \vee (|\mathfrak{a}(z)| = |\mathfrak{a}(y)| \wedge \varrho(z) > \varrho(y)))\}$$

if the right-hand side is a set, and $|x|$ is the void set if the right-hand side is a proper class. In $\mathsf{ZF}'_+$ we have, obviously, that if $x$ and $y$ are sets and $|x|$, $|y|$ are not the void set then $|x| = |y| \leftrightarrow x \approx y$. Thus, in order to prove the interpretation of $\mathsf{C}_1$ in $\mathsf{ZF}^{**}$ we have to prove that for every set $x \in M$, $|x|^{(M)}$ is non-void. Let us define in ZF**

$$\|x\| = \{y \mid y \in M \wedge y \approx^{(M)} x \wedge (\forall z \in M)(z \approx^{(M)} x \rightarrow |\mathfrak{a}(z)| > |\mathfrak{a}(y)| \vee (|\mathfrak{a}(z)| = |\mathfrak{a}(y)| \wedge \varrho(z) > \varrho(y)))\}.$$

The clause "if the right-hand side is a set, ... proper class" is now superfluous; since in ZF** $At$ is a set; hence for every $\alpha$ $\{u \mid \varrho(u) = \alpha\}$ is a set too, and therefore the right-hand side above is always a set. By 54, and the absoluteness with respect to relativization to $M$ of cardinality equalities and inequalities for finite sets, we see that for $x \in M$ if $\|x\| \in M$ then it satisfies the relativization of the definition of $|x|$ in $\mathrm{ZF}'_+$, i.e., $\|x\| = |x|^{(M)}$. Since for $x \in M$ $\|x\|$ is anyway a class of $M$, given in $M$ by the right-hand side of the definition of $|x|$, $\|x\| \in M$ will follow, by the axiom schema of subsets of $\mathrm{ZF}'_+$, once we prove that $\|x\|$ is a subset of some member of $M$. The latter statement will be proved in 60.

56. $\{x \mid x \in M \wedge \varrho(x) = \alpha \wedge \mathrm{Sp}(x) \subseteqq At_u\} \in M$.

*Proof.* This follows directly from 45 and the fact that for $\sigma \in G$ $\varrho(\sigma(x)) = \varrho(x)$.

57. For $b \in B$ we write $b = \{b_1, \ldots, b_k\}$, where $b_1 < b_2 < \cdots < b_k$. For $k \geqq 0$ let $N^k$ be the set of all $k$-tuples $\langle n_1, \ldots, n_k \rangle$ of natural numbers. For $\boldsymbol{n} \in N^k$ we put $n = \sum_{i=1}^{k} n_i$. We define, for $|b| = z$ and $\boldsymbol{n} \in N^k$, $\Delta_b(\boldsymbol{n})$ to be the set of all sets $\{u_1, \ldots, u_n\}$ where $u_i \in At$ for $1 \leqq i \leqq n$,

$$u_1 < u_2 < \cdots < u_{n_1} < b_1$$

and

$$b_i < u_{n_i+1} < u_{n_i+2} < \cdots < u_{n_{i+1}} < b_{i+1}$$

for $1 \leqq i < n$. Notice that if $n = 0$ $\Delta_b(\boldsymbol{n}) = \{0\}$. It is easily seen, by 46 and 45, that $\Delta_b(\boldsymbol{n})$ is a $b$-symmetric member of $M$.

58. For every $y \in M$ if $b = \mathfrak{a}(y)$ and $k = |b|$ then there is a function $\alpha$ on $N^k$ whose values are ordinals such that $y \approx^{(M)} \bigcup_{\boldsymbol{n} \in N^k} \alpha(\boldsymbol{n}) \times \Delta_b(\boldsymbol{n})$.

*Proof.* Let $z \in y$ and let $t = \{\sigma(z) \mid \sigma \in G_b\}$. For every $\sigma \in G_b$ we have $\sigma(z) \in \sigma(y) = y$, hence $t \subseteqq y$ and $t \subseteqq M$. Let $\beta$ be an ordinal. Consider the relation $r = \{\langle \sigma(z), \langle \beta, \sigma(\mathfrak{a}(z)) - b \rangle\rangle \mid \sigma \in G_b\}$. By 36 and since $G_b$ is a group $r$ is $b$-symmetric. By $t \subseteqq M$, 46, 36 and 20 $r \subseteqq M$. We shall now prove that $r$ is a $1-1$ function. Let $\sigma_1, \sigma_2 \in G_b$, $\sigma_1(z) = \sigma_2(z)$ then, by 48, $\sigma_1(\mathfrak{a}(z)) = \mathfrak{a}(\sigma_1(z)) = \mathfrak{a}(\sigma_2(z)) = \sigma_2(\mathfrak{a}(z))$ hence $\langle \beta, \sigma_1(\mathfrak{a}(z)) - b \rangle = \langle \beta, \sigma_2(\mathfrak{a}(z)) - b \rangle$ and $r$ is a function. Again, if $\sigma_1, \sigma_2 \in G_b$ and $r(\sigma_1(z)) = r(\sigma_2(z))$ then $\sigma_1(\mathfrak{a}(z)) - b = \sigma_2(\mathfrak{a}(z)) - b$.

Therefore we get, since $\sigma_2\sigma_1^{-1} \in G_b$,

$$\sigma_2\sigma_1^{-1}[\sigma_1(\mathfrak{a}(z)) - b] = \sigma_2\sigma_1^{-1}\sigma_1(\mathfrak{a}(z)) - \sigma_2\sigma_1^{-1}(b)$$
$$= \sigma_2(\mathfrak{a}(z)) - b = \sigma_1(\mathfrak{a}(z)) - b\,.$$

Since $\sigma_2\sigma_1^{-1}$ is order-preserving and $\sigma_1(\mathfrak{a}(z)) - b$ is finite

$$\sigma_2\sigma_1^{-1}[\sigma_1(\mathfrak{a}(z)) - b] = \sigma_1(\mathfrak{a}(z)) - b$$

entails $\sigma_2\sigma_1^{-1}(v) = v$ for every $v \in \sigma_1(\mathfrak{a}(z)) - b$, and hence, since $\sigma_2\sigma_1^{-1} \in G_b$, $\sigma_2\sigma_1^{-1}(v) = v$ for every $v \in \sigma_1(\mathfrak{a}(z))$, i.e. by 48, for every $v \in \mathfrak{a}(\sigma_1(z))$. Therefore, by definition of $\mathfrak{a}(\sigma_1(z))$ $\sigma_2\sigma_1^{-1}(\sigma_1(z)) = \sigma_1(z)$, i.e., $\sigma_2(z) = \sigma_1(z)$, and $r$ is $1-1$. We denote $(-\infty, u) = \{w \mid w \in At \wedge w < u\}$, $(u, v) = \{w \mid w \in At \wedge u < w < v\}$. Let $n_1 = |\mathfrak{a}(z) \cap (-\infty, b_1)|$ and for $1 < i \leqq k$, $n_i = |\mathfrak{a}(z) \cap (b_{i-1}, b_i)|$, and let $\boldsymbol{n} = \langle n_1, \ldots, n_k\rangle$. We claim that the range of $r$ is $\{\beta\} \times \Delta_b(\boldsymbol{n})$. If $q$ is in the range of $r$ then

$$q = \langle\beta, \sigma(\mathfrak{a}(z)) - b\rangle$$

for some $\sigma \in G_b$. Since $z \in y$ we have $\mathrm{Sp}(z) \subseteqq \mathrm{Sp}(y) \subseteqq At_b$, by 51, hence, by 52, $\mathfrak{a}(z) \subseteqq At_b$, i.e., $\mathfrak{a}(z)$ has no members beyond $b_k$. Since $\sigma$ is an order preserving map of $At$ which does not move the members of $b$, $\sigma(\mathfrak{a}(z))$ has the same number of members as $\mathfrak{a}(z)$ between $b_{i-1}$ and $b_i$, for $1 \leqq i \leqq k$ (where $b_0 = -\infty$), namely $n_i$ members, and like $\mathfrak{a}(z)$, $\sigma(\mathfrak{a}(z))$ has no members beyond $b_k$; hence $\sigma(\mathfrak{a}(z)) - b \in \Delta_b(\boldsymbol{n})$ and $q \in \{\beta\} \times \Delta_b(\boldsymbol{n})$. If, on the other hand $q = \langle\beta, d\rangle \in \{\beta\} \times \Delta_b(\boldsymbol{n})$ and $\sigma \in G_b$ is such that $\sigma(\mathfrak{a}(z) - b) = d$ then we have in the range of $r$ the pair

$$\langle\beta, \sigma(\mathfrak{a}(z)) - b\rangle = \langle\beta, \sigma(\mathfrak{a}(z) - b)\rangle = \langle\beta, d\rangle = q\,.$$

Let $T = \{\{\sigma(z) \mid \sigma \in G_b\} \mid z \in y\}$. Since, as we saw, $t \in T \rightarrow t \subseteqq y$ and since $G_b$ is a group, $T$ is a partition of $y$. For $z \in t \in T$ we have

$$\{\mathfrak{a}(x) - b \mid x \in t\} = \{\mathfrak{a}(\sigma(z)) - b \mid \sigma \in G_b\} = \{\sigma(\mathfrak{a}(z)) - b \mid \sigma \in G_b\},$$

by 48; we, essentially, saw above that this set is of the form $\Delta_b(\boldsymbol{n})$, let us denote this $\boldsymbol{n}$ with $\boldsymbol{n}(t)$. Let $g$ be a function on $T$ such that for every $\boldsymbol{n} \in N^k$ if $T_{\boldsymbol{n}} = \{t \mid t \in T \wedge \boldsymbol{n}(t) = \boldsymbol{n}\}$ then $g$ restricted to $T_{\boldsymbol{n}}$ is a one-one map of $T_{\boldsymbol{n}}$ on some ordinal, which we denote with $\alpha(\boldsymbol{n})$. The function $r$ given above is given by $\{\langle x, \langle\beta, \mathfrak{a}(x) - b\rangle\rangle \mid x \in t\}$ and thus it depends on the parameters $t$ and $\beta$, let us denote it with $r_{t,\beta}$. Let $R = \bigcup_{t \in T} r_{t,g(t)}$. We saw that $r_{t,g(t)}$ is a function whose domain is $t$, therefore $R$ is a function and its domain is $\bigcup T = y$. The range of $r_{t,g(t)}$ is, as we saw above, $\{g(t)\} \times \Delta_b(\boldsymbol{n}(t))$; for $t_1 \neq t_2$ we have $\boldsymbol{n}(t_1) \neq \boldsymbol{n}(t_2)$, in which case $\Delta_b(\boldsymbol{n}(t_1)) \cap \Delta_b(\boldsymbol{n}(t_2)) = 0$, or else $g(t_1) \neq g(t_2)$, therefore $R$ is a $1-1$ function. The range of $R$ is

$$\bigcup_{\in T} \{g(t)\} \times \Delta_b(\boldsymbol{n}(t)) = \bigcup_{\boldsymbol{n} \in N^k} \bigcup_{t \in T \wedge \boldsymbol{n}(t) = \boldsymbol{n}} \{g(t)\} \times \Delta_b(\boldsymbol{n}) = \bigcup_{\boldsymbol{n} \in N^k} \alpha(\boldsymbol{n}) \times \Delta_b(\boldsymbol{n}),$$

so 57 is proved once we show that $R \in M$. We showed above that $r_{t,g(t)}$ is a $b$-symmetric subset of $M$, hence also $R$ is a $b$-symmetric subset of $M$. As is easily seen, $\mathrm{Sp}(R)$ is the union of the supports of the domain and range of $R$, i.e., $\mathrm{Sp}(R) = \mathrm{Sp}(y) \cup \mathrm{Sp}\left(\bigcup_{\boldsymbol{n}\in N^k} \alpha(\boldsymbol{n}) \times \Delta_b(\boldsymbol{n})\right) \subseteqq At_b$, by 51; hence, by 45, $R \in M$.

59. If $y \in M$ is such that for all $t \in M$ if $t \approx^{(M)} y$ then $|\mathfrak{a}(t)| \geqq |\mathfrak{a}(y)|$, and also $\mathfrak{a}(y) = b = \{b_1, \dots, b_k\}$, $k \neq 0$ and $z \approx^{(M)} y$ then $b_k \in \mathfrak{a}(z)$.

*Proof.* By 58 $y \approx^{(M)} \bigcup_{\boldsymbol{n}\in N^k} \alpha(\boldsymbol{n}) \times \Delta_b(\boldsymbol{n})$. Since, as easily seen, $\mathfrak{a}\left(\bigcup_{\boldsymbol{n}\in N^k} \alpha(\boldsymbol{n}) \times \Delta_b(\boldsymbol{n})\right) \subseteqq b = \mathfrak{a}(y)$ and since $\mathfrak{a}(y)$ is minimal we can assume, without loss of generality, $y = \bigcup_{\boldsymbol{n}\in N^k} \alpha(\boldsymbol{n}) \times \Delta_b(\boldsymbol{n})$. If $\boldsymbol{n} \in N^k$ is such that $n_k = 0$ then let us denote $\langle n_1, \dots, n_{k-1}\rangle$ by $\boldsymbol{n}'$ and $\{b_1, \dots, b_{k-1}\}$ by $b'$ and we have, obviously, $\Delta_b(\boldsymbol{n}) = \Delta_{b'}(\boldsymbol{n}')$. If $\alpha(\boldsymbol{n}) = 0$ for every $\boldsymbol{n} \in N^k$ such that $n_k \neq 0$ then $y = \bigcup_{\boldsymbol{n}\in N^k} \Delta_{b'}(\boldsymbol{n}')$ but $\mathfrak{a}\left(\bigcup_{\boldsymbol{n}\in N^k} \Delta_{b'}(\boldsymbol{n}')\right) = b' \neq b = \mathfrak{a}(y)$, contradiction. Therefore there is an $\boldsymbol{n} \in N^k$ such that $n_k \neq 0$ and $\alpha(\boldsymbol{n}) \neq 0$. Suppose $b_k \notin \mathfrak{a}(z)$ and let $w$ be an atom such that $w > b_k$ and if $\mathfrak{a}(z)$ has members beyond $b_k$ then $w$ is between $b_k$ and the next larger member of $\mathfrak{a}(z)$. Let $\sigma \in G_{\mathfrak{a}(z)\cup\{b_1,\dots,b_{k-1}\}}$ be such that $\sigma(b_k) = w$. Let $f \in M$ be a one-one function mapping $z$ on $y$ then, since $\sigma \in G$ is an automorphism of $M$, $\sigma(f)$ is a one-one function mapping $\sigma(z) = z$ on $\sigma(y)$ which is $\bigcup_{\boldsymbol{n}\in N^k} \alpha(\boldsymbol{n}) \times \Delta_{b^*}(\boldsymbol{n})$, where $b^* = \{b_1, \dots, b_{k-1}, w\}$. Let $\boldsymbol{n}$ be such that $n_k \neq 0$ and $\alpha(\boldsymbol{n}) \neq 0$ then $\{0\} \times \Delta_{b^*}(\boldsymbol{n}) \subseteqq \sigma(y)$. Let $u_1, \dots, u_{n-1}$ be fixed atoms such that

$$u_1 < \cdots < u_{n_1} < b_1 < u_{n_1+1} < \cdots < u_{n_1+n_2} < b_2 < \cdots < b_{k-1} < \cdots < u_{n-1} < w$$

and let $s = \max(b_k, u_{n-1})$. The set $\{\langle 0, \{u_1, \dots, u_{n-1}, x\}\rangle \mid s < x < w\}$ is a subset of $\{0\} \times \Delta_{b^*}(\boldsymbol{n}) \subseteqq \sigma(y)$ and it is obviously in $M$. It is also obviously equinumerous in $M$ to $\{x \mid s < x < w\}$. Since $\sigma(y) \approx^{(M)} z \approx^{(M)} y$, $\{x \mid s < x < w\}$ is equinumerous in $M$ to some subset of $y$. Let $g \in M$ be a one-one function mapping $\{x \mid s < x < w\}$ into $y$. Since $g \in M$, $g$ is $c$-symmetric for some $c \in B$. Let $u, v \in At$ be such that $s < u < v < w$, $u, v \notin c$, and $u, v$ are not separated by any member of $c$. There is a $\sigma \in G_c$ such that $\sigma(x) = x$ for $x \leqq b_k$ and $\sigma(u) = v$. $g(u) = \{t_1, \dots, t_n\} \subseteqq At_b$, thus $\langle u, \{t_1, \dots, t_n\}\rangle \in g$ and $\langle \sigma(u), \{\sigma(t_1), \dots, \sigma(t_n)\}\rangle \in \sigma(g)$; but $\sigma(g) = g$, $\sigma(t_i) = t_i$ for $1 \leqq i \leqq n$, and $g(v) = g(\sigma(u)) = \{\sigma(t_1), \dots, \sigma(t_n)\} = \{t_1, \dots, t_n\} = g(u)$, contradicting the one-one-ness of $g$.

60. If $x \in M - At$ then $\|x\|$ is a subset of some member of $M$.

*Proof.* Given $x \in M$, $\|x\|$ is a non-void subset of $M$, let $y_0 \in \|x\|$. If also $y \in \|x\|$ then $\varrho(y) = \varrho(y_0)$ and, by 59, either $\mathfrak{a}(y) = \mathfrak{a}(y_0) = 0$ or else $\mathfrak{a}(y)$ and $\mathfrak{a}(y_0)$ have the same last member; in either case $At_{\mathfrak{a}(y)} = At_{\mathfrak{a}(y_0)}$. By 51 we have $\mathrm{Sp}(y) \subseteqq At_{\mathfrak{a}(y)} = At_{\mathfrak{a}(y_0)}$, hence $\|x\| \subseteqq \{y \mid y \in M \wedge \varrho(y) = \varrho(y_0) \wedge \mathrm{Sp}(y) \subseteqq At_{\mathfrak{a}(y_0)}\}$, where the right-hand side is in $M$ by 56.

61. The relativizations to $M$ of the instances of schema 42 are provable.

*Proof.* Let $z \in M$ and assume that for all sets $x, y \in M$ $\tau^{(M)}(z, x) = \tau^{(M)}(z, y)$ if and only if $x \approx {}^{(M)}y$. Since $z \in M$ there is a $u \in At$ such that $\mathrm{Sp}(z) \subseteqq At_u$. Let $v$ be an atom $> u$, let $r, s$ be arbitrary atoms such that $u < r \leqq v < s$, and let $\xi$ be a permutation of $At$ which interchenges $r$ and $s$ and is the identity otherwise. By 12 and 26 $\xi$ is a permutation of $M$ which preserves membership and the property of being an atom. Therefore, as in the proof of 31, $\xi$ preserves all the formulas relativized to $M$ which do not contain the symbol $\mathfrak{a}$ and $\xi$ commutes with all such terms; in particular, for $\tau$'s as in schema 42

$$p, q \in M \rightarrow \xi(\tau^{(M)}(p, q)) = \tau^{(M)}(\xi(p), \xi(q)).$$

By 27 $\xi(z) = z$. Since, obviously, $\xi(At_v) = (At_v - \{r\}) \cup \{s\} \approx {}^{(M)}At_v$ we have $\tau^{(M)}(z, At_v) = \tau^{(M)}(z, \xi(At_v)) = \tau^{(M)}(\xi(z), \xi(At_v)) = \xi(\tau^{(M)}(z, At_v))$. Therefore, if $r \in \mathrm{Sp}(\tau^{(M)}(z, At_v))$ then also

$$s = \xi(r) \in \xi(\mathrm{Sp}(\tau^{(M)}(z, At_v))) = \mathrm{Sp}(\xi(\tau^{(M)}(z, At_v))) = \mathrm{Sp}(\tau^{(M)}(z, At_v)),$$

by 23; but $s$ is an arbitrary atom $> v$ hence $\mathrm{Sp}(\tau^{(M)}(z, At_v)) \supseteqq At - At_v$, contradicting $\tau^{(M)}(z, At_v) \in M$. Thus we have proved that if $u < r \leqq v$ then $r \notin \mathrm{Sp}(\tau^{(M)}(z, At_v))$. Suppose that for some $r > v$ $r \in \mathrm{Sp}(\tau^{(M)}(z, At_v))$, let $s$ be an atom $> r$ and let $\sigma \in G$ be the identity on $At_v$ and $\sigma(r) = s$, then we get

$$\begin{aligned} s = \sigma(r) \in \sigma(\mathrm{Sp}(\tau^{(M)}(z, At_v))) &= \mathrm{Sp}(\sigma(\tau^{(M)}(z, At_v))) \\ &= \mathrm{Sp}(\tau^{(M)}(\sigma(z), \sigma(At_v))) \\ &= \mathrm{Sp}(\tau^{(M)}(z, At_v)) \end{aligned}$$

hence, $\mathrm{Sp}(\tau^{(M)}(z, At_v)) \supseteqq At - At_r$, contradiction. Therefore we have $\mathrm{Sp}(\tau^{(M)}(z, At_v)) \subseteqq At_u$. Now, let $w > v$ and let $\sigma \in G$ be a map which is the identity on $At_u$ and maps $v$ on $w$. Then, by 27 and

$$\mathrm{Sp}(\tau^{(M)}(z, At_v)) \subseteqq At_u,$$

we get

$$\tau^{(M)}(z, At_v) = \sigma(\tau^{(M)}(z, At_v)) = \tau^{(M)}(\sigma(z), \sigma(At_v)) = \tau^{(M)}(z, At_w)$$

and hence, by our assumption, $At_v \approx {}^{(M)}At_w$. Let $f \in M$ be a $b$-symmetric function which maps $At_v$ on $At_w$. There is a $\sigma \in G_b$ which is the identity on $At_v$ and $\sigma(t) \neq t$ for some $t < w$; thus if $\langle r, t\rangle \in f$ where $r < v$ then also $\langle r, \sigma(t)\rangle = \langle \sigma(r), \sigma(t)\rangle \in \sigma(f) = f$. Since $\sigma(t) \neq t$ this contradicts our assumption that $f$ is a function.

## § 4. $|x|$ Can be Undefinable while it is Relatively Definable

As mentioned in § 1 we shall prove here that if ZF, or ZF′, is consistent so is the theory obtained from ZF′ by adding the axiom

62. $\exists z\, \forall x\, \forall y (\tau'(z, x) = \tau'(z, y) \leftrightarrow x \approx y)$

where $\tau'(z, x)$ is a particular formula of ZF′ which will be given in 71 below, and the schema

63. $\neg\, \forall x\, \forall y (\tau(x) = \tau(y) \leftrightarrow x \approx y)$

where $\tau(x)$ is any term of ZF with no free variables other than $x$.

As in § 3, if ZF is consistent so is the theory ZF** of § 3. Within ZF** we construct a model of ZF′ which is, again, very similar to that of Mostowski [6]. Since there are $\aleph_0$ atoms we can write them as $(n, r)$, where $n < \omega$ and $r$ is a rational number. Let $G$ be the group of all permutations $\sigma$ of $At$ such that $\sigma(n, r) = (n, \sigma_1(r))$, where $\sigma_1$ is an order-preserving permutation of the rationals; let $B$ be the set of all finite sets of atoms; and let $G_b$, for $b \in B$, be the subgroup of $G$ consisting of all $\sigma$'s such that $\sigma(u) = u$ for each $u \in b$. The notions of an element $x$ being *b-symmetric*, *symmetric* and *hereditarily symmetric* are as in 44. $M$ is taken to be the class of all hereditarily symmetric elements $x$ such that $\mathrm{Sp}(x)$ contains atoms $(n, r)$ for only finitely many $n$'s. We put $A_n = \{(n, r) \mid r \in R\}$, where $R$ is the set of all rational numbers, and $\mathfrak{b}(x) = \{n \mid \mathrm{Sp}(x) \cap A_n \neq 0\}$. For $x \in M$ $\mathfrak{b}(x)$ is a finite set.

64. Let $\lambda$ be a permutation of $\omega$, then the permutation $\xi$ of $At$ given by $\xi(n, r) = (\lambda(n), r)$ is an automorphism of $M$, and for every element $x$ we have $\mathfrak{b}(\xi(x)) = \lambda(\mathfrak{b}(x))$.

It is easy to prove all the relativizations to $M$ of the axioms of ZF′. For the axiom schema of replacement one proves first, using 64, the analogue of 34 and 53, namely:

65. Let $\psi(x_1, \ldots, x_n, y)$ be a formula of ZF′ with no free variables other than indicated. If $x_1, \ldots, x_n \in M$ and for these $x_1, \ldots, x_n$ there is exactly one $y$ such that $\psi^{(M)}(x_1, \ldots, x_n, y)$ then $\mathfrak{b}(y) \subseteq \bigcup_{i=1}^{n} \mathfrak{b}(x_i)$.

Now we proceed to show that 62 holds when relativized to $M$. We define a function $\eta$ on all elements $x$ by recursion on the membership relation as follows:

66. $\eta(n, r) = \{\{n, (0, r)\}\}$, and if $y$ is a set $\eta(y) = \{\eta(x) \mid x \in y\}$.

67. For every $x$ $\eta(x)$ is a set which contains no atoms.

*Proof.* It is clear from 66 that $\eta(x)$ is always a set, hence if $y$ is a set $\eta(y) = \{\eta(x) \mid x \in y\}$ contains no atoms. $\eta(n, r)$ does obviously contain no atoms.

68. For all elements $x, y$ $\eta(x) = \eta(y) \leftrightarrow x = y$, and $\eta(x) \in \eta(y) \leftrightarrow x \in y$.

*Proof.* We have, of course, to prove only the left to right direction. First suppose that $y$ is an atom, say $y = (n, r)$, then $\eta(y) = \{\{n, (0, r)\}\}$. If $\eta(x) = \eta(y)$ then if $x$ is an atom $(n', r')$ then $\eta(y) = \eta(x) = \{\{n', (0, r')\}\}$ and we have $n = n'$, $r = r'$, i.e., $x = y$; $x$ cannot be a set since then $\eta(x) = \eta(y) = \{\{n, (0, r)\}\}$ would entail that for some $z \in x$ $\eta(z) = \{n, (0, r)\}$ contradicting 67. For no $x$ can we have $\eta(x) \in \eta(y)$ since then $\eta(x) = \{n, (0, r)\}$, again contradicting 67. We shall now carry out the proof by transfinite induction on $\alpha = \max(\varrho(x), \varrho(y))$ (see 54). Assume $\eta(x) = \eta(y)$. If one of $x$ and $y$ is an atom then, as we have seen, $x = y$. If both are sets then let $z \in x$ then $\eta(z) \in \eta(x) = \eta(y)$, hence $\eta(z) = \eta(u)$ for some $u \in y$. $z \in x$ and $u \in y$ entail $\varrho(z) < \varrho(x)$, $\varrho(u) < \varrho(y)$ and therefore by the induction hypothesis $\eta(z) = \eta(u)$ implies $z = u$, i.e., $z \in y$. Thus we proved $\forall z(z \in x \rightarrow z \in y)$, i.e., $x \subseteqq y$ and in the same way we get also $y \subseteqq x$. Now assume $\eta(x) \in \eta(y)$, then, as we saw, $y$ is not an atom, therefore $\eta(x) = \eta(u)$ for some $u \in y$ $\varrho(u) < \varrho(y)$ hence $\varrho(x), \varrho(u) \leqq \alpha$ and by the induction hypothesis and the part we have already proved we have $x = u$, i.e., $x \in y$.

69. For every $x$, $\mathfrak{b}(\eta(x)) = \{0\}$.

*Proof.* By induction on the membership relation.

70. $x \in M \wedge \sigma \in G \rightarrow \eta(\sigma(x)) = \sigma(\eta(x))$.

*Proof.* By induction on the membership relation

71. If $x \in M$ also $\eta(x) \in M$ und if $x$ is a set then $\eta(x) \approx^{(M)} x$.

*Proof.* By induction on the membership relation. If $x \in At$ then $\eta(x) \in M$, by 36 and 20. If $x$ is a set then, by the induction hypothesis, $y \in x \rightarrow \eta(y) \in M$, hence, by 20, $\langle y, \eta(y)\rangle \in M$. Let $f = \{\langle y, \eta(y)\rangle \mid y \in x\} \subseteqq M$. Since $y \in M$ $y$ is $b$-symmetric for some $b \in B$. Let $\sigma \in G_b$; for $y \in x$ we have $\sigma(y) \in x$ and $\sigma(\langle y, \eta(y)\rangle) = \langle \sigma(y), \sigma(\eta(y))\rangle = \langle \sigma(y), \eta(\sigma(y))\rangle \in f$, by 70. Thus we have shown $\sigma \in G_b \rightarrow \sigma(f) \subseteqq f$ and hence, by 14, $\sigma(f) = f$. $\mathfrak{b}(f) = \mathfrak{b}(\mathfrak{D}(f)) \cup \mathfrak{b}(\mathfrak{R}(f)) = \mathfrak{b}(x) \cup \mathfrak{b}(\eta(x)) = \mathfrak{b}(x) \cup \{0\}$, by 69. Therefore, by the analogue of 17 and 45, $f \in M$ and $\eta(x) = \{\eta(y) \mid y \in x\} \in M$. $f$ is a one-one function by 68, hence $x \approx^{(M)} \eta(x)$.

72. We take now $\tau'(z, x)$ to be the term "*the set of all sets $y$ of least rank among the sets $y \approx x$ whose supports are included in $z$*". It is easily shown in ZF′ that for every set $z$ and ordinal $\alpha$, the lements $y$ whose support is included in $z$ and whose rank is $\alpha$ constitute a set. Therefore we have, in ZF′, as easily seen: If $x$ and $y$ are equinumerous to sets supported by subsets of $z$ then $\tau'(z, x) = \tau'(z, y) \leftrightarrow x \approx y$. Since in our

interpretation every set $x$ is equinumerous to the set $\eta(x)$ supported by $A_0$ (by 68 and 70), we have

$$\forall x\, \forall y\,(x, y \in M \to \tau'(A_0, x) = \tau'(A_0, y) \leftrightarrow x \approx^{(M)} y).$$

73. The relativization of 63 to $M$ holds.

*Proof.* Assume that $\tau^{(M)}(x) = \tau^{(M)}(y) \leftrightarrow x \approx y$. Let $u \in R$ and let $A_{n,u} = \{(n, r) \mid r \leqq u\}$. Obviously $A_{n,u} \in M$. For some fixed $u$ let $m \in \mathfrak{b}(\tau^{(M)}(A_{0,u}))$. Let $p$ be any member of $\omega$, let $\lambda$ be a permutation of $\omega$ such that $\lambda(m) = p$, and let $\xi$ be a permutation of $At$ given by $\xi(n, r) = (\lambda(n), r)$. Since $m \in \mathfrak{b}(\tau^{(M)}(A_{0,u}))$ we have by 64,

$$\begin{aligned} p = \lambda(m) \in \lambda(\mathfrak{b}(\tau^{(M)}(A_{0,u}))) &= \mathfrak{b}(\xi(\tau^{(M)}(A_{0,u}))) \\ = \mathfrak{b}(\tau^{(M)}(\xi(A_{0,u}))) &= \mathfrak{b}(\tau^{(M)}(\xi(A_{\lambda(0),u})). \end{aligned}$$

For every $k < \omega$ we have, as easily seen, $A_{k,u} \approx^{(M)} A_{0,u}$ hence $\tau^{(M)}(A_{\lambda(0),u}) = \tau^{(M)}(A_{0,u})$ and $p \in \mathfrak{b}(\tau^{(M)}(A_{\lambda(0),u})) = \mathfrak{b}(\tau^{(M)}(A_{0,u}))$. Thus we have shown $p < \omega \to p \in \mathfrak{b}(\tau^{(M)}(A_{0,u}))$, i.e., $\omega \subsetneqq \mathfrak{b}(\tau^{(M)}(A_{0,u}))$, contradicting $\tau^{(M)}(A_{0,u}) \in M$. Therefore we have $\mathfrak{b}(\tau^{(M)}(A_{0,u})) = 0$, hence $\mathrm{Sp}(\tau^{(M)}(A_{0,u})) = 0$. Let $\sigma_1$ be a permutation of $R$ such that $\sigma_1(u) = v \neq u$, let $\sigma(n, r) = (n, \sigma_1(r))$, $\sigma \in G$. Since $\mathrm{Sp}(\tau^{(M)}(A_{0,u})) = 0$ we have, by 27, $\sigma(\tau^{(M)}(A_{0,u})) = \tau^{(M)}(A_{0,u})$. On the other hand, $\sigma(\tau^{(M)}(A_{0,u})) = \tau^{(M)}(\sigma(A_{0,u})) = \tau^{(M)}(A_{0,v})$, therefore $A_{0,u} \approx^{(M)} A_{0,v}$, where $u \neq v$; but this leads to a contradiction, as at the end of the proof of 61.

## References


1. Asser, G.: Theorie der logischen Auswahlfunktionen. Z. math. Logik Grundlagen Math. **3**, 30–68 (1957).
2. Fraenkel, A. A., and H. Bar-Hillel: Foundation of set theory. Amsterdam: North Holland Publishing Co. 1958.
3. Lévy, A.: The interdependence of some consequences of the axiom of choice. Fundamenta Mathematicae **54**, 135–157 (1964).
4. Mendelson, E.: The idenpendence of a weak axiom of choice. J. Symbolic Logic **21**, 350–366 (1956).
5. — A semantic proof of the eliminability of descriptions. Z. math. Logik Grundlagen Math. **6**, 199–200 (1960).
6. Mostowski, A.: Über die Unabhängigkeit des Wohlordnungssatzes vom Ordnungsprinzip. Fundamenta Mathematicae **32**, 201–252 (1939).
7. Scott, D.: Definitions by abstraction in axiomatic set theory (abstract). Bull. Amer. Math. Soc. **61**, 442 (1955).
8. Specker, E.: Zur Axiomatik der Mengenlehre (Fundierungs- und Auswahlaxiom). Z. math. Logik Grundlagen Math. **3**, 173–210 (1957).
9. Suppes, P.: Introduction to logic. Princeton: D. Van Nostrand Co. 1957.
10. Tarski, A.: General principles of induction and recursion in axiomatic set theory (abstract). Bull. Amer. Math. Soc. **61**, 442–443 (1955).

# The Use of Symbolic Logic in Proving Mathematical Theorems by Means of a Digital Computer

BERNARD MELTZER

## Introduction

Unbeknown to most mathematicians and others, the last decade has been marked by the initiation of a major revolution in the development of mathematics and implicitly of the sciences and other disciplines which depend on it. For the advent of large-storage and high-speed digital computers has now made the implementation of LEIBNIZ's proposal for a universal calculus for carrying out mathematical proofs by rules partially, as we shall see, realizable.

That this was on the agenda of the development of science was already clear round about the beginning of this century, when in RUSSELL and WHITEHEAD's "Principia Mathematica" [1] and other systems the raw material for the formalization of mathematics had been provided in detailed fashion. For when these systems had been converted into pure syntactical axiomatic structures by HILBERT and his school, the way was open to mechanize their syntactical rules of transformation (inference) for the proof of theorems — and it only required an enormous increase in power of data-processing compared with that of human beings to apply these methods to interesting mathematical theorems, instead of only to simple and often rather trivial theorems of the pure predicate calculus, as done in "Principia Mathematica" and elsewhere. This has been one of the functions of the digital electronic computer.

The following is a small arbitrary selection of some of the theorems that have already been proved in the last decade by *systematic* computer programmes:

1) The existence of a right identity element in any algebra closed under a binary associative operation having left and right solutions.

2) Any group, the squares of whose elements are equal to the identity, is commutative.

3) The square root of a prime number is irrational.

4) In a ring $x \cdot 0 = 0$ for every $x$.

5) In a ring $-x \cdot -y = x \cdot y$ for every $x$ and $y$.

Although these theorems are rather elementary, it will be seen that the successful machine proof of more complex theorems is limited —

given the power and capacities of present-day computers — only by the efficiency and strategies of the systematic proof procedures which have been developed so far. The latter, it will be seen, have even in the short space of time of the last decade improved spectacularly and there is little reason to suppose that this process will not continue.

## The Basis of Proof Procedures

Most published work in this field has used the first-order predicate calculus. Thereby, it is all wholly based on two major results of logic associated with the name of the man whose anniversary is being celebrated in this symposium.

GÖDEL's completeness theorem [2] for the predicate calculus provides the assurance that every logically valid theorem will have a proof in the formal system. Furthermore most of the successful programmes have been based on HERBRAND's Theorem, which emerged from the work of SKOLEM [3], GÖDEL [2], and HERBRAND [4].

The form in which HERBRAND's Theorem is applied is as follows:

The — explicit or implicit — proper axioms and special hypotheses $A_1, A_2, \ldots A_n$ and the desired conclusion $C$, are expressed in terms of atomic predicates, and the formula

$$A_1 \,\&\, A_2 \ldots A_n \,\&\, \bar{C}$$

is cast into conjunctive normal form, with existential quantifiers eliminated by means of Skolem functions. The resulting conjoined disjuncts of negated and unnegated atomic predicates are usually refered to as *clauses*. Theorem-proving programmes are designed to prove the unsatisfiability of this conjunction of clauses.

We form the Herbrand universe by taking the set of all constants appearing in the clauses and all terms which can be formed from constants and the function symbols appearing in the clauses. Then HERBRAND's Theorem states that the set of clauses is unsatisfiable if, and only if, some *finite* set of instantiations of the clauses over this universe is truth-functionally contradictory.

## Ground Level Methods

These methods, which were the earliest ones used, consisted merely in generating larger and larger sets of instantiated or *ground* clauses, testing at each stage for truth-functional contradiction. The programme schema is essentially that shown in the figure, which is self-explanatory. The first proposed method, that of GILMORE [5] was to convert the set at every stage into disjunctive normal form, for which the test for contradiction consisted merely in checking whether in each conjunct two complementary predicates, i.e. predicates which are negations of each other, occurred. The amount of symbol manipulation involved in this conversion is so large that on an IBM 704 GILMORE's procedure was unsuccessful even when applied to the following theorem of the pure predicate calculus:

$$(Ex)(Ey)(z)\{[F(x,y) \rightarrow (F(y,z) \mathbin{\&} F(z,z))] \\ \mathbin{\&} [(F(x,y) \mathbin{\&} G(x,y)) \rightarrow (G(x,z) \mathbin{\&} G(z,z))]\}$$

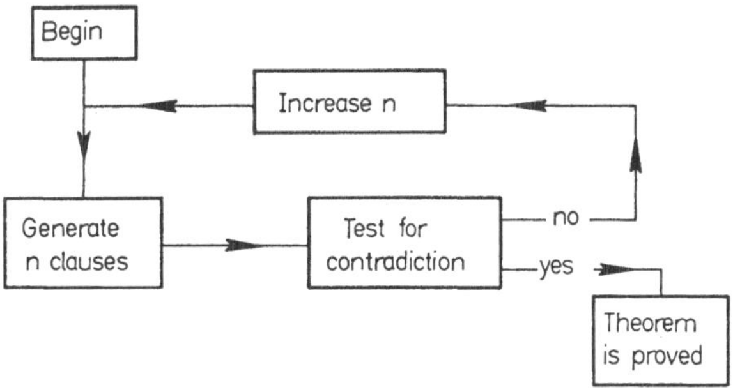


Flow chart for theorem-proving by use of ground clauses

An enormous increase in efficiency was effected by DAVIS and PUTNAM [6], who avoided the conversion to disjunctive normal form by devising a set of rules (three of them) which could be applied to a set of ground clauses, and involved eliminating predicates and clauses in a systematic manner, so that one arrived at two clauses which consisted of single predicates complementary to each other — if the set tested was truth-functionally contradictory. These rules follow from the theorem:

If $S$ is a set of ground clauses and $P$ a particular atomic predicate, and one forms first the set $R_1$, which results from eliminating from $S$ all clauses containing $P$ and all occurrences of $\bar{P}$, and the set $R_2$, which results from eliminating from $S$ all clauses containing $\bar{P}$ and all occurrences of $P$, then $S$ is contradictory if, and only if, $R_1$ and $R_2$ are both contradictory.

On the Gilmore example the Davis and Putnam procedure converged after a mere thirty minutes of *hand* computation.

In spite of this improvement, an investigation by ROBINSON [7] showed that for quite elementary theorems in group theory, for example, even the Davis-Putnam procedure would quite soon have to call on astronomically large sets of clauses. The rather irritating feature of the situation was that even though by the systematic methods used one had to generate these vast quantities of ground clauses, the actual contradiction arose only from a relatively very small set of them, maybe only four or ten. ROBINSON pointed out that this was the key issue in further development of these methods: namely, how to select the crucial contradictory set without going through the "British Museum" process of turning out vast numbers of irrelevant clauses.

## Prawitz's Idea

In a rather obscure paper [8] PRAWITZ indicated that the selection of the contradictory set of ground clauses could be effectively carried out without the systematic generation of large numbers of clauses one after the other. The idea was, given a set of non-instantiated clauses $S$, one would form a relatively limited number of replicas of $S$, each differing from $S$ only by the replacement of variables by other variables. If now one could find out what substitutions of terms for variables in this set of replicas of $S$ would produce such identifications of atomic predicates in the various clauses as to yield a truth-functional contradiction, the job of proving the theorem was done.

To illustrate this consider the theorem 1) mentioned above in the Introduction:

"In any algebra closed under a binary associative operation having left and right solutions $x$ and $y$ for all equations $x \cdot a = b$ and $a \cdot y = b$, where $a$ and $b$ are elements of the algebra, there is an element which is a right identity."

If $P(x, y, z)$ is equivalent to stating that $x \cdot y = z$, then the set $S$ of clauses which have to be shown to be unsatisfiable is the following:

(1) $\bar{P}(x, y, u)\, \bar{P}(y, z, v)\, \bar{P}(x, v, w)\, P(u, z, w)$,

(2) $\bar{P}(x, y, u)\, \bar{P}(y, z, v)\, \bar{P}(u, z, w)\, P(x, v, w)$,

(3) $P(g(x, y), x, y)$,

(4) $P(x, h(x, y), y)$,

(5) $P(x, y, f(x, y))$,

(6) $\bar{P}(k(x), x, k(x))$.

(1) and (2) express the associativity of the operation, (3) the existence of a right solution, (4) the existence of a left solution, (5) the property of closure and (6) the denial of the existence of a right identity element.

It is incidentally interesting that while, as we shall see, only four ground clauses are required to show the unsatisfiability, ROBINSON [7] calculated that in the most straightforward way of generating the ground clauses from the Herbrand universe approximately $6 \times 10^{11}$ ground clauses would have to be generated!

The only Herbrand constants actually required are

$$a, \qquad h(a, a), \qquad k(h(a, a)), \qquad g(a, k(h(a, a))),$$

and the four ground clauses required to produce a contradiction are the following

$$(1') \quad \bar{P}(g(a, k(h(a, a))), a, k(h(a, a)))\, \bar{P}(a, h(a, a), a)$$
$$\bar{P}(g(a, k(h(a, a))), a, k(h(a, a)))\, P(k(h(a, a)), h(a, a), k(h(a, a))),$$
$$(3') \quad P(g(a, k(h(a, a))), a, k(h(a, a))),$$
$$(4') \quad P(a, h(a, a), a),$$
$$(6') \quad \bar{P}(k(h(a, a)), h(a, a), k(h(a, a))).$$

It is easy to see what substitutions needed to have been made in the original set $S$ to have provided the four replicas of it from which this contradictory set of four clauses has been selected.

It will be noted for example how by appropriate substitutions the single atomic predicate of (3) in $S$ has been identified with the first atomic predicate appearing in (1) of $S$.

PRAWITZ's idea has been discussed in a very interesting way by ROBINSON [9] and his suggestion is amenable to exploitation to a greater degree than has yet occurred.

## J. A. Robinson's Resolution Method

Guided by PRAWITZ's insight J. A. ROBINSON [10] has developed a method, based on what he calls a machine-oriented logic, which seems to be the first practical method of proving theorems on the basis of HERBRAND's theorem without actually resorting to the examination of ground clauses — it operates entirely at the free-variable level.

This method is based on a "unification algorithm" which essentially is a procedure for making the identification of predicates required by PRAWITZ. In addition a single rule of inference is used for the effective test of unsatisfiability.

Resolution operates on two clauses at the free-variable level: if by the algorithm a substitution is found which transforms a set of predicates in one of the clauses into a single predicate, and transforms a set of predicates in the other clause into the negation of the same predicate, then a new clause is inferred or generated, which consists of the transform (under the same substitution) of all the remaining predicates in the two parent clauses. ROBINSON's basic result is that the original set is unsatisfiable if and only if, when resolution is carried out on the clauses as they are generated, the empty clause is produced.

A good deal of work has now been done in many institutions, using the resolution principle — a lot of it unpublished as yet. Subsidiary strategies, some of them very greatly improving the efficiency of the basic procedure, have been developed.

ROBINSON [10] himself discussed one or two in his original paper.

WOS, G. A. ROBINSON and CARSON [11, 12] have used what they call the unit preference and set of support strategies. The former is based on using in resolution preferentially single-predicate clauses and going to progressively longer and longer clauses only when required. The set of support strategy is based, roughly speaking, on using resolutions whose ancestry can be traced back to precisely the special hypotheses and the denial of the theorem rather than only to the basic axioms of the theory.

Later, ROBINSON [13] developed a more powerful method which he termed "hyper-resolution", which processes more than two clauses and is based on the following remarkable discovery:

If a set $S$ of clauses is unsatisfiable then the empty clause can be generated by a series of resolutions in every one of which one parent consists entirely of positive, i.e. non-negated, predicates. ROBINSON termed deductions based on such resolutions P1-deductions. MELTZER [14] showed that P1-deduction was a special case of a large number of deduction types, an optimum choice from which should be possible for a given problem. He also indicated an alternative approach to automatic deduction based on the renaming of predicates by means of their negations.

It is interesting to note that the resolution method in its original form, using ROBINSON's so-called "purity" and "replacement" supplementary strategies [10], when applied to the Gilmore example given above, converged in 6 stages; however, only 2 "hyper-resolutions" were required.

## Final Remarks

The computer programs for theorem proving are, most of them, written not in conventional arithmetic-oriented languages but in list-processing languages such as LISP invented by JOHN MCCARTHY [15] and his group at M. I. T. In fact some of the motivation for the development of these languages came indeed from the requirements of theorem-proving and logic — it is interesting that MCCARTHY incorporated CHURCH's [16] lambda notation used in the theory of recursive functions.

Finally it may be pointed out that the procedures which have been described above have the potential of partially fulfilling LEIBNIZ's dream of a universal mechanical calculus for mathematics. They are complete procedures in the sense that, when confronted with a proposition that is a theorem, they are in principle capable of showing that it is a theorem — and only limitations of computing speed and storage capacity could prevent this. But of course, CHURCH's theorem [17] assures us that they are not universally effective procedures in that, when confronted with a non-theorem they will not in general be able to demonstrate this.

## References

1. Russell, B., and A. N. Whitehead: Principia mathematica, Vols. I—III. Cambridge: University Press 1910—1913.
2. Gödel, K.: Die Vollständigkeit der Axiome des logischen Funktionenkalküls. Monatsh. Math. Phys. **37**, 349—360 (1930).
3. Skolem, T.: Über einige Grundlagenfragen der Mathematik. Skrifter utgibt av Det Norske Videnskaps-Academi i Oslo, I. Matematisk-naturvidenskapelig klasse, Number 4 (1929).
4. Herbrand, J.: Recherches sur la theorie de la demonstration. Travaux de la Societé des Sciences et des Letters de Varsovie, Classe III sciences mathematiques et physiques, Number 33 (1930).
5. Gilmore, P. C.: A proof method for quantification theory. I.B.M. J. Res. Develop. **4**, 28—35 (1960).
6. Davis, M., and H. Putnam: A computing procedure for quantification theory. J. Ass. Comp. Mach. **7**, 201—215 (1960).
7. Robinson, J. A.: Theorem-proving on the computer. J. Ass. Comp. Mach. **10**, 163—174 (1963).
8. Prawitz, D.: An improved proof procedure. Theoria **26**, 102—139 (1960).
9. Robinson, J. A.: A review of automatic theorem-proving. Proceedings of Symposia in Applied Mathematics, Mathematical Aspects of Computer Science, Vol. 19. American Mathematical Society, 1967
10. — A machine-oriented logic based on the resolution principle. J. Ass. Comp. Mach. **12**, 23—41 (1965).
11. Wos, L., D. Carson, and G. Robinson: The unit preference strategy in theorem proving. A.F.I.P.S. Conference Proceedings 26, pp. 615—621. Washington, D.C.: Spartan Books 1964.
12. — G. A. Robinson, and D. F. Carson: Efficiency and completeness of the set of support strategy in theorem proving. J. Ass. Comp. Mach. **12**, 536—541 (1965).
13. Robinson, J. A.: Automatic deduction with hyper-resolution. Int. J. Computer Math. **1**, 227—234 (1965).
14. Meltzer, B.: Theorem-proving for computers: some results on resolution and renaming. Computer J. **8**, 341—343 (1966).
15. McCarthy, J., et al.: LISP I.5 Programmer's Manual. Cambridge, Mass.: M.I.T. Press 1962.
16. Church, A.: The calculi of lambda-conversion. Princeton: University Press 1951.
17. Kleene, S. C.: Introduction to metamathematics. Amsterdam: North-Holland Publishing Co. 1964.

## Note Added in Proof

There has been considerable progress in the development of this discipline since the present review was written more than two years ago. Two of the highlights are the following:

The first reported contribution of a computer theorem-proving program to the solution of an open mathematical problem (in lattice theory) is given in „CRT-aided semi-automated mathematics“, a report by J. H. Bennett et al., Applied Logic Corporation, Princeton, USA, 1967.

A very powerful generalisation of resolution, and a most illuminating conceptual framework for considering the problems of automatic theorem-proving, are presented in „The generalised resolution principle“ by J. A. Robinson, published in Machine Intelligence **3**, edited by D. Michie, Edinburgh University Press, 1968.

# A New Approach to the Foundations of Probability

Herman Rubin*

In the history of the theory of probability, until recently there was no suitable foundational system on which probability calculations could be based which would yield satisfactory results. The first formulation which was adequate for general probability computations, and the only one now in wide use[1], is that of Kolmogoroff. However, we claim that it is not altogether satisfactory, and we shall give an alternative formulation, and show the relationships and differences of the two approaches.

In the "prehistory" of probability, events were usually considered as sentences. Since sentences do not have the formal properties of a system adequate for probability, Kolmogoroff naturally turned to the types of systems in mathematical use which possess many of the properties of sentences, namely, set algebra and Boolean algebra. While Boolean algebra seems to handle events reasonably well, random variables have not been satisfactorily treated from a Boolean algebra approach. The other possibility, that of set algebra, was the one exploited by Kolmogoroff.

However, there are philosophical difficulties and even slight mathematical difficulties with this approach. Briefly, let us review the Kolmogoroff formalization. First, one starts with a sample space. It is only recently that authors have seriously allowed the possibility that the sample space may not be unique; as we shall see, its non-uniqueness makes some of the definitions imprecise. An *event* is defined to be a subset of the sample space; it is hard to see how, for example, the certain event, which is defined to be the whole point-set of the sample space, can be the same if sample spaces have different point-sets, as they usually do when different sample spaces are considered. Also, a *random variable* is defined to be a measurable function[2] defined on a sample space; that is, it is a fixed function on a fixed space, and is consequently neither random nor a variable. Aside from these philosophical difficulties, there is a mathe-

* Research supported by ONR Contract Nonr-2587(02).

[1] An exception to this can be found in Lehmann [4].

[2] Kolmogoroff defines a random variable to be real-valued. In probability literature, various extensions are made — in fact, authors feel free to define random variables with any desired type of range in any particular argument. We follow Blackwell and Girshick [1] in placing no restriction on the range of a random variable.

matical difficulty. In many cases, it is desired to use a random variable independent of a given set of random variables. If one does not exist, it is customary to utilize a product space, with the new random variable a function of the new coordinate, and the original random variables function of the original coordinates. Since the domain of definition of the random variables is now changed[3], the "original" random variables are not even functions on the new sample space. We shall attempt to give a formulation of probability theory in the von Neumann-Bernays-Gödel set theory[4] which avoids all of these difficulties, and preserves all the theorems of the Kolmogoroff formulation. We shall obtain our formulation by obtaining properties from the Kolmogoroff formulation which will then serve as an axiom system, and add one additional axiom.

If we consider a given sample space, the events form, by assumption, a $\sigma$-Boolean algebra. Consequently, we postulate.

**Axiom 1.** *The operations of disjunction and conjunction are defined on certain subsets of the class of all events, including all countable subsets, the result yielding an event, and negation carries events into events. There is a certain event $C$ and an impossible event $M$.*

**Axiom 2.** *Every set of events is contained in a $\sigma$-Boolean algebra of events, with the supremum operation corresponding to disjunction, the infimum operation to conjunction, the complementation operation to negation, the zero to $M$, and the unit to $C$.*

Next we consider random variables. Since a random variable is a measurable function in the Kolmogoroff formulation, to each random variable there is a field. Also for any set in the field, the statement that the random variable is in that set is an event, and this correspondence between sets and events is a complete homomorphism[5] of the field into the class of events. Formally

**Axiom 3.** *There exists a function whose domain is the class of random variables. For each random variable $X$, the class of sets $F(X)$ is a $\sigma$-Boolean algebra with unit.*

**Axiom 4.** *If $X$ is a random variable and $A$ is an element of $F(X)$, there is an associated event $[X \in A]$. If $\mathscr{A}$ is a class of elements of $F(X)$ whose union $B$ is an element of $F(X)$, the disjunction of the events $[X \in A]$, $A \in \mathscr{A}$, is $[X \in B]$, and the negation of $[X \in A]$ is $[X \in S \sim A]$, where $S$ is the unit of $F(X)$.*

---

[3] It is possible to concoct examples in which the domain is not changed; in those cases the function is changed.

[4] For an exposition of this system see Gödel [2].

[5] A complete homomorphism preserves suprema and infima not only of finite or countable collections of elements, but even of arbitrary collections of elements.

It follows from the properties of Boolean algebras that the corresponding requirement for conjunction is automatically satisfied.

Another property of random variables is that a measurable function of a random variable is a random variable, and the induced distribution is related to the inducing distribution by the customary measure-theoretic relation. Formally

**Axiom 5.** *If $X$ is a random variable, $\mathscr{C}$ is a $\sigma$-Boolean algebra with unit, and $f$ is an $F(X) - \mathscr{C}$ measurable function, there is a random variable $Y$ such that $F(Y) = \mathscr{C}$, and for all $A \in \mathscr{C}$, $[Y \in A] = [X \in f^{-1}[A]]$. In this case we say $Y = f(X)$.*

We have already considered one correspondence between random variables and events, namely, to each measurable subset of the range of a random variable there is an event. The characteristic function (indicator) of a set classically gives a correspondence between events and random variables. Formally,

**Axiom 6.** *There is a function $K$ such that for each event $E$, $K(E)$ is a random variable with field $\{\emptyset, \{0\}, \{1\}, \{0, 1\}\}$. The event $E$ is equal to $[K(E) \in \{1\}]$.*

Axioms 5 and 6 point out a difficulty which needs some care to resolve. In Axiom 6, if $K$ is the impossible event, the range of $K(E)$ would just be $\{0\}$ and if $K(E)$ is the certain event, 0 is not an element of the range of $K$. Furthermore, if we define $J(E)$ to be the real-valued random variable "equal to" $K(E)$, the field $F(J(E))$ would be the BOREL sets of the reals. Also, the events $[J(E) = 1]$ and $[J(E) > 1/2]$ are the same. We see that the "range" of a random variable need not be the unit element of its field, and that random variables with totally different fields can be equal. To achieve this, we associate to every set of random variables a set of functions called the range of that set of random variables and postulate its natural properties.

**Axiom 7.** *There is a function $R$ such that for every set $\mathscr{X}$ of random variables $R(\mathscr{X})$ is a set of functions on $\mathscr{X}$. Furthermore, if $\mathscr{Y} \subseteqq \mathscr{X}$, $R(\mathscr{Y})$ is the restriction to $\mathscr{Y}$ of the elements of $R(\mathscr{X})$ and if $X \in \mathscr{X}$, $Y \in \mathscr{X}$, $Y = f(X)$, and $S \in R(\mathscr{X})$, $S(Y) = f(S(X))$. In addition, $A \cap R\{X\} = B \cap R\{X\}$ if and only if $[X \in A] = [X \in B]$.*

Another desirable property of the Kolmogoroff model is that random variables can be combined. We may accomplish this end by

**Axiom 8.** *Let $\mathscr{X}$ be a set of random variables. Then there is a random variable $Y$ and a function $G$ such that for all $X \in X$, $\mathscr{X} = G(X)(Y)$.*

We have now found enough properties to show that we have achieved an axiomatization of the Kolmogoroff model. We have not yet introduced probability, but the probability structure in the classical model is that of a measure on a measurability space, and everything except the actual value of probability merely involves the measurability concepts.

**Theorem 1.** *Let $\mathscr{E}$ be a set of events and $\mathscr{X}$ be a set of random variables. Then there is a random variable $Y$ such that to each event $E$ of $\mathscr{E}$ there is associated an element $E'$ of $F(Y)$, to each random variable $X$ of $\mathscr{X}$ there is associated an $F(Y) - F(X)$ measurable function $X'$ as in Axiom 6, and the correspondence between the events $[Y \in A]$ and the sets $A \in F(Y)$ is a complete Boolean isomorphism.*

The proof of this theorem is an immediate consequence of the axioms, with the complete homomorphism properties postulated in Axiom 8 enabling the complete isomorphism to be obtained. Thus we see that any set of random variables and events has a Kolmogoroff representation. Furthermore, if we characterize probability by

**Axiom 9.** *Probability is a non-negative, countably additive realvalued function $P$ defined on the class of events such that $P(C) = 1$, this representation also extends to the inclusion of probability.*

So far, we have achieved nothing more than an abstract axiomatization of a system which has a Kolmogoroff representation, at least when sets are considered. This is of the same type as the formal axiomatization of Desarguian projective geometry, which can be shown to be equivalent to that of homogeneous linear spaces over a division ring. However, no geometer would *define* Desarguian projective geometry by this characterization. But we can do more. In the customary manner, we may introduce the concept of independence. Then we can postulate

**Axiom 10.** *Let $\mathscr{E}$ be a set of events, and let $(Z, \mathscr{B}, m)$ be a measure space with $m(Z) = 1$. Then there is a random variable $X$ independent of each element of $\mathscr{E}$, with field $\mathscr{B}$, such that for every $A \in \mathscr{B}$, $P([X \in A]) = m(A)$.*

**Theorem 2.** *From Axioms 1—10, we can conclude that the events form a proper class.*

That this is so follows from the fact that if $m$ and $A$ are chosen so that $m(A)$ is neither 0 nor 1, $[X \in A]$ cannot be in the set $\mathscr{E}$ of Axiom 10.

This possibility is not present in the Kolmogoroff formulation as all events there must correspond to elements of a fixed set of sets.

We have shown that Axioms 1—10 imply the existence of proper classes. We now show that the axioms are consistent with von Neumann-Bernays-Gödel set theory.

**Theorem 3.** *In any model of von Neumann-Bernays-Gödel set theory, there exist classes $\mathscr{A}$ (events) and $\mathscr{V}$ (random variables), operations $\vee$*

*(disjunction),* $\wedge$ *(conjunction), and* $'$ *(negation), sets* $C$ *and* $M$, *functions* $F$, $R$, $K$, $P$, *and* [ ], *satisfying Axioms* 1—10.

*Proof*. Let $\mathcal{N}$ be the class of all probability spaces, i.e., triples $(Z, \mathcal{B}, m)$ with $\mathcal{B}$ a $\sigma$-Boolean algebra of subsets of $Z$ with unit $Z$ and $m$ a countably additive function on $\mathcal{B}$. Let $\mathcal{W}$ be the collection of sub*sets* of $\mathcal{N}$ containing a given $N \in \mathcal{N}$. For each $W \in \mathcal{W}$, we form the product set, product $\sigma$-Boolean algebra, and product measure. For any $W_1, W_2 \in \mathcal{W}$, $W_1 \subsetneqq W_2$, and any subset $B$ of the product set of $W_1$, there is a natural extension of $B$ to a subset of the product set of $W_2$, and any function on the product set of $W_1$ can be extended to a function on the product set of $W_2$. Now let $\mathcal{A}$ be the collection of all pairs $(W, B)$, where $B$ is an element of the product $\sigma$-algebra of $W$ which is not an extension from a smaller element of $\mathcal{W}$, and let $\mathcal{V}$ be the set of all pairs $(f, \mathcal{F})$ where $f$ is a $\mathcal{C} - \mathcal{F}$ measurable function on the product set $H$ of some element $W$ of $\mathcal{W}$ with product $\sigma$-algebra $\mathcal{C}$, and $f$ is not a natural extension for some smaller element of $\mathcal{W}$. Also set $M = (\{N\}, \emptyset)$ and $C = (\{N\}, S)$, where $S$ is the space of $N$. The operation $\vee$ ($\wedge$), is the natural one obtained by extending, if necessary, all sets to an appropriate element of $\mathcal{W}$ and noting if the set obtained by union (intersection) is an extension of a set in a product $\sigma$-field of a (possibly) smaller element of $\mathcal{W}$. The operation $'$ corresponds to complementation in the appropriate space. The function $F$ is defined by $F(f, \mathcal{F}) = \mathcal{F}$. The function $K$ is defined to be the indicator function in the appropriate space. The event $[(f, \mathcal{F}) \in A]$, where $f$ is defined on the product set of $W$, is the minimal pair $(W^*, B)$ such that $f^{-1}(A)$ is the extension of $B$ to the product set of $W$. The function $K$ is the natural range obtained by extending the functions to a suitable space, and $P(W, B)$ is the $W$-product measure of $B$.

It is easy to verify that this construction yields a model of Axioms 1—10 in the given model of set theory.

There are other features of this formulation which are at best difficult in the traditional model. For example, one can impose the restriction that all distributions are perfect. Also, one can consider situations in which the operations of conjunction and disjunction have smaller ranges, as in quantum mechanics or where there is a choice of experiments.

## References

1. BLACKWELL, D., and M. A. GIRSHICK: Theory of games and statistical decisions. New York: Wiley 1954.
2. GÖDEL, K.: The consistency of the continuum hypothesis. Ann. Math. Studies No. 3, Princeton: 1940.
3. KOLMOGOROFF, A.: Grundbegriffe der Wahrscheinlichkeitsrechnung. Ergebn. Mathematik **2**, 3 (1933).
4. LEHMANN, E. L.: Testing statistical hypotheses. New York: Wiley 1959.

# Measure-Theoretic Uniformity

Gerald E. Sacks *, 1

Here we present the principal ideas and results of [5] with some indications of proof. The general notion of uniformity is difficult to harvest; nonetheless, various offshoots of it have borne fruit in all fields of mathematical logic. In this paper we introduce the notion of measure-theoretic uniformity, and we describe its use in recursion theory, hyperarithmetic analysis, and set theory. In recursion theory we show that the set of all sets $T$ such that the ordinals recursive in $T$ are the recursive ordinals has measure 1. In set theory we obtain all of Cohen's independence results [1, 2]. Solovay [8, 9] has extended Cohen's method by forcing statements with closed, measurable sets of conditions rather than finite sets of conditions; in this manner he exploits the concepts of forcing and genericity to prove: if $ZF$ is consistent, then $ZF +$ "there exists a translation-invariant, countably additive extension of Lebesgue measure defined on all sets of reals" + "the dependent axiom of choice" is consistent. Solovay's result is also a consequence of the notion of measure-theoretic uniformity.

We begin with the simplest possible example of measure-theoretic uniformity. Let $T$ be an arbitrary set of natural numbers, and let $P$ be the power set of the natural numbers. We think of $P$ as the product of countably many copies of a two-point set $\{a, b\}$. We assign the unbiased measure: $m(\{a, b\}) = 1$, $m(\{a\}) = m(\{b\}) = \frac{1}{2}$, and $m(\varphi) = 0$. We give $P$ the induced product measure denoted by $u$. Of course $u$ is merely Lebesgue measure for the unit interval in a thin disguise.

Let $R(T, x, y)$ be a recursive predicate of the set-variable $T$ and the number variables $x$ and $y$. A familiar uniformity can be expressed as follows: if for some given $T$ we have $(x)(Ey)R(T, x, y)$, then there exists a function $f$ recursive in the given $T$ such that $(x)R(T, x, f(x))$; in addition, there is a method of computing $f$ from $T$ which is independent of $T$. Before we introduce the measure-theoretic counterpart of this

* The preparation of this paper was supported by U.S. Army Contract ARO-D-373. The author wishes to thank Professor Anil Nerode for many helpful conversations on uniformity and definability in set theory.

[1] This paper is a slightly amplified version of a paper of the same name reprinted with permission of the publisher from the Bulletin of the American Mathematical Society; pp. 169—174; 

uniformity, we must shift our point of view from Skolem functions to bounding functions in order to make the measure come out right: if for some given $T$ we have $(x)(Ey)R(T, x, y)$, then there exists a function $f$ recursive in $T$ such that $(x)(Ey)_{y \leqq f(x)} R(T, x, y)$. Note that the existence of a Skolem function is equivalent to the existence of a bounding function. It is not hard to verify: if $\{T \mid (x)(Ey)R(T, x, y)\}$ has measure 1, then $\{T \mid (Ef)(f \text{ recursive } \& (x)(Ey)_{y \leqq f(x)} R(T, x, y))\}$ has measure 1. Thus the restriction of the bounding function $f$ to the recursive functions resulted merely in a restriction of $T$ to a set of measure 1. Theorem 1 is our simplest, tidy example of measure-theoretic uniformity.

**Theorem 1.** *Let $B(T, x, y)$ be arithmetical. Then the set of all $T$ satisfying the following condition has measure 1: if $(x)(Ey)B(T, x, y)$, then there exists an arithmetical function $f$ such that $(x)(Ey)_{y \leqq f(x)} B(T,x,y)$.*

**Corollary 2.** *Let $B(T)$ be arithmetical. If the set $\hat{T}B(T)$ has positive measure, then $B(A)$ holds for some arithmetical $A$* [2].

The proof of Theorem 1 turns on the fact that the measure of the set $\hat{T}B(T, x, y)$ is an arithmetical function of $x$ and $y$.

Let $T_0, T_1, T_2, \ldots$ be an arbitrary sequence of sets of natural numbers. For all results below prior to Theorem 8, we define $\mathcal{M}(T_0, T_1, T_2, \ldots)$ as follows: let $\mathcal{M}_0(T_0, T_1, T_2, \ldots)$ be the set of all sets arithmetical in some finite subsequence of $T_0, T_1, T_2, \ldots$; for each recursive ordinal $\alpha > 0$, let $\mathcal{M}_\alpha(T_0, T_1, T_2, \ldots)$ be the set of all sets "hyperarithmetic" in some finite subsequence of $T_0, T_1, T_2, \ldots$ with the set-quantifiers of the hyperarithmetic definitions restricted to

$$\bigcup \{\mathcal{M}_\beta(T_0, T_1, T_2, \ldots) \mid \beta < \alpha\};$$

finally, let

$$\mathcal{M}(T_0, T_1, T_2, \ldots) = \bigcup \{\mathcal{M}_\alpha(T_0, T_1, T_2, \ldots) \mid \alpha < \omega_1\},$$

where $\omega_1$ is the least non-recursive ordinal. We put a probability measure on sequences of sets of natural numbers by putting the product measure on the product of countably many copies of $P$, the power set of the natural numbers. FEFERMAN [3] proves that if $T_0, T_1, T_2, \ldots$ is a generic sequence, then $\mathcal{M}(T_0, T_1, T_2, \ldots)$ is a model of the $\Sigma_1^1$-axiom of choice (and hence of the hyperarithmetic comprehension axiom). The set of all generic sequences has measure 0.

**Theorem 3.** *With probability 1: $\mathcal{M}(T_0, T_1, T_2, \ldots)$ is a model of the $\Sigma_1^1$-axiom of choice and no $T_i$ is hyperarithmetic in any finite sequence of the remaining $T_j$'s.*

---

[2] (added in proof) Corollary 2 was independently obtained by H. TANAKA [11].

SPECTOR [10] proved the existence of two incomparable hyperdegrees by observing that the set $\{(T_0, T_1) \mid T_0$ and $T_1$ are hyperarithmetically incomparable$\}$ has measure 1. Theorem 3 is a consequence of the measure-theoretic uniformity expressed by Theorem 4. The origin of Theorem 4 is a result of KREISEL [4]: if $B(T, x, y)$ is $\pi_1^1$, and if for some given $T$ we have $(x)(Ey)B(T, x, y)$, then there exists a function $f$ hyperarithmetic in the given $T$ such that $(x)B(T, x, f(x))$.

**Theorem 4.** *Let $B(T, x, y)$ be $\pi_1^1$. The set of all $T$ satisfying the following condition has measure* 1: *if $(x)(Ey)B(T, x, y)$, then there exists a hyperarithmetic function $f$ such that $(x)(Ey)_{y \leqq f(x)} B(T, x, y)$.*

**Corollary 5.** *Let $B(T)$ be $\pi_1^1$. If the set $\hat{T} B(T)$ has positive measure, then $B(H)$ holds for some hyperarithmetic $H$*[3].

**Corollary 6.** *With probability* 1, *the ordinals recursive in $T$ are the recursive ordinals.*

**Corollary 7.** *For every $T$, if the set $\{X \mid T$ is hyperarithmetic in $X\}$ has positive measure, then $T$ is hyperarithmetic.*

Now let $\mathcal{M}$ be a countable, initial segment of the constructible sets which is a model of $ZF$. (It appears necessary that $\mathcal{M}$ be countable in order to apply the notion of measure-theoretic uniformity. On the other hand, forcing with infinite sets of conditions can be applied to uncountable models.) For each $T$, let $\mathcal{M}(T)$ be the set of all sets constructible from $T$ by means of the ordinals of $\mathcal{M}$. Let $\mathcal{M}_0(T) = \{T\}$; for each $\alpha > 0$ in $\mathcal{M}$, let $\mathcal{M}_\alpha(T)$ be the set of all subsets of $\bigcup\{\mathcal{M}_\beta(T) \mid \beta < \alpha\}$ definable by means of some first-order formula of $ZF$ with constants in, and quantifiers restricted to, $\bigcup\{\mathcal{M}_\beta(T) \mid \beta < \alpha\}$. Then

$$\mathcal{M}(T) = \bigcup\{\mathcal{M}_\beta(T) \mid \beta \in \mathcal{M}\}.$$

FEFERMAN [3] provides a language $\mathcal{L}$ suitable for dissecting $\mathcal{M}(T)$. $\mathcal{L}$ includes the usual logical symbols, the $\in$-symbol for membership, numerals for all the ordinals in $\mathcal{M}$, a $\mathcal{T}$-symbol to denote the set $T$, abstraction symbols, and restricted quantifier symbols $(Ex)_{x<\alpha}$ and $(x)_{x<\alpha}$. The intended meaning of "$(x)_{x<\alpha}$" in $\mathcal{M}(T)$ is "for all $x$ in $\bigcup\{\mathcal{M}_\beta(T) \mid \beta < \alpha\}$". Each member of $\mathcal{M}(T)$ corresponds in a natural manner to some formula of $\mathcal{L}$, with all quantifiers bounded, which defines it.

Let $f$ be a function from the ordinals of $\mathcal{M}$ into the ordinals of $\mathcal{M}$. We say $f$ is definable in $\mathcal{M}$ if there exists a formula $F(x, y)$ of $ZF$ with constants in $\mathcal{M}$ such that for all $\alpha, \beta \in \mathcal{M}$, we have $f(\alpha) = \beta \leftrightarrow \models_{\mathcal{M}} R(\alpha, \beta)$.

[3] (added in proof) H. TANAKA [12] has found another proof of Corollary 5 in which he quotes Corollary 6 buts avoids Theorem 4.

Theorem 8 expresses the key measure-theoretic uniformity we associate with the constructible hierarchy of $\mathscr{M}$.

**Theorem 8.** *Let $B(x, y)$ be a formula of $\mathscr{L}$. Then the set of all $T$ satisfying the following condition has measure* 1: *if $\models_{\mathscr{M}(T)} (x)(Ey) B(x, y)$, then there exists a function $f$ definable in $\mathscr{M}$ such that for all $\alpha$ in $\mathscr{M}$,*

$$\models_{\mathscr{M}(T)} (x)_{x<\alpha} (Ey)_{y<f(\alpha)} B(x, y).$$

Any instance of measure-theoretic uniformity is obtained by showing that the probability of truth for an appropriate class of statements is definable in some model associated with that class of statements. For each sentence $\mathscr{F}$ of $\mathscr{L}$, we observe that the set $\{T \mid \models_{\mathscr{M}(T)} \mathscr{F}\}$ is Borel, since $\mathscr{M}$ is countable; we define $p(\mathscr{F})$, the probability that $\mathscr{F}$ is true in $\mathscr{M}(T)$, to be the measure of this Borel set.

**Theorem 9.** *For each $k \geqq 0$, the function $p(\mathscr{F})$, restricted to sentences of $\mathscr{L}$ having at most $k$ unbounded quantifiers, is definable in $\mathscr{M}$.*

An intuitive word about the proof of Theorem 9. Imagine an observer A at large in the universe V who computes $p(\mathscr{F})$. From the point of view of A, the set $\{T \mid \models_{\mathscr{M}(T)} \mathscr{F}\}$ is Borel and its measure is computed by regarding it as the union or intersection of an expanding or contracting, countable sequence of Borel sets of lower rank. An observer B, trapped in $\mathscr{M}$, can with some pain regard $\{T \mid \models_{\mathscr{M}(T)} \mathscr{F}\}$ as the union or intersection of an expanding or contracting, transfinite sequence of sets of lower rank. A computes measure by taking limits of monotone, countable sequences. B computes measure by taking limits of monotone, transfinite sequences. A transfinite induction shows A and B obtain the same answer, and at the same time, defines $p(\mathscr{F})$.

Theorem 8 follows from Theorem 9 easily. Let

$$p(\alpha, \beta) = p((x)_{x<\alpha} (Ey)_{y<\beta} B(x, y)).$$

By Theorem 8, $p(\alpha, \beta)$ is definable in $\mathscr{M}$. For each $\alpha$, $p(\alpha, \beta)$ is a non-decreasing, bounded function of $\beta$. But then for each $\alpha$, there is a $\gamma$ such that $p(\alpha, \gamma) =$ least upper bound of

$$\{p(\alpha, \beta) \mid \beta \in \mathscr{M}\} = p((x)_{x<\alpha} (Ey)\, \beta(x, y));$$

let the least such $\gamma$ be $f(\alpha)$.

It follows from Theorem 8 that the replacement axiom holds in $\mathscr{M}(T)$ with probability 1. Both COHEN [1, 2] and SOLOVAY [8, 9] show the replacement axiom holds in their models by observing that the forcing relation, restricted to sentences with at most $k$ unbounded quantifiers ($k \geqq 0$), is definable in $\mathscr{M}$. We use Theorem 9 instead.

**Theorem 10.** *With probability one, $\mathscr{M}(T)$ is a model of $ZF + V \neq L$ and the cardinals of $\mathscr{M}(T)$ are the same as those of $\mathscr{M}$.*

Let $T_0, T_1, T_2, \ldots$ be an arbitrary sequence of sets of natural numbers. $\mathscr{M}(T_0, T_1, T_2, \ldots)$ is defined like $\mathscr{M}(T)$, save for some slight complications designed to maximize the symmetry of $\mathscr{M}(T_0, T_1, T_2, \ldots)$. Let $\mathscr{L}$ be augmented by symbols $\mathscr{T}_0, \mathscr{T}_1, \mathscr{T}_2, \ldots$; Theorems 8 and 9 remain true. FEFERMAN [3] showed that if $T_0, T_1, T_2, \ldots$ is a generic sequence, then $\mathscr{M}(T_0, T_1, T_2, \ldots)$ is a model of $ZF$ in which the Boolean algebra of all sets of natural numbers has no non-principal, maximal ideals.

**Theorem 11.** *With probability* 1, $\mathscr{M}(T_0, T_1, T_2, \ldots)$ *is a model of* $ZF$ *in which the Boolean algebra of all sets of natural numbers has no non-principal, maximal ideals.*

The proof of Theorem 11 uses only one trick not used to prove Theorem 10. We formulate this trick as Lemma 12. Let $\tau$ be an arithmetical function from $\omega^2$ into $\{0, 1\}$; $\tau$ induces a transformation of the sentences of $\mathscr{L}$ as follows: "$n \in \mathscr{T}_j$" is replaced by

$$\text{"}(n \in \mathscr{T}_j \;\&\; \tau(n, j) = 0) \vee (n \notin \mathscr{T}_j \;\&\; \tau(n, j) = 1)\text{"}.$$

In a similar fashion, $\tau$ transforms all sequences of sets of natural numbers.

**Lemma 12.** $$\{(T_0, T_1, T_2, \ldots) \mid \models_{\mathscr{M}(T_0, T_1, T_2, \ldots)} \tau(\mathscr{F})\} = \tau(\{T_0, T_1, T_2, \ldots) \mid \models_{\mathscr{M}(T_0, T_1, T_2, \ldots)} \mathscr{F}\}),$$

*for each arithmetical function* $\tau$.

Lemma 12 is FEFERMAN's transformation lemma [3] with "forcing" replaced by "truth" and "generic sequence" replaced by "arbitrary sequence". A useful consequence of Lemma 12 is Lemma 13.

**Lemma 13.** *Let* $\mathscr{F}$ *be a sentence of* $\mathscr{L}$ *whose only* $\mathscr{T}_i$*-symbols are* $\mathscr{T}_0, \mathscr{T}_1, \ldots, \mathscr{T}_n (n \geqq 0)$. *Let* $T_0, T_1, \ldots, T_n$ *be an arbitrary finite sequence of sets of natural numbers. Then the conditional probability that* $\mathscr{F}$ *is true in* $\mathscr{M}(\mathscr{T}_0, \mathscr{T}_1, \mathscr{T}_2, \ldots)$, *given that* $\mathscr{T}_i = T_i$ *for all* $i \leqq n$, *is either* 0 *or* 1.

Let $X^\omega$ be a variable of $\mathscr{L}$ restricted to sets of natural numbers in $\mathscr{M}(T_0, T_1, T_2, \ldots)$, and let $B(X^\omega)$ be a formula of $\mathscr{L}$ with all quantifiers bounded. For each formula $B(X^\omega)$ and each sequence $T = (T_0, T_1, T_2, \ldots)$, we define the *absolute measure* of $\hat{X}^\omega B(X^\omega)$, denoted by $\mu_a^T(X^\omega B(X^\omega))$, as follows: let $\mathscr{T}_B$ be some $\mathscr{T}_i$-symbol not occurring in $B(X^\omega)$; then $\mu_a^T(X^\omega B(X^\omega))$ is the conditional probability that $B(\mathscr{T}_B)$ is true, given that $\mathscr{T}_i = T_i$ for every $\mathscr{T}_i$-symbol occurring in $B(X^\omega)$. We say $\mathscr{M}(T_0, T_1, T_2, \ldots)$ is *measure-complete* if for every pair of formulas $B_0(X^\omega)$ and $B_1(X^\omega)$, if $\hat{X}^\omega B_0(X^\omega) = \hat{X}^\omega B_1(X^\omega)$ is true in $\mathscr{M}(T_0, T_1, T_2, \ldots)$, then $\mu_a^T(\hat{X}^\omega B(X^\omega)) = \mu_a^T(\hat{X}^\omega B_1(X^\omega))$.

Thus, in a measure-complete model, we can associate absolute measure with sets rather than definitions of sets.

**Lemma 14.** *With probability one, $\mathcal{M}(T_0, T_1, T_2, \ldots)$ is a measure-complete model of ZF in which the absolute measure is definable.*

SOLOVAY [9] has results virtually identical with Theorem 11 and Lemma 13 save that $T_0, T_1, T_2, \ldots$ is always a sequence generic in his sense.

**Lemma 15.** *With probability one, the absolute measure agrees with Lebesgue measure on all Lebesgue-measurable sets, is translation-invariant, and is countably additive.*

**Lemma 16.** *With probability one, the dependent axiom of choice holds in $\mathcal{M}(T_0, T_1, T_2, \ldots)$.*

The result of SOLOVAY [9] we quoted on the first page of this paper follows from Lemmas 14, 15, and 16. The translation-invariance of the absolute measure follows from some elementary facts about independent, uniformly distributed, random variables. The countable additivity of the absolute measure follows from the measure-completeness and the countable additivity of Borel measure. In short, the sets of reals in a countable, measure-complete model form a good copy of a certain countable family of Borel sets in the "real" world. We wish to stress that this last fact has no vital connection with the ideas of forcing and genericity.

Let $\mathcal{M}$ be a countable initial segment of the constructible sets which is a model of $ZF$. We define the fundamental equivalence type of a one-element extension of $\mathcal{M}$, namely $\mathrm{FET}_1(\mathcal{M})$, with the aid of the $0-1$ law. Let $\mathcal{F}$ be a sentence of $ZF$. Then the Borel set $\{T \mid \models_{\mathcal{M}(T)}\}$ has measure 0 or 1 by the $0-1$ law, since $\mathcal{F}$ has no constants[4]. Let $\mathrm{FET}_1(\mathcal{M})$ consist of all $\mathcal{F}$ such that $\{T \mid \models_{\mathcal{M}(T)} \mathcal{F}\}$ has measure 1; as we saw above, $\mathrm{FET}_1(\mathcal{M})$ includes $ZF + V \neq L$. All Cohen-generic extensions $\mathcal{M}(T)$ have the same elementary equivalence type, and that type is not $\mathrm{FET}_1(\mathcal{M})$. In [6] we will define the fundamental equivalence types of general, countable models, and we will discuss measure-theoretic uniformity in a wider context.

Call $T$ fundamental if $\mathcal{M}(T)$ has the fundamental equivalence type. Several persons have observed that if $T$ is fundamental, then there exists a Solovay-generic $S$ such that $\mathcal{M}(T) = \mathcal{M}(S)$. Another way of expressing this fact is: a set is fundamental if and only if it has the same degree of non-constructibility as some Solovay-generic set. When we pass from sets to degrees, we pass from forcing to measure-theoretic

[4] Professor SIMON KOCHEN suggested the use of the $0-1$ law here in place of Lemma 13.

uniformity. In [7] Scott and Solovay provide an interesting general theory which makes possible an accurate comparison between the forcing approach and the measure-theoretic uniformity approach.

## References


1. Cohen, P. J.: The independence of the continuum hypothesis I. Proc. nat. Acad. Sci. **50**, 1143—1148 (1963).
2. — The independence of the continuum hypothesis II. Proc. nat. Acad. Sci. **51**, 105—110 (1964).
3. Feferman, S.: Some applications of the notions of forcing and generic sets. Fund. Math. **56**, 325—345 (1965).
4. Kreisel, G.: The axiom of choice and the class of hyperarithmetic functions. Indag. Math. **24**, 307—319 (1962).
5. Sacks, G. E.: Measure-theoretic uniformity in recursion theory and set theory. Trans. Amer. Math. Soc., to appear.
6. — On the fundamental equivalence type of a countable model. In preparation.
7. Scott, D., and R. Solovay: Boolean-valued models and forcing. To appear.
8. Solovay, R.: The measure problem, Abstract 65T-62. Not. Amer. Math. Soc. **12**, 217 (1965).
9. — The measure problem. To appear.
10. Spector, C.: Measure-theoretic construction of incomparable hyperdegrees. J. Symb. Log. **23**, 280—288 (1958).
11. Tanaka, H.: Some results in the effective descriptive set theory. Publ. RIMS, Kyoto Univ. Ser A, 3 II—52 (1967).
12. — A basis result for $\Pi_1^1$— sets of positive measure. Mimeographed notes, Hosei University 1967.

# On the Cardinality of $\mathbf{\Sigma}_2^1$ Sets of Reals[1]

ROBERT M. SOLOVAY[2]

What are the possible cardinalities of subsets of the reals? $\boldsymbol{R}$ itself has power $2^{\aleph_0}$; it is trivial to exhibit subsets of the integers of any power $\leq \aleph_0$. The Continuum Hypothesis of CANTOR conjectures that these examples exhaust the possible cardinalities of subsets of $\boldsymbol{R}$.

GÖDEL [7] and COHEN [4] have shown that the truth or falsity of the continuum hypothesis cannot be decided on the basis of the usual axioms of set-theory, including the axiom of choice. Using techniques introduced by LEVY and SOLOVAY [12], one can show in a systematic way that even if the usual axioms are enlarged by various sorts of "large cardinal" axioms, the continuum hypothesis remains undecidable.

It is natural to expect that the cardinality of fairly simple subsets of $\boldsymbol{R}$ will be easier to determine. In fact, this is the case. Recall that a set of reals, $K$, is *perfect* if it is compact, non-void, and every point of $K$ is an accumulation point of $K$. CANTOR showed that every perfect set has power $2^{\aleph_0}$. Moreover, the Cantor-Bendixon theorem asserts that every closed subset of $\boldsymbol{R}$ is countable or the union of a perfect set and a countable set. Thus no closed set has a cardinal strictly between $\aleph_0$ and $2^{\aleph_0}$.

We now review the results of classical descriptive set theory on the cardinality of projective sets of reals. We shall use the modern notation of ADDISON [2] and SHOENFIELD [18] for the levels of the analytic hierarchy. Recall, in particular, that lightface (e.g. $\Sigma_1^1$) means "without parameters"; bold face (e.g. $\mathbf{\Sigma}_1^1$) means "parameters allowed". Thus, e.g., there are $\aleph_0$ $\Sigma_1^1$ sets of reals, but $2^{\aleph_0}$ $\mathbf{\Sigma}_1^1$ sets of reals.

The classical results are as follows ([13]): Every uncountable $\mathbf{\Sigma}_1^1$ set of reals contains a perfect subset. Every $\mathbf{\Pi}_1^1$ set of reals is the disjoint union of $\aleph_1$ Borel sets. It follows that every $\mathbf{\Sigma}_1^1$ set has power $\leq \aleph_0$ or $= 2^{\aleph_0}$. Since Borel sets are $\mathbf{\Sigma}_1^1$, it then follows that every $\mathbf{\Pi}_1^1$ set of reals has cardinality $\leq \aleph_1$ or $= 2^{\aleph_0}$. However, classical descriptive set theory was unable to answer the following question: Is there an uncountable

[1] The main results of this paper (Theorem 1 through 3) were obtained independently a few months later by R. MANSFIELD.

[2] The author is a Sloan Foundation fellow. This research was partially supported by National Science Foundation grant GP-5632.

$\mathbf{\Pi}^1_1$ set without perfect subsets? (A negative answer would rule out $\mathbf{\Pi}^1_1$ sets with power between $\aleph_0$ and $2^{\aleph_0}$; a positive answer would yield a $\mathbf{\Pi}^1_1$ set whose power is $\aleph_1$, even if $2^{\aleph_0} > \aleph_1$.)

It has turned out that there were deep reasons for the failure of classical descriptive set theory to settle this question. Let $ZF^*$ be Zermelo-Fraenkel set theory (including the axiom of choice).

1) GÖDEL [6] showed that it is consistent with $ZF^*$ to assume a) the existence of an uncountable $\Pi^1_1$ set without perfect subsets and b) $2^{\aleph_0} = \aleph_1$.

2) Using the models constructed by COHEN [4], VOPENKA et al. [23] and the author showed the consistency of $ZF^*$ with the assumption 1) $2^{\aleph_0} > \aleph_1$, 2) there is an uncountable $\Pi^1_1$ set without perfect subsets (of power $\aleph_1$).

3) LEVY [11] constructed a model of $ZF^*$ in which every projective set of reals (or more generally, every set of reals which is ordinal-definable from a single real) is countable or has power $2^{\aleph_0}$, while $2^{\aleph_0}$ can have "any possible value" (e.g., $\aleph_2$).

4) The author has shown [22] that in LEVY's model, every uncountable projective set of reals (or more generally, every uncountable set of reals ordinal-definable from a single real) contains a perfect subset.

We now state the main results of this paper.

**Theorem 1.** *The following three propositions are equivalent:*

1) *Every uncountable* $\mathbf{\Pi}^1_1$ *set of reals contains a perfect subset.*

2) *Every uncountable* $\mathbf{\Sigma}^1_2$ *set of reals contains a perfect subset.*

3) *For each real s, only countable many reals are constructible from s.*

GAIFMAN [5] and ROWBOTTOM [17] have shown that part 3) of Theorem 1 follows from the existence of measurable cardinals. Hence we have the following corollary to Theorem 1.

**Theorem 2.** *If there is a measurable cardinal, then every uncountable* $\mathbf{\Sigma}^1_2$ *set of reals contains a perfect subset (and thus has power* $2^{\aleph_0}$*).*

The following theorem sheds light on the structure of countable $\mathbf{\Sigma}^1_2$ sets of reals.

**Theorem 3**[3]**.** *Assume that one (and hence all) of the equivalent propositions of Theorem 1 is valid. Then every countable* $\Sigma^1_2$ *set of reals consists of constructible reals. More generally, if the code (see § 1) for the* $\mathbf{\Sigma}^1_2$ *set A is* $\alpha \in N^N$, *then if A is countable, every element of A is constructible from* $\alpha$. *(Thus the reals constructible from* $\alpha$ *form the maximum countable* $\mathbf{\Sigma}^1_2$ *set with code recursive in* $\alpha$.*)*

The remainder of this paper is organized as follows. In § 1, we recall some conventions about coding various families of sets and give an

[3] This theorem was suggested by a conversation with STEVE GARLAND.

effective version of the decomposition theorem for $\mathbf{\Pi}_1^1$ sets. § 2 contains the heart of the argument. We show, assuming 3) of Theorem 1, that if a $\mathbf{\Pi}_1^1$ set $A$ contains no perfect subset, then every element of $A$ is constructible from each code for $A$. The key idea is that 3) of Theorem 1 allows us to do Cohen extensions of $L[s]$, for any real $s$. ($L[s]$ is the class of sets constructible from $s$. See [18], for example.) § 3 deduces Theorems 1—3 from the result in § 2 by means of the Kondo-Addison theorem[4].

We close this introduction with a few remarks.

1) Mansfield [14] has used Theorems 1 and 3 to prove the following theorem (assuming measurable cardinals). Let $A$ be the universal $\Sigma_1^1$ set in $\boldsymbol{R} \times \boldsymbol{R}$. (Thus every $\mathbf{\Sigma}_1^1$ set $B$ of $\boldsymbol{R}$ is obtainable as $\{t^s \langle s, t\rangle \in A\}$ for some real $s$.) Then $A$ does not lie in the $\sigma$-ring generated by rectangles whose sides are $\mathbf{\Sigma}_2^1$. This result settles a problem of Ulam.

2) Let $M$ be the proposition "there is a measurable cardinal". Then one can show, using recent work of Silver [20] that the results contained in Theorems 2 and 3 are optimal for the theory $ZF^* + M$. In fact it is relatively consistent with $ZF^* + M$ to assume that 1) there is a $\Pi_2^1$ set of reals of power $\aleph_1$ which contains no perfect subset and either 2a) $2^{\aleph_0} = \aleph_1$ or 2b) $2^{\aleph_0} > \aleph_1$. The results of Levy and the author mentioned previously also generalize to $ZF^* + M$, using the techniques of [12].

3) It seems very likely to the author that by using stronger large cardinal axioms, Theorem 2 will be considerably extended. To state the appropriate conjecture, let $L(P(\omega))$ be the smallest transitive class which is a model of $ZF$ (without choice) and which contains all ordinals and all subsets of $\omega$. (Here $\omega$ is the least infinite ordinal.) We say that a set of reals, $A$, is quasiprojective if $A \in L(P(\omega))$. (In particular, every projective set of reals is quasi-projective.) For the notion of a supercompact cardinal, we refer the reader to [15]. We conjecture that the following assertion is a theorem of $ZF^*$: If supercompact cardinals exist, every uncountable quasi-projective set of reals contains a perfect subset. Undoubtedly, Theorem 3 should also generalize to higher levels of the projective hierarchy, but the correct formulation of the generalization is still obscure.

## § 1. An effective form of the decomposition theorem for $\mathbf{\Pi}_1^1$ sets

As mentioned in the introduction, if $A$ is a $\mathbf{\Pi}_1^1$ set, then there is a more-or-less canonical decomposition of $A$ into $\aleph_1$ Borel sets:

$$A = \bigcup_{\lambda < \aleph_1} A_\lambda .$$

[4] I am extremely grateful to Shoenfield and Moschovakis for patiently expounding this theorem and its proof to me.

We wish to study to what extent this decomposition is effective. In order to discuss this, we use functions from the non-negative integers, $N$ into itself, to code up descriptions of $\mathbf{\Pi}_1^1$ sets, countable ordinals, and Borel sets. The main result of this paragraph states that if $\alpha \in N^N$ (the set of functions from $N$ into $N$) codes the $\mathbf{\Pi}_1^1$ set $A$ and $\gamma \in N^N$ codes the countable ordinal $\lambda$, then there is a code, $\beta$, for $A$, as a Borel set, which is uniformly recursive in $\alpha$ and $\gamma$.

We do not consider this result as new. It is simply a byproduct of the classical proof of the decomposition theorem, obtained by worrying about recursiveness.

1.1. Let $N$ be given the discrete topology, and let $N^N$ be given the product topology. Then $N^N$ is a complete separable metric space. Because of the canonical homeomorphism of

$$N^N \cong N^N \times N^N,$$

it is more convenient to take $N^N$ as our basic space rather than $\boldsymbol{R}$. All the results previously quoted from descriptive set theory concerning $\boldsymbol{R}$ remain valid for $N^N$. Similarly, we shall see below that the $\boldsymbol{R}$-versions of Theorems 1 through 3 follow easily from the $N^N$-versions.

1.2. A basic open set $U$ of $N^N$ is determined by a finite sequence from $N$, $\langle n_0, \ldots, n_{k-1}\rangle$ via $U = \{f \in N^N\colon f(i) = n_i,\ 0 \leqq i < k\}$. We fix a recursive enumeration of the finite sequences of elements of $N$, $s_0, s_1, \ldots$, and let $U_n$ be the basic open set determined by $s_n$.

**Definition.** *We say that $\alpha \in N^N$ codes the closed set $F \subseteqq N^N$ if 1) $\alpha\colon N \to \{0, 1\}$; 2) $\alpha(n) = 1$ iff $U_n$. is disjoint from $F$.*

It is clear that in this case there is an unique $\alpha \in N^N$ that codes a given closed set $F$.

1.3. We now introduce codes for the countable ordinals $\geqq \omega$. We say that $\alpha \in N^N$ codes the ordinal $\lambda$ if the set

$$R_\alpha = \{\langle r, s\rangle \in N \times N\colon \alpha(2^r \cdot 3^s) = 0\}$$

is a well-ordering of $N$ of order type $\lambda$.

We let $W = \{\alpha\colon \alpha \text{ codes some ordinal}\}$. $W$ is a $\Pi_1^1$ subset of $N^N$.

1.4. For codes for the Borel subsets of $N^N$, we follow [22]. We refer the reader to [22] for a detailed study of this coding. We define the relation "$\alpha$ codes $B$" recursively as follows:

1) $\alpha$ codes $U_n$ if $\alpha(0) \equiv 0 \pmod 3$ and $\alpha(1) = n$.

2) Suppose $\alpha_i$ codes $B_i$, $i = 0, 1, 2, \ldots$; then $\alpha$ codes $\bigcup_i B_i$ if $\alpha(0) \equiv 1 \pmod 3$ and $\alpha(2^a(2b+1)) = \alpha_a(b)$.

3) Suppose $\beta$ codes $B$, $\alpha(0) \equiv 2 \pmod 3$, and $\alpha(n+1) = \beta(n)$, for all $n \in N$. Then $\alpha$ codes the complement of $B$.

4) $\alpha$ codes $B$ only as required by 1)—3).

1.5. We now introduce codes for $\mathbf{\Pi}_1^1$ subsets of $N^N$ as follows: According to KLEENE [9], every $\mathbf{\Pi}_1^1$ set $A \subseteqq N^N$ has the form

$$\{\alpha \in N^N\colon\ (\gamma)(\exists n)\, T'(\bar{\alpha}(n), \bar{\beta}(n), \bar{\gamma}(n))\}.$$

Here, for $\alpha \in N^N$, and $n \in N$,

$$\bar{\alpha}(n) = \prod_{i<n} p_i^{\alpha(i)+1};$$

i.e. $\bar{\alpha}(n)$ is a sequence number that codes the first $n$ values of $\alpha$. $T'$ is a variant of KLEENE's $T$ predicate, and $\beta \in N^N$ is a suitable parameter. We let $\beta$ be the code for $A$. $A$ is $\Pi_1^1$ just in case it has a recursive code, $\beta$.

In a similar way, one can introduce codes for $\mathbf{\Sigma}_2^1$ subsets of $N^N$.

1.6. The following result is a slight reformulation of KLEENE's classical result [10] on the connection between $\mathbf{\Pi}_1^1$ sets and well-orderings.

**Lemma.** *There is a Gödel number $e \in N$ with the following properties:*

*1) Let $\{e\}^{\alpha,\beta}$ be the function partial recursive in the functions $\alpha, \beta \in N^N$ with Gödel number e. For any $\alpha$ and $\beta$ in $N^N$, $\{e\}^{\alpha,\beta}$ is everywhere defined, and maps $N$ into $\{0, 1\}$.*

*2) The relation*

$$T_{\alpha,\beta} = \{\langle r, s\rangle\colon\ \{e\}^{\alpha,\beta}(2^r \cdot 3^s) = 0\}$$

*is always a linear ordering of $N$ with last element 0.*

*3) $T_{\alpha,\beta}$ is a well-ordering iff*

$$(\gamma)(\exists n)\, T'(\bar{\alpha}(n), \beta(n), \bar{\gamma}(n)).$$

Thus if $\beta$ codes the $\mathbf{\Pi}_1^1$ subset $A$, of $N^N$ then $\alpha \in A$ if $\{e\}^{\alpha,\beta} \in W$.

We remark that since 0 is the last element of $\langle N; T_{\alpha,\beta}\rangle$, $T_{\alpha,\beta}$, if it is a well-ordering, has an order-type of the form $\lambda + 1$.

For each ordinal $\lambda < \aleph_1$, we put

$$A_\lambda = \{\alpha \in A\colon\ \{e\}^{\alpha,\beta} \text{ has order type } \lambda + 1\}.$$

Thus we get a decomposition of $A$ into $\aleph_1$ sets, $\{A_\lambda\}$, (depending on the code $\beta$ for $A$).

1.7. We shall now prove the following lemma.

**Lemma.** *$A_\lambda$ is BOREL. If $\gamma$ is a code for $\lambda$, and $\beta$ is a code for $A$ qua $\mathbf{\Pi}_1^1$ set, then there is a code $\delta \in N^N$ for $A_\lambda$ which is uniformly recursive in $\beta$ and $\gamma$.*

*Proof.* 1. Let $R$ be a linear ordering of $N$. If $j \in N$, we let $\mathrm{seg}_R(j) = \{i \in N\colon i R j\}$. We consider $\mathrm{seg}_R(j)$ as ordered by the restriction of $R$.

We define a linear ordering of $N$, $S$, as follows: 1) 0 will be the $S$-last member of $N$; 2) $(n+1) S (m+1)$ if $\gamma(2^n \cdot 3^m) = 0$. Since $\gamma$ codes $\lambda$, $\mathrm{seg}_S(0) \cong \lambda$. (Throughout this proof, $\cong$ means "is order-isomorphic to".)

If $\alpha \in N^N$, let $T_\alpha$ be the linear ordering determined by $\{e\}^{\alpha, \beta}$:

$$T_\alpha = \{\langle s, t\rangle \in N \times N\colon \{e\}^{\alpha, \beta}(2^s \cdot 3^t) = 0\}.$$

2. We put

$$B_{i,j} = \{\alpha \in N^N\colon \operatorname{seg}_{T_\alpha}(i) \cong \operatorname{seg}_S(j)\}.$$

Note that $B_{0,0} = A_\lambda$. Thus to complete the proof it suffices to establish the following

**Sublemma 1.** *There is a Gödel number $e_1$ (not depending on $\gamma$ or $\beta$) such that if*

$$f\colon N \times N \times N \to N$$

*is recursive in $\gamma$ and $\beta$ with Gödel number $e_1$, then for each $i, j \in N$ the function (using* CHURCH's *lambda notation),*

$$\lambda k f(i, j, k)\colon N \to N$$

*is a code for $B_{i,j}$ qua Borel set. (in particular, each $B_{i,j}$ is* BOREL.*)*

3. Let $\hat{S}$ be $N$ considered as an ordered set via the ordering $S$. We are going to define the function

$$f(i, j, k)$$

by transfinite induction on $j \in \hat{S}$. We begin by stating a lemma of ROGERS [16] in a form suitable for our use.

If $f\colon N\times N\times N \to N$, and $j \in N$, we define $f_j\colon N\times N\times N \to N$ as follows: If $j' S j$, $f_j(i, j', k) = f(i, j', k)$; otherwise $f_j(i, j', k) = 0$.

Thus $f_j$ describes the restriction of $f$ to $N \times \operatorname{seg}_S(j) \times N$.

**Sublemma 2.** *Suppose that $f\colon N\times N\times N \to N$. Put $g_j(i, k) = f(i, j, k)$. Suppose further that $g_j$ is recursive in $j$, $f_j$, $\beta$, $\gamma$ with a Gödel number $e_2$ not depending on $j$, $\beta$, or $\gamma$. Then $f$ is recursive in $\beta$ and $\gamma$ with Gödel number $e_1$. Moreover, $e_1$ can be effectively computed from $e_2$.*

4. The following sublemma is the effective version of the key step in the proof that $A_\lambda$ is BOREL.

**Sublemma 3.** *There is a Gödel number $e_2$ with the following property:*

*Let $j \in N$. Suppose that $h\colon N\times N\times N \to N$ is such that 1) $h(i, j', k) = 0$ unless $j' S j$; 2) if $i \in N$, and $j' S j$, then*

$$\lambda k h(i, j', k)$$

*is a Borel code for $B_{i,j'}$.*

*Let $g\colon N\times N \to N$ be the function (possibly partial) recursive in $j$, $h$, $\beta$, $\gamma$ with Gödel number $e_2$. Then $g$ is total. Moreover, for each $i \in N$,*

$$\lambda k g(i, k)$$

*is a Borel code for $B_{i,j}$.*

Sublemma 3 will suffice to complete the proof of the lemma. For let $f: \boldsymbol{N}\times\boldsymbol{N}\times\boldsymbol{N} \to \boldsymbol{N}$ be defined by transfinite induction on $j \in \hat{S}$ so that the hypotheses of Sublemma 2 hold. Then using Sublemma 3, we see that $f$ is defined everywhere and that for each $\langle i, j\rangle \in \boldsymbol{N}\times\boldsymbol{N}$,

$$\lambda k f(i, j, k)$$

is a code for $B_{i,j}$. By Sublemma 2, $f$ is uniformly recursive in $\beta$ and $\gamma$ with Gödel number $e_1$. Thus Sublemma 1 holds, and the lemma is proved.

5. We turn to the proof of sublemma 3. Note that $\alpha \in B_{i,j}$ just in case the following two assertions are true:

1) For all $i' \in \boldsymbol{N}$, if $i' T_\alpha i$, then for some $j' \in \boldsymbol{N}$, we have $j' S j$ and $\operatorname{seg}_{T_\alpha}(i') \cong \operatorname{seg}_S(j')$.

2) For all $j' \in \boldsymbol{N}$, if $j' S j$, then there is an $i' \in \boldsymbol{N}$, such that $i' T_\alpha i$ and $\operatorname{seg}_{T_\alpha}(i') \cong \operatorname{seg}_S(j')$.

Indeed, clause 1 states that each proper initial segment of $\operatorname{seg}_{T_\alpha}(i)$ is well-ordered of order type less than that of $\operatorname{seg}_S(j)$. Thus $\operatorname{seg}_{T_\alpha}(i)$ is well-ordered and

$$|\operatorname{seg}_{T_\alpha}(i)| \leqq |\operatorname{seg}_S(j)|\,.$$

Granted clause 1) holds, clause 2) amounts to the reverse inequality

$$|\operatorname{seg}_{T_\alpha}(\mathrm{i})| \geqq |\operatorname{seg}_S(j)|\,.$$

Using clause 1) and 2), we can express $B_{i,j}$ as a complicated $\sigma$-algebraic expression in the $B_{i',j'}$'s with $j'$ $S$-less than $j$. Taking advantage of the fact that infinite unions and complements of Borel sets correspond to certain recursive operations on the codes, it is now straightforward to prove Sublemma 3. We leave details to the reader.

## § 2. $\boldsymbol{\Pi}_1^1$ Sets without Perfect Subsets

2.1. We use $P$ as a shorthand notation for the following proposition: For every real $s$, there are only countably many reals constructible from $s$.

**Lemma (P).** *Let $s$ be a real. Let $\lambda$ be a countable ordinal. Then $\lambda$ has only countably many subsets in $L[s]$.*

*Proof.* Pick a real $t$ from which both $s$ and some code for $\lambda$ are constructible. Then $\lambda$ clearly has at least as many subsets in $L[t]$ as in $L[s]$. Since $\lambda$ is countable in $L[t]$, there is in $L[t]$ a $1-1$ correspondence between reals constructible from $t$ and subsets of $\lambda$ lying in $L[t]$. Thus the lemma follows from our assumption $P$.

2.2. **Theorem 4 (P).** *Let $A$ be a $\boldsymbol{\Pi}_1^1$ subset of $\boldsymbol{N}^{\boldsymbol{N}}$ with code $\alpha \in \boldsymbol{N}^{\boldsymbol{N}}$. Then either every element of $A$ is constructible from $\alpha$ or $A$ contains a perfect subset.*

The proof of Theorem 4 will be a proof by contradiction. So we assume, from now on, that $A$ contains elements not constructible from $\alpha$ but contains no perfect subset.

Let

$$A = \bigcup_{\lambda < \aleph_1} A_\lambda$$

be the canonical decomposition of $A$ into $\aleph_1$ Borel sets. Note that each $A_\lambda$ is countable. Otherwise, there would be a perfect set $P$ such that

$$P \subseteqq A_\lambda \subseteqq A .$$

We pick $\lambda_0 < \aleph_1$ such that $A_{\lambda_0}$ has members not constructible from $\alpha$.

The remainder of the proof, in outline, is the following. Let $\gamma$ be a code for $\lambda_0$. We show that

(1) $$A_{\lambda_0} \subseteqq L[\alpha, \gamma] .$$

We next observe that for certain codes $\gamma$, which we call "generic", $L[\alpha, \gamma]$ is a Cohen extension of $L[\alpha]$. Let $\mathscr{L}$ be a ramified language appropriate to describing $L[\alpha, \gamma]$ from inside $L[\alpha]$. Using (1), we construct a term $x$ of $\mathscr{L}$ which is forced to denote a member of $A_{\lambda_0}$ not constructible from $\alpha$. To finally get our contradiction, we construct a generic $\gamma$ in which $x$ denotes a real not in $A_{\lambda_0}$.

We turn to the detailed proof of Theorem 4.

2.3. **Lemma** *Let $B$ be a countable Borel set and let $\beta$ be a code for $B$. Then every member of $B$ is constructible from $\beta$. Let $\beta' \in N^N$ be such that $\beta$ is constructible from $\beta'$. Let $B'$ be the Borel set coded by $\beta$ in $L[\beta']$. Then $B' = B$.*

*Proof.* Our proof will be based on the results of [22, § 3]. The first such result states that

$$B' = B \cap L[\beta'] .$$

Thus if every element of $B$ is constructible from $\beta$, the equality $B = B'$ will follow, for arbitrary $\beta'$. Thus we may as well assume that $\beta' = \beta$. We may also assume that $B \neq \emptyset$.

Next, according to [22, § 3], $B$ is countable (resp. non-void) iff $B'$ is countable (resp. non-void) in $L[\beta]$. Thus let

(2) $$\{b_0, b_1, b_2, \ldots\}$$

be an enumeration of $B'$ (possibly with repititions) *inside* $L[\beta]$.

Let $\gamma \in L[\beta] \cap N^N$ be such that $\gamma$ codes the set $B'$ in both $L[\beta]$ and the real world. ($\gamma$ expresses $B'$ as a countable union of singletons.) The key result of [22, § 3] now states that since $\gamma$ and $\beta$ code the same Borel set in $L[\beta]$, they code the same Borel set in the real world. I.e., $B = B'$. It follows that every element of $B$ is constructible from $\beta$ and the proof is complete.

**2.4. Lemma.** *Let $F: N \to \lambda$ be a bijection. Then there is a code $\gamma$ for $\lambda$ such that*

$$L[F] = L[\gamma].$$

Moreover, we have

$$A_\lambda \subseteqq L[\alpha, \gamma].$$

*Proof.* Let $\gamma \in N^N$ be defined by

$$\gamma(2^r 3^s) = 0 \quad \text{if} \quad F(r) < F(s);$$

otherwise $\gamma(n) = 1$. Then clearly $\gamma$ is constructible from $F$.

Put

$$R = \{\langle r, s\rangle : \gamma(2^r 3^s) = 0\}.$$

Then $\langle N; R\rangle$ is a well-ordering, and $F$ is an order isomorphism between $\langle N; R\rangle$ and $\langle \lambda, <\rangle$. Thus $\gamma$ codes $\lambda$.

Now clearly $R \in L[\gamma]$, and $R$ is still a well-ordering inside $L[\gamma]$. Since $L[\gamma]$ is a model for $ZF$, there is an ordinal $\lambda'$ and an order isomorphism

(3) $$F': \langle N, R\rangle \cong \langle \lambda', <\rangle,$$

inside $L[\gamma]$. But clearly $F'$ must still give an order isomorphism in the real world. By the uniqueness of the isomorphism $F$, we have $F = F'$, $\lambda = \lambda'$. Thus $F \in L[\gamma]$. So

$$L[F] = L[\gamma].$$

The last sentence is clear from Lemma 1.7 and Lemma 2.3.

2.5. We are going to use COHEN's forcing method [4] to construct well behaved bijections

(4) $$F: N \cong \lambda_0.$$

Our ground model will be $L[\alpha]$. Inside $L[\alpha]$ we construct a ramified language $\mathscr{L}$ appropriate for extensions of the form $L[\alpha, F]$, where $F$ is as in (4). (It is worth remarking that $\lambda_0 \geqq \omega$ since $A_{\lambda_0}$ is non-empty).

We next describe the set of conditions $\mathscr{P}$. $\mathscr{P}$ consists of all functions

$$f: \text{domain}(f) \to \lambda_0$$

such that $f$ is an injection and domain$(f)$ is a finite initial segment of $N$. We order $\mathscr{P}$ by $\subseteqq$.

If $P \in \mathscr{P}$, then we say that $P$ forces '$\langle n, \lambda\rangle \in F$'' just in case $P(n) = \lambda$. We define the forcing relation for all sentences of $\mathscr{L}$ in the usual way. As discussed in [22, § 2] we may do this so that for any statement $\varphi$ of $\mathscr{L}$, either $P \Vdash \varphi$ or for some $P' \supseteqq P$, $P' \Vdash \neg\varphi$. (Here "$\Vdash$" means "forces".)

We now recall the notion of a dense subset of $\mathscr{P}$ and of a complete sequence of elements of $\mathscr{P}$: A subset $X$ of $\mathscr{P}$ is *dense* if: (1) For all $P \in \mathscr{P}$, there exists $Q \supseteqq P$ such that $Q \in X$; (2) if $P \in X$, $Q \in \mathscr{P}$, and $P \subseteqq Q$, then $Q \in X$.

An increasing sequence of conditions

$$P_0 \subseteqq P_1 \subseteqq P_2 \ldots$$

is *complete* (over $L[\alpha]$) if for each dense subset $X$ of $\mathscr{P}$, *lying in* $L[\alpha]$, we have $P_n \in X$ for all sufficiently large $n$.

It follows easily from Lemma 2.1 that there are only countably many subsets of $\mathscr{P}$ lying in $L[\alpha]$. Thus it is trivial to construct complete sequences starting at any $P \in \mathscr{P}$.

If $\{P_n\}$ is a complete sequence, then $\bigcup_n P_n$ is easily seen to be a bijection $F\colon N \cong \lambda_0$. We call an $F\colon N \cong \lambda_0$ which arises in this way a generic collapsing map for $\lambda_0$.

As discussed in [22, § 2], we can view $L[\alpha, F]$ as a Cohen extension of $L[\alpha]$. Thus if $F$ is generic, $\{P_n\}$ is a complete sequence giving rise to $F$, and $\varphi$ is a statement of $\mathscr{L}$, then $\varphi$ is true in $L[\alpha, F]$ iff $P_n \mid\vdash \varphi$ for all sufficiently large $n$.

2.6. We follow the general convention that if $x$ is some object, $\underset{\sim}{x}$ is a term of the ramified language $\mathscr{L}$ that denotes $x$. We let

1) $\gamma$ be the code for $\lambda_0$ determined by $F$. (Thus $\gamma\colon \boldsymbol{N} \to \{0, 1\}$; $\gamma(2^r 3^s) = 0$ if $F(r) < F(s)$, $\gamma(n) = 1$ otherwise.)

2) $\beta$ be the code for $A_{\lambda_0}$ determined from $\gamma$ and $\alpha$ via Lemma 1.7.

3) $<$ be the well-ordering of the elements of $\boldsymbol{N}^{\boldsymbol{N}}$ in $L[\alpha, F]$ in order of their construction from $\langle \alpha, F \rangle$. (The reader may supply sufficient conventions to precisely specify $<$.)

4) $x$ be the $<$-least element of $\boldsymbol{N}^{\boldsymbol{N}}$ in the Borel set coded by $\beta$ in $L[a, F]$ which is not constructible from $\alpha$.

(It is important to see that $x$ "always exists". By Lemma 2.3 the Borel set coded by $\beta$ in $L[a, F]$ is $A_{\lambda_0}$ itself. But by assumption, $A_{\lambda_0}$ has members not constructible from $\alpha$.)

It is clear from the definition of $x$ that the following two statements hold for every generic choice of $F$ (and so are forced by every $P \in \mathscr{P}$).

1) $\underset{\sim}{x}$ is not constructible from $\underset{\sim}{\alpha}$;

2) $\underset{\sim}{x}$ lies in the Borel set coded by $\underset{\sim}{\beta}$.

2.7. Let $a \in \boldsymbol{N}^{\boldsymbol{N}}$ and let $P$ be a condition. We say that $P$ "makes $x$ unequal to $a$" if for some $m, n \in \boldsymbol{N}$, we have $a(m) = n$ and $P$ forces "$\underset{\sim}{x}(\underset{\sim}{m}) \neq \underset{\sim}{n}$". Because of the connection between forcing and truth, it is clear that if 1) $F\colon \boldsymbol{N} \to \lambda_0$ is a generic collapsing map, 2) $P$ makes $x$ unequal to $a$, and 3) $P \subseteqq F$, then in $L[\alpha, F]$, $\underset{\sim}{x}$ does not denote $a$. (Note that $a$ need not lie in $L[\alpha]$.)

**Lemma.** *Let $P \in \mathscr{P}$, and let $a$ be in $\boldsymbol{N}^{\boldsymbol{N}}$. Then $P$ has an extension $Q$ which makes $x$ unequal to $a$.*

*Proof.* Let $T = \{\langle m, n\rangle \in \boldsymbol{N} \times \boldsymbol{N} : P \Vdash \langle \underset{\sim}{m}, \underset{\sim}{n}\rangle \in \underset{\sim}{x}\}$. Clearly $T$ is a function with domain a subset of $\boldsymbol{N}$, taking values in $\boldsymbol{N}$. Since forcing is "expressible in the ground model", $T$ lies in $L[\alpha]$. Clearly, $P$ forces "$\underset{\sim}{T} \subseteqq \underset{\sim}{x}$".

If domain $(T) = \boldsymbol{N}$, then clearly, $\underset{\sim}{T} \subseteqq \underset{\sim}{x}$ implies $\underset{\sim}{T} = \underset{\sim}{x}$. This contradicts the fact that $P$ forces "$\underset{\sim}{x}$ is not constructible from $\alpha$". Thus domain $(T) \neq \boldsymbol{N}$. Let $m \in \boldsymbol{N}$-domain $(T)$. Let $n = a(m)$. Then clearly, $P$ does not force $\underset{\sim}{x}(\underset{\sim}{m}) = \underset{\sim}{n}$. By a fundamental property of forcing (recalled in § 2.5), $P$ has an extension $Q$ which forces $\underset{\sim}{x}(\underset{\sim}{m}) \neq \underset{\sim}{n}$. The lemma is proved.

2.8. Let $a_0, a_1, a_2, \ldots$ be an enumeration of the countable set $A_{\lambda_0}$. We are now going to construct a complete sequence $\{P_n\}$ which keeps $\underset{\sim}{x}$ out of the countable set $A_{\lambda_0}$. This will contradict the fact that $\underset{\sim}{x}$ is forced to lie in $A_{\lambda_0}$, and so complete the proof of Theorem 4.

Let $\{X_n\}$ be an enumeration of the dense subsets of $\mathscr{P}$ lying in $L[\alpha]$. Let $P_0$ be some arbitrary member of $\mathscr{P}$. Suppose $P_0 \subseteqq P_1 \subseteqq \cdots \subseteqq P_n$ have been defined. Since $X_n$ is dense, there is a $Q_n \supseteqq P_n$, $Q_n \in X_n$. By Lemma 2.7, there is an extension $P_{n+1}$ of $Q_n$ which makes $\underset{\sim}{x}$ unequal to $a_n$.

By construction the sequence $\{P_n\}$ is complete. Let $F$: $\boldsymbol{N} \to \lambda_0$ be the bijection determined by $\{P_n\}$. Then if $x$ is the element of $\boldsymbol{N}^{\boldsymbol{N}}$ denoted by $\underset{\sim}{x}$ in $L[\alpha, F]$, we have

$x$ lies in the Borel set coded by $\beta$. (Cf. end of § 2.6.)

Thus $x = a_n$, for some $n$, by Lemma 2.3. But this is absurd since we insured, when constructing $P_n$, that $x$ would not be equal to $a_n$. This contradiction completes the proof of Theorem 4.

## § 3. Proofs of Theorems 1—3

3.1. We shall first prove the $\boldsymbol{N}^{\boldsymbol{N}}$ versions of Theorems 1 through 3. Later we shall show how the corresponding results for $\boldsymbol{R}$ are immediate corollaries.

3.2. We now state the Kondo-Addison theorem [1]. (For a proof, see the section on basis results in [19, Chapter 7].)

**Lemma.** *Let $A$ be a $\boldsymbol{\Sigma}^1_2$ subset of $\boldsymbol{N}^{\boldsymbol{N}}$ with code $\alpha$. Then there is a $\boldsymbol{\Pi}^1_1$ subset, $B$, of $\boldsymbol{N}^{\boldsymbol{N}} \times \boldsymbol{N}^{\boldsymbol{N}}$ with code $\beta$ such that:*

1) $A = \{s \in \boldsymbol{N}^{\boldsymbol{N}}: (\exists t \in \boldsymbol{N}^{\boldsymbol{N}})(\langle s, t\rangle \in B)\}$;

2) *For each $s \in A$, there is exactly one $t \in \boldsymbol{N}^{\boldsymbol{N}}$ such that $\langle s, t\rangle \in B$;*

3) *$\beta$ is uniformly recursive in $\alpha$.*

(Note that 1) and 2) state that the restriction of the projection map to $B$ is a bijection $\pi$: $B \to A$.)

3.3. *Proof of Theorem 1.*

3) $\rightarrow$ 1). Let $A$ be an uncountable $\mathbf{\Pi}_1^1$ set, with code $\alpha$. By 3), there is $\gamma \in A$ with $\gamma$ not constructible from $\alpha$. By Theorem 4, which is valid in the presence of 3), $A$ contains a perfect subset.

1) $\rightarrow$ 2). Let $A$ be an uncountable $\mathbf{\Sigma}_2^1$ subset of $\mathbf{N}^{\mathbf{N}}$. We apply the Kondo-Addison theorem, getting a $B \subseteqq \mathbf{N}^{\mathbf{N}} \times \mathbf{N}^{\mathbf{N}}$ which is $\mathbf{\Pi}_1^1$ and such that the projection map restricts to a bijection

$$\pi\colon B \rightarrow A .$$

Since $\pi$ is onto, $B$ is uncountable. Since $\mathbf{N}^{\mathbf{N}} \times \mathbf{N}^{\mathbf{N}}$ is homeomorphic to $\mathbf{N}^{\mathbf{N}}$, we can apply 1), getting a perfect subset $K$ of $B$. Let $K' = \pi[K]$. Then as the one-one continuous image of a perfect set, $K'$ is perfect. Since $K' \subseteqq A$, we are done.

2) $\rightarrow$ 3). We shall prove the contrapositive. Thus suppose that $s$ is real such that uncountably many reals are constructible from $s$. We shall construct an uncountable $\mathbf{\Sigma}_2^1$ set $s$ without perfect subsets.

Since the reals constructible from $s$ are in $1 - 1$ correspondence with $\aleph_1^{L[s]}$, we must have

$$\aleph_1 = \aleph_1^{L[s]}.$$

We now let $<$ be the canonical $\mathbf{\Sigma}_2^1$-ordering of $\mathbf{N}^{\mathbf{N}} \cap L[s]$ (cf. [3]). Within $L[s]$, $<$ is $\mathbf{\Delta}_2^1$. Thus the set:

$$W' = \{\alpha \in \mathbf{N}^{\mathbf{N}} \cap L[s] \colon \alpha \text{ codes an ordinal but for each } \beta < \alpha, \\ \beta \text{ does not code the ordinal coded by } \alpha\}$$

is easily seen to be $\mathbf{\Delta}_2^1$ inside $L[\alpha]$. Moreover, since $\aleph_1 = \aleph_1^{L[s]}$, $W'$ contains exactly one code for each countable ordinal $\geqq \omega$. Let $S(x)$ be a $\mathbf{\Sigma}_2^1$ predicate such that

$$W' = \{\alpha \in \mathbf{N}^{\mathbf{N}} \cap L[s] \colon\ L[s] \models S(\alpha)\} .$$

By a theorem of Shoenfield, $\mathbf{\Sigma}_2^1$ predicates relativise to $L[s]$ (if their parameters lie in $L[s]$). (Cf. [18].) Thus $W' = \{\alpha \in \mathbf{N}^{\mathbf{N}} \cap L[s] \colon S(\alpha)\}$.

A theorem of Gödel ([6], cf. also [3]) states that there is a $\mathbf{\Sigma}_2^1$ predicate $T(x)$ such that

$$(\alpha \in \mathbf{N}^{\mathbf{N}}) (\alpha \in L[s] \Leftrightarrow T(\alpha)) .$$

Thus $W' = \{\alpha \mid T(\alpha) \wedge S(\alpha)\}$ so $W'$ is $\mathbf{\Sigma}_2^1$. Since $W'$ contains a code for each infinite countable ordinal, $W'$ is uncountable.

To complete the proof, we show that $W'$ contains no perfect subset. Suppose that $K \subseteqq W'$, $K$ perfect. Then $K \subseteqq W$, and a classical theorem (whose proof is recalled in [21]) states that there is a countable ordinal $\lambda_1$ such that each $\gamma \in K$ codes an ordinal $\leqq \lambda_1$. But $W'$ contains exactly one code for each countable ordinal. Thus $K$ is countable, which is absurd, since $K$ is perfect.

3.4. *Proof of Theorem 3.*

Let $A$ be a countable $\boldsymbol{\Sigma}_2^1$ subset of $\boldsymbol{N}^{\boldsymbol{N}}$ with code $\alpha$. We apply the Kondo-Addison theorem to $A$, getting a $\boldsymbol{\Pi}_1^1$ subset $B$ of $\boldsymbol{N}^{\boldsymbol{N}} \times \boldsymbol{N}^{\boldsymbol{N}}$ which projects one-one onto $A$, and whose code $\beta$ is recursive in $\alpha$.

Since $A$ is countable, so is $B$. By Theorem 4, each element of $B$ is constructible from $\beta$ and therefore from $\alpha$. Now if $s$ is in $A$, then for some $t \in \boldsymbol{N}^{\boldsymbol{N}}$, $\langle s, t\rangle \in B$. Therefore, $\langle s, t\rangle$ is constructible from $\alpha$; a fortiori, $s$ is constructible from $\alpha$.

3.5. We now show how the $\boldsymbol{R}$-versions of Theorems 1 through 3 follow from the $\boldsymbol{N}^{\boldsymbol{N}}$ versions. We first introduce codes for reals and for the $\boldsymbol{\Pi}_1^1$ subsets of $\boldsymbol{R}$ as follows: Let $\{q_i\colon i \in \boldsymbol{N}\}$ be a recursive enumeration of the rationals without repetitions. We say that $\alpha \in \boldsymbol{N}^{\boldsymbol{N}}$ codes the real $r \in \boldsymbol{R}$ if $\alpha(n) = 1 \Leftrightarrow q_n < r$, and otherwise $\alpha(n) = 0$. Clearly each real $r$ is coded by exactly one $\alpha$; call this $\alpha$, $\alpha_r$. If $\beta \in \boldsymbol{N}^{\boldsymbol{N}}$ codes the $\boldsymbol{\Pi}_1^1$ set $A \subseteqq \boldsymbol{N}^{\boldsymbol{N}}$, we shall also say that $\beta$ codes $\{r \in \boldsymbol{R}\colon \alpha_r \in A\}$. In this way, we introduce codes for the $\boldsymbol{\Pi}_1^1$ subsets of $\boldsymbol{R}$.

It is clear that the $\boldsymbol{R}$-versions of Theorems 1 through 3 are equivalent to the (0, 1)-versions. (Use the homeomorphism

$$x \rightarrow \frac{1}{2} + \frac{1}{\pi} \operatorname{arc\ tan} x$$

of $(-\infty, \infty)$ with (0, 1).) Next, let $J$ be

$$\{r \in (0, 1)\colon r \text{ is irrational}\}.$$

Since $J$ omits only countably many points of (0, 1), and each point omitted is constructible, the (0, 1)-versions of Theorems 1—3 are equivalent to the $J$-versions.

But $J$ is canonically homeomorphic to $\boldsymbol{N}^{\boldsymbol{N}}$ via the continued fraction expansion:

$$r = \cfrac{1}{n_0 + \cfrac{1}{n_1 + \dots}}$$

(see [13, p. 5]). We write $\beta_r$ for the function giving the continued fraction expansion of $r$. Then $\beta_r$ and $\alpha_r$ are recursive in one another, uniformly in $r$. (This is easy to see.) Using this fact, it is not hard to prove the following. Let $\gamma_0$ code the $\boldsymbol{\Pi}_1^1$ set of reals, $A$, and let

$$B = \{\beta_r\colon r \in A\}.$$

Then $B$ is a $\boldsymbol{\Pi}_1^1$ subset of $\boldsymbol{N}^{\boldsymbol{N}}$, and we can find a code for $B$ uniformly recursive in $\gamma_0$. Similarly, given a $\boldsymbol{\Pi}_1^1$ subset $B$ of $\boldsymbol{N}^{\boldsymbol{N}}$, with code $\gamma_1$, and putting

$$A = \{r \in J\colon \beta_r \in B\},$$

we can find a code $\gamma_0$ for $A$, qua $\mathbf{\Pi}_1^1$ set, uniformly recursive in $\beta$. Finally, totally analogous results hold for $\mathbf{\Sigma}_2^1$ sets.

Using the results of the preceding paragraph, it is easy to see that the $J$-versions of Theorems 1 through 3 are equivalent to their $\boldsymbol{R}$-versions. Thus we have succeeded in proving the $\boldsymbol{N}^{\boldsymbol{N}}$-version and $\boldsymbol{R}$-version of our theorem equivalent.

In a similar manner, one could show that the $2^{\boldsymbol{N}}$-versions are also equivalent to the $\boldsymbol{N}^{\boldsymbol{N}}$-versions.

3.6. Suppose now that $P$ holds (cf. § 2.1). We show that if $X$ is a complete separable metric space, and $A$ is an uncountable $\mathbf{\Sigma}_2^1$ subset of $X$, then $A$ contains a perfect subset.

Let $I$ be the unit interval, [0, 1]. Let $I^{\boldsymbol{N}}$ be the cartesian product of countably many copies of $I$. Then there is a homeomorphism into:

$$h\colon X \to I^{\boldsymbol{N}}$$

(cf. [8, p. 125]). Moreover, the image of $h$ is a $G_\delta$ in $I^{\boldsymbol{N}}$ ([8, Problem 6K]). It follows that $h(A)$ is an uncountable $\mathbf{\Sigma}_2^1$ subset of $I^{\boldsymbol{N}}$. Now our proof of Theorem 4 applies word-for-word to subsets of $I^{\boldsymbol{N}}$, and the same is true of the Kondo-Addison theorem. Thus we may conclude that $h(A)$ contains a perfect subset. But $h$ was a homeomorphism into; therefore, $A$ must also contain a perfect subset.

3.7. Mansfield's proof of Theorem 1 yields the following extra information. If $A$ is an uncountable $\mathbf{\Sigma}_2^1$ set with code $\alpha$, then $A$ contains a perfect subset $K$ with code (qua closed set) constructible in $\alpha$.

We wish to show that MANSFIELD's sharpened results can be deduced from Theorem 1. As usual, by the Kondo-Addison theorem, we may reduce matters to the case when $A$ is $\mathbf{\Pi}_1^1$. But now

$$\{\beta \mid \beta \text{ codes, qua closed set, a perfect subset of } A\}$$

is easily seen to be $\mathbf{\Pi}_1^1$ with code recursive in $\alpha$. But a theorem of SHOENFIELD [18] now asserts that there is a $\beta$ in this set with $\beta$ constructible from $\alpha$. (In fact, using the basis theorem for $\mathbf{\Pi}_1^1$ sets, we can find such a $\beta$ $\Delta_2^1$ in $\alpha$.)

## Postscript I

The author has recently constructed a model of $ZF + AC$ in which $2^{\aleph_0} = \aleph_2$ and every subset of $\boldsymbol{R}$ of power $\aleph_1$ is $\mathbf{\Pi}_1^1$. This result will appear elsewhere.

## Postscript II

(October 17, 1967.) In a recent letter, MANSFIELD has pointed out that the proof of Theorems 1 to 3 can be used to prove the following:

**Theorem.** Let $A$ be $\mathbf{\Sigma}_2^1$, and suppose that $A$ contains a real not constructible from the code for $A$. Then $A$ contains a perfect subset.

*Proof.* By the Kondo-Addison theorem, we may reduce to the case that $A$ is $\mathbf{\Pi}_1^1$. Let $\alpha$ be a code for $A$. We suppose the theorem false, and define $A_{\lambda_0}$ as in § 2.2. We now consider a Boolean valued model $V^{(\mathscr{B})}$ (cf. a forthcoming paper of Scott and the author entitled "Boolean-valued models for set-theory") in which $\aleph_1^V$ is countable. Lemma 2.4, and Lemmas 1.7 and 2.3 show that $A_{\lambda_0}^{(\mathscr{B})}$ is countable in $V^{(\mathscr{B})}$ and contains a real not constructible from the code for $A^{(\mathscr{B})}$. ($A^{(\mathscr{B})}$ is the $\mathbf{\Pi}_1^1$ set coded in $V^{(\mathscr{B})}$ by the code $\alpha$ for $A$; $A_{\lambda_0}^{(\mathscr{B})}$ is the corresponding constituent computed in $V^{(\mathscr{B})}$.) We can now apply the proof of Theorem 4 inside $V^{(\mathscr{B})}$ to get a contradiction.

**Corollary** Let $\aleph_1 > \aleph_1^{L[\alpha]}$. Then every uncountable $\mathbf{\Sigma}_2^1$ set with code constructible from $\alpha$ contains a perfect subset.

*Proof.* Let $A$ be $\mathbf{\Sigma}_2^1$ with code constructible from $\alpha$. If $A$ is uncountable, and $\aleph_1 > \aleph_1^{L[\alpha]}$, then $A$ contains reals not constructible from $\alpha$, and the theorem applies.

We wish to emphasize that the theorem and corollary are provable in $ZF^*$ (and do not require a large cardinal assumption).

## References


1. Addison, J. W.: Hierarchies and the axiom of constructibility. Summaries of talks at the Summer Institute of Symbolic Logic in 1957 at Cornell University **3**, 355—362 (1957).
2. — Separation principles in the hierarchies of classical and effective descriptive set theory. Fund. Math. **46**, 123—135 (1958).
3. — Some consequences of the axiom of constructibility. Fund. Math. **46**, 337—357 (1959).
4. Cohen, P. J.: The independence of the continuum hypothesis, Parts I, II. Proc. nat. Acad. Sci. **50**, 1143—1148 (1963); **51**, 105—110 (1964).
5. Gaifman, H.: Self extending models, measurable cardinals and the constructible universe. Mimeographed notes.
6. Gödel, K.: The consistency of the axiom of choice and of the generalized continuum-hypothesis. Proc. nat. Acad. Sci. **24**, 556—557 (1938).
7. — The consistency of the axiom of choice and of the generalized continuum-hypothesis with the axioms of set theory. Ann. Math. Studies no. 3, second printing. Princeton 1951.
8. Kelley, J. L.: General topology. New York: Van Nostrand 1955.
9. Kleene, S. C.: Arithmetical predicates and function quantifiers. Trans. Amer. Math. Soc. **79**, 312—340 (1955).
10. — On the forms of the predicates in the theory of constructive ordinals (second paper). Amer. J. Math. **77**, 405—428 (1955).
11. Levy, A.: Independence results in set theory by Cohen's method, I, III, IV. Notices Amer. Math. Soc. **10**, 592—593 (1963).
12. —, and R. M. Solovay: Measurable cardinals and the continuum hypothesis. Israel J. Math. **5**, 234—248 (1967).

13. LJAPUNOW, A. A.: Arbeiten zur deskriptiven Mengenlehre. Berlin: VEB Deutscher Verlag der Wissenschaften 1955.
14. MANSFIELD, R.: The solution of one of Ulam's problems concerning analytic rectangles. Preprint.
15. REINHARDT, W., and R. M. SOLOVAY: Strong axioms of infinity and elementary embeddings. To appear.
16. ROGERS jr., H.: Recursive functions over well ordered partial orderings. Proc. Amer. Math. Soc. **10**, 847—853 (1959).
17. ROWBOTTOM, F.: Doctoral dissertation. Madison: Univ. of Wisconsin 1964.
18. SHOENFIELD, J. R.: The problem of predicativity. Essays on the foundations of mathematics. Jerusalem 1961, pp. 132—139.
19. — Mathematical logic. Addison-Wesley Publishing Co. 1967.
20. SILVER, J.: The consistency of the generalized continuum hypothesis with the existence of a measurable cardinal. Notices Amer. Math. Soc. **13**, 721 (1966).
21. SOLOVAY, R. M.: Some consequences of the axiom of determinateness. To appear.
22. — The measure problem. I. A model of set theory in which all sets of reals are Lebesgue measurable. To appear.
23. VOPENKA, P., and L. BUKOVSKY: The existence of a PCA-set of cardinality $\aleph_1$. Commentationes Mathematicae Universitatis Caroline (Prague) **5**, 125—128 (1964).

# The Universe of Set Theory[1]

Gaisi Takeuti

Since Cohen's discovery of forcing, many problems in set theory have been proved to be independent of ZF-set theory just as in the case of the parallel postulate in plane geometry. In plane geometry, only the independence of the parallel postulate was considered, but in set theory it seems that infinitely many problems can be proved to be mutually independent. The consideration of many set theories might not be of advantage to us because set theory is a basis of mathematics and working mathematicians cannot believe that both "yes" and "no" are equally reasonable answers to their problems in natural numbers, real numbers or Hilbert spaces.

Now consider again the independence of the parallel postulate from plane geometry. This independence means that the parallel postulate is not a problem inside plane geometry but a problem outside of plane geometry. For example, the problem whether the space in which we live satisfies the parallel postulate or not is a problem of physics but not a problem of plane geometry. Similarly, a problem which is proved to be independent of ZF is not a problem inside ZF but a problem outside of ZF. Then what does "outside of ZF" mean? I am quite sure that "outside of ZF" does not mean physics. I think that it means the metamathematics of how to build up a stronger set theory.

We should remember that Gödel already predicted the independence of the continuum hypothesis in [9] and stated his opinion as follows:

> "It is to be noted, however, that even if one should succeed in proving its undemonstrability as well, this would (in contradistinction, for example, to the proof for the transcendency of $\pi$) by no means settle the question definitively. Only someone who (like the intuitionist) denies that the concepts and axioms of classical set theory have any meaning (or any well-defined meaning) could be satisfied with such a solution, not someone who believes them to describe some well-determined reality. For in this reality Cantor's conjecture must be either true or false, and its undecidability from the axioms as known today can only mean that these axioms do not contain a complete description of this reality; and such a belief is by no means chimerical, since it is possible to point out ways in which a decision of the question, even if it is undecidable from the axioms in their present form, might nevertheless be obtained."

[1] Part of this work was supported by *NSF* GP-4616. The outline of this work was discussed in a symposium on the Current Status of Set Theory at a joint session of *ASL* and *APA* held on December 28, 1965.

Then our problem is how to build up stronger set theories. A natural and easy way to strengthen ZF is to introduce type theory of ZF. Since the basic idea of type theory is the same as the basic idea of set theory and also since the consistency of the set theory of lower type can be proved in the set theory of higher type, it is very natural to introduce type theory into set theory. In [28], we have introduced a transfinite type set theory TT. In this paper, we shall define a transfinite type theory over set theory in the same spirit but in a different formulation.

Now, in order to consider our problem more systematically, we have to consider what the universe of sets is. The universe of sets is obtained by the transfinitely iterated applications of the operation "set of." Therefore the main character of our universe should be expressed by the answers to the following two questions:

1) What is the construction of "set of", i. e., the construction of subsets of some fixed domain $a$? This question should be expressed by the following similar questions: How many sets are there in $P(a)$? How wide is $P(a)$? How many new sets can we get by the construction of $R(\alpha + 1)$ from $R(\alpha)$? How wide is $R(\alpha)$? These questions are called the question of "how wide our universe is".

2) How far can we keep constructing $R(\alpha)$? We can express this in the following ways. How far can we construct the ordinal numbers? How long is On (the class of all ordinal numbers)? These questions are called the question of "how long our universe is".

Since the main idea of Cantor's set theory is the creation of more and more new sets, what we should do about On is to keep constructing the ordinal numbers endlessly. At this moment, we have a more precise intuitive feeling about the problem "how long" than about "how wide". If we formalize the fact that we can go until some point, then it is usually very easy to state that we can go further. Therefore we shall first consider the length of On (Chapter I). The axioms on length are usually called the axioms of (strong) infinity. There are two elegant plausible principles here. One is Mahlo's principle ([17], [18] and [19]) and the other is Lévy's reflection principle in the first order language ([12]). Bernays [3] presented an elegant generalization of both of them. We shall generalize Bernays' reflection principle using transfinite type theory.

For this purpose, we shall introduce the notion of "node". Using the notion of node, we shall formalize a set theory NTT with a generalization of both the reflection principle in transfinite type theory and Mahlo's principle. The theory NTT will be developed in Chapter I. We shall prove that $V = L$ is relatively consistent with NTT and that TT can be interpreted in NTT.

In Chapter II we shall consider the problem of the width of the universe. Since our intuition on width is not yet mature enough, we cannot present at this moment many axioms about width as we can in the case of length. Therefore what we should do is to mature our intuition about the width of the universe. The best way to do this is to define a rather complete classification of widths and to investigate the mathematical implications of each case of the classification.

Since the problem of how wide our universe is is the problem of how many sets exist, what we need in order to classify width are scales to measure the numbers of sets. Our scales are the following:

1) Classification of the next cardinal.

2) Classification of the functions $f$ such that $\overline{\overline{2^{\aleph_\alpha}}} = \aleph_{f(\alpha)}$.

3) Degree of undefinability of $2^{\aleph_\alpha}$.

1) *Classification of the next cardinal.* Usually the scale to measure how many sets are in a given set is the cardinality of the set. However, the cardinality is not adequate for out purpose for the reason that, since it is expressed by $\overline{\overline{a}} = \mu\alpha\,\exists f(f:\alpha \xrightarrow{\text{onto}} a)$, the notion of $\exists f$ (hence the notion of width) is involved in the notion of $\overline{\overline{a}}$. In general, if we quantify over sets in defining some notion, such a notion itself depends on how wide the universe is. Therefore what we need is a scale which is defined independently of the width of our universe. Such a notion must be defined by a notion concerning only ordinal numbers. For the development of such a notion, the example of recursive function theory of natural numbers is very promising. Recursive functions, arithmetical functions and hyperarithmetical functions of ordinal numbers are obtained in a similar way to the corresponding notions on natural numbers (cf. [16], [23], [25], [26] and [28]).

Now consider the notion of cardinals. What should be considered is the classification of how large the next cardinal is. Though there are infinitely many cases, we classify them into the following three groups:

i) There exists an arithmetical function $f$ such that $a^+ \leqq f(a)$, where $a^+$ is the first cardinal which is greater than $a$.

ii) There exists no arithmetical function $f$ such that $a^+ \leqq f(a)$, i. e., the next cardinal cannot be reached by means of arithmetical functions.

iii) There exists no hyperarithmetical function $f$ such that $a^+ \leqq f(a)$, i. e., the next cardinal cannot be reached by means of hyperarithmetical functions.

ii) and iii) are called *Transcendency of cardinals* (TC) ([26]) and *Hypertranscendency of cardinals* (HTC) ([28]), respectively. TC and HTC mean that the notion of cardinal is essentially analytical, and neither arithmetical nor hyperarithmetical. Roughly speaking, we can show (i) TC and HTC are consistent with ZF assuming some axiom of infinity, (ii) TC is equivalent to $\forall\alpha((R(\aleph_\alpha) \cap L)$ is an elementary submodel of $L)$

and (iii) TC implies $\exists x \subseteqq \omega (x \notin L)$. Both TC and HTC mean that our universe is fairly wide. The author thinks that TC and HTC are plausible.

2) *Classification of $f$ such that* $\overline{\overline{2^{\aleph_\alpha}}} = \aleph_{f(\alpha)}$. After we classify the notion of cardinality, "how wide $\overline{\overline{2^{\aleph_\alpha}}}$ is" can be expressed roughly by the function $f$ such that $\overline{\overline{2^{\aleph_\alpha}}} = \aleph_{f(\alpha)}$. EASTON [5] showed that $\overline{\overline{2^{\aleph_\alpha}}} = \aleph_{f(\alpha)}$ is consistent with ZF for almost any function $f$ satisfying certain necessary conditions.

3) *Degree of undefinability of* $2^{\aleph_\alpha}$. One $f$ in 2) is fixed, our problem becomes "how complex $2^{\aleph_\alpha}$ is". First we define the first degree of undefinability of $2^{\aleph_\alpha}$ as the least ordinal $\beta$ such that

$$\exists h((h\colon \beta \overset{\text{into}}{\longrightarrow} 2^{\aleph_\alpha}) \wedge \exists g(g \in L(h) \wedge (2^{\aleph_\alpha} = g''\aleph_{f(\alpha)}))),$$

where $L(h)$ is the smallest complete model of ZF such that $\mathrm{On} \subseteqq L(h)$ and $h \in L(h)$. This means that if we assume the knowledge of the $\beta$ functions which map $\aleph_\alpha$ into 2, then we have enough knowledge of $2^{\aleph_\alpha}$. Therefore, the smaller the $\beta$ is, the simpler $2^{\aleph_\alpha}$ is. It turns out by using a result of HAJNAL [10] that the $\beta$ can take one of the following values.

Case 1) $\beta = 0$. This means that $2^{\aleph_\alpha} \in L$.

Case 2) $\beta = 1$. This means that there exists a function $g$ in $2^{\aleph_\alpha}$ such that $2^{\aleph_\alpha} \in L(g)$.

In both case 1) and case 2) we shall show that we can imitate ADDISON's work ([1] and [2]) and KURATOWSKI's work ([11]) and obtain many results of projective set theory.

Case $\infty - 1$) $\beta = \aleph_\gamma$, where $f(\alpha) = \gamma + 1$ and $\aleph_\gamma$ is a singular cardinal. This is a very singular case.

Case $\infty$) $\beta = \aleph_{f(\alpha)}$. The further classification of this case should be investigated.

I would like to thank very sincerely the referee of this article, who read it painstakingly and made many suggestions for improvement. Unfortunately, however, there was not time to take account of all of them.

# Chapter I

# Nodal Transfinite Type Set Theory

## § 1. Introduction of Typed Class Variables to Set Theory

We begin with giving an intuitive idea about typed variables. Let $\mathbb{L}$ be a language of type theory on set theory. Let $R(\alpha)$ be $\bigcup_{\gamma<\alpha} P(R(\gamma))$. When we try to interpret formulas in $\mathbb{L}$ in a first order theory, letting first order variables range over $R(\alpha)$, we must consider formulas in $\mathbb{L}$ of the forms $\forall X(\ldots)$ and $\exists X(\ldots)$, where $X$ is a second order variable, are translated in formulas of the forms $\forall x(x \in R(\alpha + 1) \to \cdots)$ and

$\exists x(x \in R(\alpha + 1) \wedge \cdots)$, respectively; in general, formulas of the forms $\forall \xi(\ldots)$ and $\exists \xi(\ldots)$, where $\xi$ is a variable of any type, in formulas of the forms $\forall x(x \in R(\beta) \rightarrow \cdots)$ and $\exists x(x \in R(\beta) \wedge \cdots)$, respectively, where $\beta$ is an ordinal $\geq \alpha$, and depends on the type of $\xi$. In this sense, we can consider that a type is determined by a certain relation $A(\alpha, \beta)$ on ordinal numbers $\alpha$ and $\beta$ satisfying

$$\forall \alpha \, \exists ! \beta \, A(\alpha, \beta) \wedge \forall \alpha \, \forall \beta (A(\alpha, \beta) \rightarrow \alpha \leqq \beta)$$

(denoted by $C_A$ in the following).

In order to carry out this idea formally, we define typed variables and extend the notion of formulas inductively by introducing "degree" of (typed) variables and formulas.

1.1 We start with the first order language of set theory which consists of set variables $a, b, c, \ldots$, and $x, y, z, \ldots$, one predicate constant $\varepsilon$ and logic symbols. We shall use the notations in [7] mostly without mention; for example $\alpha, \beta, \gamma, \ldots$ stand for ordinal numbers. We define the degree of a first order variable $a$ (denoted $\deg(a)$) to be 0; and the degree of a first order formula $\varphi$ (denoted $\deg(\varphi)$) to be 0. For every formula (of first order) $A(\alpha, \beta)$ having exactly two free variables $\alpha$ and $\beta$, variables of type $A$ are introduced and denoted by capital italics with superscripts $A$ like $X^A$, $Y^A, \ldots$. (In particular, when $A(\alpha, \beta)$ is $\beta = \alpha + 1$, we omit the superscript in writing the variables of type $A$; those are second order variables in the ordinary sense.) We define $\deg(X^A)$ to be 1 and extend the notion of formula by adjoining variables of degree 1 and quantifiers over the typed variables to our starting language. The degree of any formula in this language is defined to be the maximum of the degrees of the variables in it. Assume that we have introduced variables of degree $n$ and formulas of degree $n$ in this way. Then for every formula $A(\alpha, \beta)$ of degree $n$ having $\alpha$ and $\beta$ as its sole free variables, we introduce variables of type $A$. The degree of a variable of type $A$ where $A(\alpha, \beta)$ is of degree $n$ is defined to be $n + 1$. The notion of formula is extended by adjoining variables of degree $n + 1$ and quantifiers with respect to variables of degree $n + 1$ to the language. The degree of a formula is defined to be the maximum of the degrees of the variables in it. To sum up: atomic formulas of our language are of the form $a \in b$, $X^A \in b$, $a \in X^B$, $X^A \in X^B$, where $a$ and $b$ are first order variables: $X^A$ and $X^B$ are variables of type $A$ and of type $B$, respectively. Formulas are built up from atomic formulas as usual, including universal and existential quantifiers over type variables.

**Lemma 1.** *Let $A(\alpha, \beta)$ be a formula having $\alpha$ and $\beta$ as the only free variables. Then $X^A$ is defined and*

$$\deg(X^A) = \deg(A) + 1 = \max(\deg(Y^B)) + 1,$$

*where $Y^B$ ranges over the variables in $A$, including the first order variables.*

1.2. The relativization of formulas to $R(\alpha)$. *For an arbitrary formula* $\varphi$ *we shall define* $\varphi^{R(\alpha)}$.

1.2.1. If $\varphi$ is an atomic formula, then $\varphi^{R(\alpha)}$ is $\varphi$ itself.

1.2.2. $(\varphi \wedge \psi)^{R(\alpha)}$ and $(\neg \varphi)^{R(\alpha)}$ are $\varphi^{R(\alpha)} \wedge \psi^{R(\alpha)}$ and $\neg \varphi^{R(\alpha)}$ respectively.

1.2.3. $(\forall x\, \varphi(x))^{R(\alpha)}$ is $\forall x \in R(\alpha)\, \varphi^{R(\alpha)}(x)$.

1.2.4. $(\forall X^A\, \varphi(X^A))^{R(\alpha)}$ is

$$(C_A \wedge \exists \beta (A(\alpha, \beta) \wedge \forall x \in R(\beta)\, \varphi^{R(\alpha)}(x))) \vee (\neg C_A \wedge \forall x \in R(\alpha)\, \varphi^{R(\alpha)}(x)),$$

where $C_A$ denotes $\forall \alpha\, \exists! \beta A(\alpha, \beta) \wedge \forall \alpha\, \forall \beta (A(\alpha, \beta) \rightarrow \alpha \leqq \beta)$.

**Lemma 2.** *Let* $\varphi$ *be a formula whose degree is greater than* 0 *and whose only free variables are of degree* 0. *Then* $\deg(\varphi^{R(\alpha)}) < \deg(\varphi)$.

*Proof.* This can easily be proved by induction on the number of logical symbols in $\varphi$. Notice that if $\varphi$ is of the form $\forall X^A\, \psi(X^A)$ and $\deg(\varphi) = n + 1$, then $\deg(A) \leqq n$, i.e., $\deg(C_A) \leqq n$.

As an easy corollary of Lemma 2, we have

**Lemma 3.** *If* $\varphi$ *is a formula whose degree is less than or equal to* 1 *and whose only free variables are of first order, then* $\varphi^{R(\alpha)}$ *is a formula of the first order language.*

Those lemmas will be made use of mostly without mention.

We can generalize the reflection principle using our language. Our principle is that there exists an ordinal $\alpha_0$ such that $R(\alpha_0)$ is an $\mathbb{L}$-submodel of the whole universe $V$ for every finite or transfinite type theory $\mathbb{L}$. Namely our reflection principle is

1.3. $\forall x \in R(\alpha_0)(\varphi(x) \leftrightarrow \varphi^{R(\alpha_0)}(x))$,

where $\varphi(x)$ has no free variables other than $x$ and is an arbitrary formula of the transfinite type theory which we have defined above. In [28], we defined the transfinite type set theory TT whose main part is the first order formulation of this principle 1.3. Since we are going to introduce the notion of node and to generalize 1.3 by using it, we will not introduce the constant $\alpha_0$ to our language or take up 1.3 as an axiom of our theory.

## § 2. The Presentation of NTT

There are two different, equally convincing standpoints for the absolute set theory.

A. There exists a fixed absolute universe of all sets.

B. The creation of sets is endless and so we cannot consider the totality of all sets. The notion "every set" should be interpreted without referring to the totality of all sets.

Now we shall explain the latter. As we explained in the introduction, one of two basic principles of our set theory is the creation of ordinals.

From the beginning of the set theory, we have the following difficult question about ordinal numbers. Is there any fixed absolute universe of all ordinal numbers? We cannot answer "yes" for this question. For, if so, we have BURALI-FORTI's paradox. Therefore, we have to choose one of the following possibilities.

1) The class of all ordinals of our absolute universe of sets is not maximal, namely our absolute universe of sets is rather a small universe.

2) Our creation of ordinals is endless and therefore there is no fixed absolute universe of our set theory.

It seems to us that 1) contradicts our basic idea of set theory i.e. the creation of as many sets as possible. One might say that our universe is maximal in the sense that it is maximal among the universes which satisfies a beautiful set theory e.g. $ZF$ set theory and an informal axiom of foundation. However, this situation is very implausible and even if this is the case, we are interested in the maximal universe more than a beautiful set theory. Therefore we believe that only 2) is possible. However in this case, there are so many possible universes of set theory i.e. $R(\alpha)$'s. Every $R(\alpha)$ provides us with a candidate of our set theory i.e. the set theory which is satisfied in $R(\alpha)$. Is there an absolute set theory? If so, what is its definition? We believe that there exists an absolute set theory and that the following is its definition. Let us fix a language of set theory e.g. the first order language with one constant $\varepsilon$. Let $\mathfrak{T}_\alpha$ be the set theory of $R(\alpha)$ w.r.t. this language. Since the creation of ordinals is endless and the cardinality of the set of all possible theories in our fixed language is bounded, some theory must appear endlessly many times among $\mathfrak{T}_\alpha$'s. We believe that only one theory will appear overwhelmingly densely among them. We wish to define the absolute set theory to be this theory. Our intuitive notion "overwhelmingly dense" is very close to "almost everywhere" in measure theory. For each $\mathfrak{A} \subseteqq \mathrm{On}$, $\mathfrak{A}$ is said to be nodal, if $\mathfrak{A}$ is overwhelmingly dense in On. "$\mathfrak{A}$ is nodal" is written as $\mathscr{N}(\mathfrak{A})$. An ordinal $\alpha$ is called a node of a closed formula $\varphi$ if $\varphi^{R(\alpha)}$ and $\alpha$ is called a node of a function $\mathfrak{F}$ if

$$\forall x \in R(\alpha)(\mathfrak{F}`x \in R(\alpha)).$$

In [28], we explained that the class of all nodes of $\mathfrak{F}$ is nodal. In our standpoint, "$\varphi$ is true" is equivalent to "the class of all nodes of $\varphi$ is nodal". In other words, the construction of ordinals is so endless that nodes of any true statement and nodes of any function $\mathfrak{F}$ should occur overwhelmingly densely[2]. Using the notion of nodes, we are going to intro-

[2] Let $A$ and $B$ be the sets of all multiples of 3 and 7 respectively. If our construction of natural numbers is up to 19, then $A$ and $B$ are disjoint and the densities of $A$ and $B$ are completely different. However if we construct natural numbers endlessly (i. e. up to $\omega$), then the members of $A$ and $B$ occur equally densely in the set of natural numbers so that even $A \cap B$ is as dense as $A$ and $B$.

duce an generalization of BERNAYS' reflection principle (cf. Introduction) and call it NTT.

The symbols of NTT are the predicate constant $\varepsilon$, the logical symbols, all the typed variables which have been defined in § 1 and a new symbol $\mathcal{N}$. Formulas are defined as usual, $\mathcal{N}(X), \mathcal{N}(Y), \ldots$ being accepted as prime formulas if $X, Y, \ldots$ are of second order. $\mathcal{N}(X)$ reads "$X$ is a nodal class". Evidently the formulas with degree constitute a subclass of the formulas of NTT.

The axioms of NTT are those of ZF with the local version of the axiom of choice and those given by the following 2.1—2.6. In particular the subsystem of NTT which is obtained from NTT by restricting the variables to those of degree less than or equal to 1 is called $\mathrm{NTT}^1$.

The axioms of NTT.

2.1. $\forall X(\mathcal{N}(X) \rightarrow 0 \neq X \subseteqq \mathrm{On})$.

2.2. $\forall X\, \forall Y(\mathcal{N}(X) \wedge X \subseteqq Y \subseteqq \mathrm{On} \rightarrow \mathcal{N}(Y))$.

2.3. $\forall X\, \forall a(\mathcal{N}(X) \rightarrow \mathcal{N}(X - a))$.

2.4. $\forall X(\forall x\, \mathcal{N}(X_x) \rightarrow \mathcal{N}(\{\alpha \mid \forall x \in R(\alpha)(\alpha \in X_x)\}))$,

where $X_x$ denotes $X``\{x\}$.

2.5. $\forall X(\mathcal{N}(\{\alpha \mid \forall x \in R(\alpha)(\varphi(x, X) \leftrightarrow \varphi^{R(\alpha)}(x, X \cap R(a)))\}))$

for an arbitrary formula $\varphi$ with degree (that is $\mathcal{N}$ does not occur in $\varphi$) whose only free variables are $x$ and $X$.

2.6. Infinite induction

$$\frac{\varphi(i) \text{ for each numeral } i}{\forall i(i \in \omega \rightarrow \varphi(i)).}$$

*Remark.* Throughout this paper, the class notion $\{x/\varphi(x)\}$ shall be used only for the formulas $\varphi$ which have no $\mathcal{N}$-symbol unless mentioned otherwise. $\mathcal{N}(\{\alpha/\varphi(\alpha)\})$ should be understood to be the abbreviation of $\forall X(\forall \alpha(\alpha \in X \leftrightarrow \varphi(\alpha)) \rightarrow \mathcal{N}(X))$.

## § 3. Elementary Properties of NTT

The properties of $\mathcal{N}$ are similar to the properties of $\mathcal{G}^0$ in [24]. Therefore many proofs of the following propositions are taken from [24].

**Proposition 1.** $\mathcal{N}(\mathrm{On})$.

**Proposition 2.** $\mathcal{N}(X) \rightarrow \mathcal{N}(X \cap \{\alpha \mid a \in R(\alpha)\})$.

*Proof.* Since $\exists \beta(X \cap \{\alpha \mid a \in R(\alpha)\} \supseteqq X - \beta)$, this follows from 2.2 and 2.3.

**Proposition 3.** $\mathcal{N}(\{\alpha \mid \forall x \in R(\alpha)(F`x \in R(\alpha))\})$.

*Proof.* We have

$$\mathcal{N}(\{\alpha \mid (\forall x\, \exists y(F`x = y) \wedge \forall \beta\, \exists \gamma(\gamma = \beta + 1))$$
$$\leftrightarrow (\forall x \in R(\alpha)\, \exists y \in R(\alpha)((F \cap R(\alpha))`x = y)) \wedge \alpha \in K_{\mathrm{II}})\}),$$

which implies

$$\mathcal{N}(\{\alpha \mid \forall x \in R(\alpha)(F`x \in R(\alpha))\}).$$

**Proposition 4.** $a \neq 0 \wedge \forall x \in a \mathcal{N}(X_x) \rightarrow \mathcal{N}(\bigcap_{x \in a} X_x)$.

*Proof.* Let $Y$ be defined by $Y_x = X_x$ for $x \in a$; $Y_x = \mathrm{On}$ otherwise. Then $\forall x \mathcal{N}(Y_x)$. By 2.4 and Proposition 2, we have

$$\mathcal{N}(\{\alpha \mid \forall x \in R(\alpha)(\alpha \in Y_x)\} \cap \{\alpha \mid a \in R(\alpha)\}).$$

However since

$$\{\alpha \mid \forall x \in R(\alpha)(\alpha \in Y_x)\} \cap \{\alpha \mid a \in R(\alpha)\} \subseteqq \{\alpha \mid \forall x \in a(\alpha \in X_x)\} = \bigcap_{x \in a} X_x,$$

the proposition follows from 2.2.

**Proposition 5.** $\mathcal{N}(X) \wedge \mathcal{N}(Y) \rightarrow \mathcal{N}(X \cap Y)$.

*Proof.* Define

$$\tilde{X} = \{x \mid \exists y((y \in X \wedge x = \langle y\, 0\rangle) \vee (y \in Y \wedge x = \langle y\, 1\rangle))\}.$$

Then $\tilde{X}_0 = X$ and $\tilde{X}_1 = Y$. Therefore the proposition follows from Proposition 4 by substituting $\tilde{X}$ and 2 for $X$ and $a$ respectively.

**Lemma.** *For every closed formula $\psi$ without $\mathcal{N}$*

$$\psi \leftrightarrow \mathcal{N}(\{\alpha \mid \psi^{R(\alpha)}\}).$$

*Proof.* Let $\mathfrak{A} = \{\alpha \mid \psi \leftrightarrow \psi^{R(\alpha)}\}$, $\mathfrak{B} = \{\alpha \mid \neg\psi \leftrightarrow \psi^{R(\alpha)}\}$, $\mathfrak{C} = \{\alpha \mid \psi^{R(\alpha)}\}$ and $\mathfrak{D} = \{\alpha \mid \neg \psi^{R(\alpha)}\}$. By 2.5, $\mathcal{N}(\mathfrak{A})$ and $\mathcal{N}(\mathfrak{B})$. If $\psi$, then $\psi^{R(\alpha)}$ for every $\alpha \in \mathfrak{A}$, i.e. $\mathcal{N}(\mathfrak{C})$. Likewise, if $\neg\psi$, then $\mathcal{N}(\mathfrak{D})$. Since $\mathfrak{C} \cap \mathfrak{D} = 0$, this implies $\neg\mathcal{N}(\mathfrak{C})$: i.e. $\mathcal{N}(\mathfrak{C})$ implies $\psi$.

**Proposition 6.** Let $\varphi(x_1, \ldots, x_m, X^{A_1}, \ldots, X^{A_n}, x)$ be an arbitrary formula without $\mathcal{N}$ having no other free variables than $x_1, \ldots, x_m$, $X^{A_1}, \ldots, X^{A_n}, x$. Then

(1) $\forall\alpha \forall x_1 \ldots \forall x_m \forall X^{A_1} \ldots \forall X^{A_n} \forall u \in R(\alpha) \exists y \in R(\alpha) \forall x \in R(\alpha)$
$(x \in y \leftrightarrow \varphi(x_1, \ldots, x_m, X^{A_1}, \ldots, X^{A_n}, x) \wedge x \in u)$.

*Proof.* By induction on $\deg(\varphi(x_1, \ldots, x_m\, X^{A_1}, \ldots, X^{A_n}, x))(= d)$. If $d = 0$, then (1) is ZF-provable. In the following let $d > 0$ and let $\forall\alpha \forall x_1 \ldots \forall x_m \mathfrak{B}(\alpha, x_1, \ldots, x_m)$ be short for the formula (1). Suppose our assertion is true for every formula whose degree $< d$ and let

$$\deg(\varphi(x_1, \ldots, x_m, X^{A_1}, \ldots, X^{A_n}, x) = d.$$

We remark that every occurrence of variables in

$$\varphi(x_1, \ldots, x_m, X^{A_1}, \ldots, X^{A_n}, x)$$

is bound in $\forall\alpha \forall x_1 \ldots \forall x_m \mathfrak{B}(\alpha, x_1, \ldots, x_m)$ and that

$$\deg(\varphi^{R(\beta)}(x_1, \ldots, x_m, y_1, \ldots, y_n, x)) < d.$$

Thus by the inductive hypothesis

$$(2)\qquad \forall\beta\ \forall x_1 \ldots \forall x_m\ \forall y_1 \ldots \forall y_n\ \forall u \in R(\alpha)\ \exists y \in R(\alpha)\ \forall x \in R(\alpha)$$
$$(x \in y \leftrightarrow \varphi^{R(\beta)}(x_1, \ldots, x_m, y_1, \ldots, y_n, x) \wedge x \in u)$$

is provable for every $\alpha$. By Lemma in order to prove (1), it suffices to show $(\forall x_1 \ldots \forall x_m\ \mathfrak{B}(\alpha, x_1, \ldots, x_m))^{R(\beta)}$ for each pair $\alpha$ and $\beta$ such that $\alpha \leqq \beta$: hence it suffices to prove $\forall x_1 \ldots \forall x_m\ \mathfrak{B}^{R(\beta)}(\alpha, x_1, \ldots, x_m)$ for each pair $\alpha$ and $\beta$ such that $\alpha \leqq \beta$. We prove this by induction on $n$ (= the number of free typed variables in $\varphi(x_1, \ldots, x_m, X^{A_1}, \ldots, X^{A_n}, x)$). If $n = 0$, this follows from (2). Induction step: $\mathfrak{B}(\alpha, x_1, \ldots, x_m)$ is of the form

$$\forall X^A\ \mathfrak{B}_1(\alpha, x_1, \ldots, x_m, X^A).$$

$\mathfrak{B}^{R(\beta)}(\alpha, x_1, \ldots, x_m)$ is

$$(C^A \wedge \exists\gamma(A(\alpha, \gamma) \wedge \forall z \in R(\gamma)\ \mathfrak{B}_1^{R(\beta)}(\alpha, x_1, \ldots, x_m, z)))$$
$$\vee (\neg C_A \wedge \forall z \in R(\beta)\ \mathfrak{B}_1^{R(\beta)}(\alpha, x_1, \ldots, x_m, z)).$$

where $\forall x_1 \ldots \forall x_m\ \forall z\ \mathfrak{B}_1^{R(\beta)}(\alpha, x_1, \ldots, x_m, z)$ is provable by the inductive hypothesis on $n$. Thus $\forall x_1 \ldots \forall x_m\ \mathfrak{B}^{R(\beta)}(\alpha, x_1, \ldots, x_m)$ is also provable.

Proposition 6 implies

**Proposition 7.**

$$\forall X^{A_1} \ldots \forall X^{A_n}\ \forall u\ \exists y\ \forall x(x \in y \leftrightarrow \varphi(X^{A_1}, \ldots, X^{A_n}, x) \wedge x \in u),$$

where $\varphi$ and $X^{A_1}, \ldots, X^{A_n}$ satisfy the same conditions as in Proposition 6.

**Proposition 8.** (Comprehension axiom of type A)

$$C_A \rightarrow \forall Y^{B_1} \ldots \forall Y^{B_n}\ \exists X^{\tilde{A}}\ \forall X^A(\varphi(Y^{B_1}, \ldots, Y^{B_n}, X^A) \leftrightarrow X^A \in X^{\tilde{A}}),$$

where $\tilde{A}(\alpha, \beta) \leftrightarrow C_A \wedge \exists\gamma(A(\alpha, \gamma) \wedge \beta = \gamma + 1)$ and $\varphi$ does not contain the symbol $\mathcal{N}$.

*Proof.* Notice that $C_A \leftrightarrow C_{\tilde{A}}$ and $C_A \rightarrow (A(\alpha, \beta) \leftrightarrow \tilde{A}(\alpha, \beta + 1))$ are provable. Let $B$ be

$$\forall Y^{B_1} \ldots \forall Y^{B_n}\ \exists X^{\tilde{A}}\ \forall X^A(\varphi(Y^{B_1}, \ldots, Y^{B_n}, X^A) \leftrightarrow X^A \in X^{\tilde{A}}),$$

By means of Lemma and 2.1, it suffices to prove

$$(1)\qquad C_A \rightarrow B^{R(\alpha)}$$

for an arbitrary $\alpha$. On the other hand, we have

$$(2)\qquad C_A \wedge A(\alpha, \beta) \rightarrow [\forall y_1 \ldots \forall y_n\ \exists y \in R(\beta + 1)\ \forall x \in R(\beta)$$
$$(\varphi^{R(\alpha)}(y_1, \ldots, y_n, x) \leftrightarrow x \in y) \rightarrow B^{R(\alpha)}].$$

Take $\varphi^{R(\alpha)}(y_1, \ldots, y_n, x)$ as the $\varphi$ in Proposition 7 and $R(\beta)$ as the $u$

there. Then Proposition 7 implies

(3) $$\exists y \in R(\beta+1)\, \forall x \in R(\beta)\, (\varphi^{R(\alpha)}(y_1, \ldots, y_n, x) \leftrightarrow x \in y).$$

(1) follows from (2) and (3).

As a special case of Proposition 8 we have

$$\forall Y^{B_1} \ldots \forall Y^{B_n} \exists X\, \forall y\, (y \in X \leftrightarrow \varphi(Y^{B_1}, \ldots, Y^{B_n}, y)).$$

**Proposition 9.** $\mathcal{N}(\{\alpha \mid \forall x \in a\, \forall y \in R(\alpha)\, (Y^{\text{‘}}_x y \in R(\alpha))\})$.

*Proof.* This follows from

$$\{\alpha \mid \forall x \in a\, \forall y \in R(\alpha)\, (Y^{\text{‘}}_x y \in R(\alpha))\} = \bigcap_{x \in a} \{\alpha \mid \forall y \in R(\alpha)\, (Y^{\text{‘}}_x y \in R(\alpha))\},$$

by Proposition 3 and 4.

**Proposition 10.** $\forall F\, \forall a\, (\{F^{\text{‘}}x \mid x \in a\}$ is a set).

*Proof.* From Propositions 2 and 3 follows

$$\mathcal{N}(\{\alpha \mid \forall x \in R(\alpha)\, (F^{\text{‘}}x \in R(\alpha))\} \cap \{\alpha \mid a \in R(\alpha)\}).$$

Therefore we have $\exists \alpha\, \forall x \in a\, (F^{\text{‘}}x \in R(\alpha))$, whence follows the proposition.

**Proposition 11.** $\mathcal{N}(\{\alpha \mid \mathrm{In}(\alpha)\})$, where $\mathrm{In}(\alpha)$ means "$\alpha$ is a strongly inaccessible cardinal".

*Proof.* Let $B$ be ("Axiom of power set" $\wedge$ Proposition 10 $\wedge$ "Axiom of infinity"), where the axiom of infinity means the existence of the set $\omega$. Using a special case of 2.5, we have $\mathcal{N}(\{\alpha \mid B^{R(\alpha)}\})$, which implies the proposition.

**Proposition 12.**

$$\mathcal{N}(Y) \to \forall X\, \exists \alpha\, (\mathrm{In}(\alpha) \wedge \alpha \in Y \wedge \forall x \in R(\alpha)\, (X^{\text{‘}}x \in R(\alpha))).$$

*Proof.* Since $\mathcal{N}(\{\alpha \mid \forall x \in R(\alpha)\, (X^{\text{‘}}x \in R(\alpha))\})$, we have

$$\mathcal{N}(\{\alpha \mid \mathrm{In}(\alpha) \wedge \alpha \in Y \wedge \forall x \in R(\alpha)\, (X^{\text{‘}}x \in R(\alpha))\}),$$

whence follows the proposition.

**Proposition 13.** $\mathcal{N}(Y) \to \mathcal{N}(\{\alpha \mid \forall x \subseteqq R(\alpha)\, \exists \beta < \alpha\, (\mathrm{In}(\beta)$
$\wedge\, \beta \in Y \wedge \forall y \in R(\beta)\, (x^{\text{‘}}y \in R(\beta))\})$.

*Proof.* By applying 2.5 to Proposition 12, we have the proposition.

**Proposition 14.** (1) $\mathrm{In}(\alpha) \wedge a \in R(\alpha)$

$$\begin{aligned}&\to \forall X^{A_1} \ldots \forall X^{A_n} [\forall x \in R(\alpha)\, \forall y \in R(\alpha)\, \forall z \in R(\alpha)\, (\varphi(X^{A_1}, \ldots, X^{A_n}, y, x)\\ &\quad \wedge \varphi(X^{A_1}, \ldots, X^{A_n}, z, x) \to y = z)\\ &\to \exists x \in R(\alpha)\, \forall y \in R(\alpha)\, (y \in x \leftrightarrow \exists z \in a\, \varphi(X^{A_1}, \ldots, X^{A_n}, y, z))],\end{aligned}$$

where $\varphi$ does not contain $\mathcal{N}$.

*Proof.* By Proposition 7 there is a set $t$ such that

$$\forall z (z \in t \leftrightarrow \exists x \exists y (\varphi(X^{A_1}, \ldots, X^{A_n}, y, x) \wedge \langle y, x \rangle = z \wedge z \in R(\alpha) \times a),$$

taking $u$ in Proposition 7 as $R(\alpha) \times a$. Since $t$ is a function by our assumption and $\mathscr{D}(t) \subseteqq a \in R(\alpha)$, by a theorem of ZF, $\mathscr{W}(t) \in R(\alpha)$, q.e.d.

**Proposition 15.**

$$\forall a \forall X^{A_1} \ldots \forall X^{A_n} [\forall x \forall y \forall z (\varphi(X^{A_1}, \ldots, X^{A_n}, y, x) \wedge \varphi(X^{A_1}, \ldots, X^{A_n}, z, x) \rightarrow y = z) \rightarrow \exists x \forall y (y \in x \leftrightarrow \exists z \in a \, \varphi(X^{A_1}, \ldots, X^{A_n}, y, z))],$$

where $\varphi$ does not contain $\mathscr{N}$.

*Proof.* Call the formula $B$. It suffices to prove $B^{R(\alpha)}$ for every inaccessible number $\alpha$. But $B^{R(\alpha)}$ is a consequence of a special case of Proposition 14.

## § 4. Relative Consistency of $V = L$ with NTT

In this section we shall prove that "$V = L$" is consistent with NTT. We use the notation of [7] mostly without mention. In particular $L$ and $F$ are those in [7]. The whole argument is based on the following principle: GÖDEL's construction does not depend on ordinals or the particular well-ordering of ordinals: Given a set $x$ well-ordered by $w$ (i. e. $x \mathscr{W}_\varepsilon w$), let $|y|_w$ denote the order-type of an element $y$ of $x$ with respect to $w$ and $|x|_w$ the order-type of $x$ with respect to $w$. Then we can define a function $\tilde{F}$ on $x$ depending also on $w$ such that

$$\forall \alpha \forall y \in x (|y|_w = \alpha \rightarrow \tilde{F}`y = F`\alpha).$$

More directly we can define the function $\tilde{F}$ on $x$ simply imitating the Gödel construction along the well-ordering $w$ instead of $\in$ and define a class $\tilde{L}$ as follows:

(*) $a \in \tilde{L} \underset{\text{df}}{\leftrightarrow} \exists u \exists v (u \mathscr{W}_\varepsilon v \wedge \exists f (f \mathscr{F}n\, u \wedge$
"$f$ is defined analogous to Gödel construction along $v$"
$\wedge \exists x \in u (f`x = a)))$.

It is well-known that $a \in L \leftrightarrow a \in \tilde{L}$. In order to prove the relative consistency of $V = L$ with NTT, we wish to follow Gödel's construction in a system with typed variables. Though we have not developed the notion of ordinals for objects of higher type theory and it is not convenient to extend the notion "$a \in L$", it is rather natural and easy to extend the idea of the definition of $\tilde{L}$ to higher type theory, e.g. considering variables $X^A$ of type $A$ instead of first order variables. As will be seen below this construction can be carried out by using a finite number of types in addition to type $A$.

4.1. Definition of $\mathscr{A}^L$ and $\mathscr{L}_A$: For any formula $A$ with degree which determines a type and for any formula $\mathscr{A}$ in NTT we shall define the relativization $\mathscr{A}^L$ to $L$ and the notion $\mathscr{L}_A$, where $\mathscr{L}_A(X^{A^L})$ means $X^{A^L}$ is constructible, simultaneously by induction on the degree of a formula where the degree of a formula is defined to be the maximum of the degrees of the variables in it. For simplicity we shall write $X^{A^*}$, $Y^{A^*}, \ldots$ instead of $X^{A^L}$, $Y^{A^L}, \ldots$ .

1°. If $\deg(\mathscr{A}) = 0$, then $\mathscr{A}^L$ is defined as usual. The definition of $\mathscr{L}_A$ for $A$ with degree 0 will be seen in 3°. Assume that $\mathscr{B}^L$ and $\mathscr{L}_B(X^{B^*})$ are defined whenever $\deg(\mathscr{B}) \leqq n$ and $\deg(B) \leqq n$; and that $\deg(\mathscr{A}) = n+1$.

2°. $\mathscr{A}^L$ is defined as follows by induction on the number of logical symbols in $\mathscr{A}$ using the inductive hypothesis on degree.

$$(\mathscr{N}(X))^L \text{ is } \mathscr{N}(X);$$
$$(x \in X^B)^L \text{ is } x \in X^{B^*};$$
$$(X^B \in x)^L \text{ is } X^{B^*} \in x;$$
$$(X^B \in Y^C)^L \text{ is } X^{B^*} \in Y^{C^*};$$
$$(\forall X^C B(X^C))^L \text{ is } \forall X^{C^*}(\mathscr{L}_C(X^{C^*}) \to B^L(X^{C^*}));$$
$$(B \wedge C)^L \text{ is } B^L \wedge C^L;$$
$$(\neg B)^L \text{ is } \neg B^L;$$
$$(\forall x B(x))^L \text{ is } \forall x \in L\, B^L(x).$$

3°. Let $A(\alpha, \beta)$ be a formula with degree which determines a type. In order to define $\mathscr{L}_A(X^{A^*})$ we recall the remark at the beginning of this section. Let $A_i^L(\alpha, \beta)$ denote $C_{A^L} \wedge \exists\gamma(A^L(\alpha, \beta) \wedge \beta = \gamma + i)$ for $i < \omega$ and let $X^i$, $Y^i, \ldots$ denote the variables of type $A_i^L$. (This convention is used only in 4.1.) Using a finite number of types of the form $A_i^L$ we can define the notion $X^1 \mathscr{W}_\varepsilon Y^3$ (read "$X^1$ is well-ordered by $Y^3$") and for any pair $X^1$, $Y^3$ such that $X^1 \mathscr{W}_\varepsilon Y^3$ we can follow the Gödel construction and define a function $\tilde{F}_{X^1, Y^3}$ using $X^1$ and $Y^3$ instead of On and $\in$. The unique existence of $\tilde{F}_{X^1, Y^3}$ for arbitrary $X^1$ and $Y^3$ such that $X^1 \mathscr{W}_\varepsilon Y^3$ follows from the following facts: (i) $\mathscr{N}(\{\alpha \mid \psi \leftrightarrow \psi^{R(\alpha)}\})$ for each closed formula $\psi$ without $\mathscr{N}$ (by 2.5); (ii) if the type $B$ of each typed variable in a closed formula $\psi$ determines a type, i.e. $C_B$, then $\psi^{R(\alpha)}$ is equivalent a first order formula; (iii) by our assumption on $A(\alpha, \beta)$, $A_i^L(\alpha, \beta)$ determines a type for each $i < \omega$ and (iv) the unique existence of $F \restriction \alpha$ is provable in ZF. We define $\mathscr{L}_A(X)$ to mean there exists a pair $X^1$ and $Y^3$ such that $X^1 \mathscr{W}_\varepsilon Y^3$ and $X$ is in the range of the $\tilde{F}_{X^1, Y^3}$, i.e. $X = \tilde{F}_{X^1, Y^3}{}^{\backprime}Z$ for some $Z \in X^1$. We omit writing down this definition formally, for it is not difficult, but it is tedious.

Let us consider the predicate $\mathscr{L}_A^{R(\alpha)}(a)$. From the definition of $\mathscr{L}_A$ and its relativization to $R(\alpha)$ this reads the right side of (*) with certain restrictions on the bound variables: More precisely we have the following

4.2. **Proposition.** Let $A(\alpha, \beta)$ be an arbitrary formula without $\mathcal{N}$ whose only free variables are $\alpha$ and $\beta$.

$$(1) \qquad C_{A^L} \wedge A^L(\alpha, \beta) \wedge a \in R(\beta) \to (\mathcal{L}_A^{R(\alpha)}(a) \leftrightarrow a \in L).$$

$$(2) \qquad \neg C_{A^L} \wedge a \in R(\alpha) \to (\mathcal{L}_A^{R(\alpha)}(a) \leftrightarrow a \in L).$$

**Lemma 1.** *Let $\mathcal{A}$ be a formula without $\mathcal{N}$ and $\alpha$ be a cardinal. Then $R(\alpha)^L = L^{R(\alpha)} = R(\alpha) \cap L$ and $(\mathcal{A}^L)^{R(\alpha)} \leftrightarrow (\mathcal{A}^{R(\alpha)})^L$.*

*Proof.* We shall prove this by mathematical induction on the number of logical symbols in $\mathcal{A}$.

If $\mathcal{A}$ has no logical symbols, then $(\mathcal{A}^L)^{R(\alpha)}$ and $(\mathcal{A}^{R(\alpha)})^L$ are $\mathcal{A}$ itself.

As for the induction steps, we shall deal only with the essential cases, i.e., the cases where $\mathcal{A}$ is $\forall X^A \mathcal{B}(X^A)$ and $\forall x \mathcal{B}(x)$.

First it is obvious that $(C_A)^L \leftrightarrow C_{A^L}$ is provable and that $R(\alpha)^L = R(\alpha) \cap L$.

$$\begin{aligned} ((\forall X^A \mathcal{B}(X^A))^{R(\alpha)})^L \colon\ & (C_{A^L} \wedge \exists \beta (A^L(\alpha, \beta) \\ & \wedge \forall x \in R(\beta)\, (\mathcal{L}_A^{R(\alpha)}(x) \to (\mathcal{B}^L(x))^{R(\alpha)}))) \\ & \vee (\neg C_{A^L} \wedge \forall x \in R(\alpha)\, (\mathcal{C}_A^{R(\alpha)}(x) \to (\mathcal{B}^L(x))^{R(\alpha)})). \end{aligned}$$

$$\begin{aligned} ((\forall X^A \mathcal{B}(X^A))^{R(\alpha)})^L \colon\ & (C_{A^L} \wedge \exists \beta (A^L(\alpha, \beta) \\ & \wedge \forall x \in (R(\beta) \cap L))\, \mathcal{B}^{R(\alpha)}(x))^L)) \\ & \vee (\neg C_{A^L} \wedge \forall x \in (R(\alpha) \cap L)\, (\mathcal{B}^{R(\alpha)}(x))^L). \end{aligned}$$

These are equivalent by the inductive hypothesis, and (1) and (2) in Proposition.

$$((\forall x \mathcal{B}(x))^L)^{R(\alpha)} \colon \forall x \in (L^{R(\alpha)} \cap R(\alpha))\, (\mathcal{B}^L(x))^{R(\alpha)},$$
$$((\forall x \mathcal{B}(x))^{R(\alpha)})^L \colon \forall x \in (R(\alpha) \cap L)\, (\mathcal{B}^{R(\alpha)}(x))^L.$$

Those are equivalent by the inductive hypothesis.

If $A(\alpha, \beta)$ is $\beta = \alpha + 1$, then we simply write $\mathcal{L}(X)$ instead of $\mathcal{L}_A(X)$.

**Lemma 2.** *Let $\varphi(a, a_1, \ldots, a_m, X_1, \ldots, X_n)$ be an arbitrary first order formula having $a, a_1, \ldots, a_m, X_1, \ldots, X_n$ as the only free variables in it ($m, n \leq 0$). Then*

$$(1) \quad \forall a_1 \cdots \forall a_m \forall X_1 \cdots \forall X_n (a_1 \in L \wedge \cdots \wedge a_m \in L$$
$$\wedge \mathcal{L}(X_1) \wedge \cdots \wedge \mathcal{L}(X_n) \to \mathcal{L}(\{a \mid a \in L \wedge \varphi^L(a, a_1, \ldots, a_n, X_1, \ldots, X_n)\})).$$

*Proof.* From a special case of 2.5 together with Propositions 11 and 5 in § 3, $(1) \leftrightarrow (1)^{R(\alpha)}$ for some cardinal $\alpha$. For such an $\alpha$

$$(1)^{R(\alpha)} \leftrightarrow \forall a_1 \cdots \forall a_m \forall x_1 \cdots \forall x_n \forall z \in R(\alpha + 1)$$
$$(a_1 \in L \cap R(\alpha) \wedge \cdots \wedge a_m \in L \cap R(\alpha) \wedge x_1 \in L \cap R(\alpha + 1) \wedge \cdots$$

$$\wedge\, x_n \in L \cap R(\alpha+1)$$
$$\wedge\, \forall y \in R(\alpha)\,(y \in z \leftrightarrow y \in L \cap R(\alpha)\, \wedge$$
$$\wedge\, \varphi^{L \cap R(\alpha)}(y, a_1, \ldots, a_n, x_1, \ldots, x_n)) \to z \in L)$$

by (1) in Proposition 1 and Lemma 1: The last formula is ZF-provable.

Since GÖDEL [7] proved that $(V = L)^L$ is provable in ZF, what we must show for the relative consistency-proof of $V = L$ is the following theorem.

**Theorem.** If $\mathscr{A}(X_1^{A_1}, \ldots, X_n^{A_n})$ is provable in NTT, then

$$\mathscr{L}_{A_1}(X_1^{A_1}) \wedge \cdots \wedge \mathscr{L}_{A_n}(X_n^{A_n}) \to \mathscr{A}^L(X_1^{A_1}, \ldots, X_n^{A_n})$$

is also provable, where $X_1^{A_1}, \ldots, X_n^{A_n}$ are only free variables in $\mathscr{A}(X_1^{A_1}, \ldots, X_n^{A_n})$ and first order variables may occur among them.

*Proof.* We prove this by transfinite induction on the ordinal of a proof figure to $\mathscr{A}(X_1^{A_1}, \ldots, X_n^{A_n})$. If $\mathscr{A}(X_1^{A_1}, \ldots, X_n^{A_n})$ is obtained by some inference, then the theorem follows immediately from the inductive hypotheses. Therefore we consider only the cases that $\mathscr{A}(X_1^{A_1}, \ldots, X_n^{A_n})$ is an axiom.

Case 1) $\mathscr{A}$ is a logical axiom or an axiom of ZF.

This case follows from GÖDEL [7].

Case 2) $\mathscr{A}$ is $\forall X(\mathscr{N}(X) \to 0 \neq X \subseteqq \mathrm{On})$. Since $\mathrm{On}^L = \mathrm{On}$ and $(\beta = \alpha + 1)^L$ is $\beta = \alpha + 1$ itself, $\mathscr{A}^L$ is equivalent to

$$\forall X(\mathscr{L}(X) \wedge \mathscr{N}(X) \to 0 \neq X \subseteqq \mathrm{On}),$$

which follows from 2.1.

Case 3) $\mathscr{A}$ is $\forall X\, \forall Y(\mathscr{N}(X) \wedge X \subseteqq Y \subseteqq \mathrm{On} \to \mathscr{N}(Y))$.

The proof goes in the same way as in Case 2).

Case 4) $\mathscr{A}$ is $\forall X\, \forall a(\mathscr{N}(X \to \mathscr{N}(X - a))$, i.e.,

$$\forall X\, \forall Y\, \forall a(\mathscr{N}(X) \wedge \forall x(x \in Y \leftrightarrow x \in X \wedge \neg\, x \in a) \to \mathscr{N}(Y)).$$

$\mathscr{A}^L$ is

$$\forall X\, \forall Y\, \forall a(\mathscr{L}(X) \wedge a \in L \wedge \mathscr{N}(X) \wedge \mathscr{L}(Y) \wedge \forall x \in L(x \in Y \leftrightarrow$$
$$\leftrightarrow x \in X \wedge \neg\, x \in a) \to \mathscr{N}(Y)),$$

which follows from Lemma 2.

Case 5) $A$ is

$$\forall X(\forall y\, \mathscr{N}(X_y) \to \mathscr{N}(\{\alpha \mid \forall y \in R(\alpha)(\alpha \in X_y)\})).$$

$\mathscr{A}^L$ is

$$\forall X(\mathscr{L}(X) \wedge \forall y \in L\, \mathscr{N}(X_y) \to \mathscr{N}(\{\alpha \mid \forall y \in R(\alpha) \cap L(\alpha \in X_y)\}))$$

which follows from 2.4.

Case 6) $\mathscr{A}$ is

$$\forall X\, \mathscr{N}(\{\alpha \mid \forall y \in R(\alpha)(\varphi(y, X) \leftrightarrow \varphi^{R(\alpha)}(y, R(\alpha) \cap X))\}).$$

$\mathscr{A}^L$ is

$$\forall X(\mathscr{L}(X) \to \mathscr{N}(\{\alpha | \forall y \in R(\alpha) \cap L(\varphi^L(y, X) \leftrightarrow (\varphi^L(y, R(\alpha) \cap X))^{R(\alpha)})\})),$$

which follows from

$$\forall X \mathscr{N}(\{\alpha | \forall y \in R(\alpha)((y \in L \wedge \varphi^L(y, X)) \leftrightarrow (y \in L \wedge \varphi^L(y, R(\alpha) \cap X))^{R(\alpha)})\}),$$

i.e., the special case of 2.5.

## § 5. An Interpretation of TT in NTT[1]

In § 1, Chapter I of [28] we defined a transfinite type set theory TT developed in a first order language and having a predicate constant $\varepsilon$ and an individual constant $\alpha_0$. We shall prove that the system TT can be interpreted in the subsystem $\mathrm{NTT}^1$ of NTT which is obtained by restricting the variables to those of degree $\leqq 1$:

**Theorem.** If $\varphi_0(\alpha_0, x_1, \ldots, x_n)$ is provable in TT, then

$$\mathscr{N}(\{\alpha_1 | \forall x_1, \ldots \forall x_n \varphi_0(\alpha_1, x_1, \ldots, x_n)\})$$

is provable in $\mathrm{NTT}^1$, where the $\alpha_1$ is an arbitrary variable which ranges over ordinal numbers.

**Corollary.** If $\varphi_0$ is provable in TT, where $\varphi_0$ is a closed formula of TT in which $\alpha_0$ does not occur, then it is also provable in $\mathrm{NTT}^1$.

*Proof.* Using the theorem, we have $\mathscr{N}(\{\alpha_1 | \varphi_0\})$ is provable in $\mathrm{NTT}^1$. Since $\mathscr{N}(\{\alpha_1 | \varphi_0\})$ we have $\{\alpha_1 | \varphi_0\} \neq 0$. Since $\alpha_1$ is not free in $\varphi_0$, we have $\varphi_0$.

Since the notion of satisfaction is involved in the axiom of TT, we present a survey of satisfaction relations before we start the proof of the theorem.

1) It is well known that we can formally deal with the notions concerning formulas of a first order set theory (for example, TT) in the theory itself. We shall denote the set which corresponds to a formal object $X$ of set theory (as an object of $\mathrm{NTT}^1$ without $\mathscr{N}$ here) by $\ulcorner X \urcorner *$. $X$ may be a set, one of the logical symbols, a variable or a formula. $\ulcorner X \urcorner *$ shall be called "Gödelization" of $X$.

In so doing, we use some notational conventions, i.e., $\ulcorner \varphi \urcorner *$ denotes the Gödelization of a formula $\varphi$ and $\forall \ulcorner \varphi \urcorner *$ denotes a quantified variable which ranges over the set of Gödelizations of formulas. $\ulcorner x \urcorner *$ denotes the Gödelization of a set $x$.

We define further notations which shall be seen in the supplement to this section. Among them:

$Cf(\ulcorner \varphi \urcorner *)$ means that $\varphi$ is a closed well-formed formula;

$Nl` \ulcorner \varphi \urcorner *$ shows the complexity of $\varphi$; $\ulcorner \varphi(X) \urcorner *$ denotes Gödelization of the result of substitution of $X$ for $x_i$ in $\varphi(x_i)$.

2) Satisfaction relations.

2.1) We can define a Gödel numbering of formulas of ZF (i.e. an assignment of a natural number to each formula) in a usual manner. Let $\ulcorner\varphi\urcorner$ be the Gödel number of a formula $\varphi$. We can define a satisfaction relation $[R(\alpha), f \models \ulcorner\varphi(x_1, \ldots, x_n)\urcorner]$ for the formulas $\varphi(x_1, \ldots, x_n)$ of ZF as usual, where $f$ is a function from $\omega$ to $R(\alpha)$. For the sake of convenience, we shall denote the relation

$$\exists f(f`1 = a_1 \wedge \ldots \wedge f`n = a_n \wedge [R(\alpha), f \models \ulcorner\varphi(x_1, \ldots, x_n)\urcorner])$$

by $R(\alpha) \overset{1}{\models} \ulcorner\varphi(a_1, \ldots, a_n)\urcorner$, where $a_1, \ldots, a_n$ are arbitrary elements of $R(\alpha)$. We also use the abbreviation $\forall\ulcorner\varphi(x_i)\urcorner$, which means "for any formula $\varphi(x_i)$ having at most one argument $x_i$".

2.2) We can define another formalization of satisfaction following the Gödelization stated in 1) accepting "individual constants". Here we regard every set as an individual constant. If we write $\ulcorner\varphi(\underline{a}_1, \ldots, \underline{a}_n)\urcorner^*$, $a_i$ denotes the name of a set $\underline{a}_i$. $R(\alpha) \models \ulcorner\varphi(\underline{a}_1, \ldots, \underline{a}_n)\urcorner^*$ is defined only if $a_1 \in R(\alpha) \wedge \ldots \wedge a_n \in R(\alpha)$. This relation shall be denoted by $R(\alpha) \overset{2}{\models} \ulcorner\varphi(\underline{a}_1, \ldots, \underline{a}_n)\urcorner^*$.

These relations are formalized by well-known methods. It is also well-known that the following assertion concerning those relations can be proved.

3) Let $m$ be a natural number. $m^*(\underline{a}_1, \ldots, \underline{a}_n)$ is defined by:

$$m^*(\underline{a}_1, \ldots, \underline{a}_n) = \begin{cases} \ulcorner\varphi(\underline{a}_1, \ldots, \underline{a}_n)\urcorner^* \quad \text{if} \quad m = \ulcorner\varphi(x_1, \ldots, x_n)\urcorner \\ \text{where } \varphi(x_1, \ldots, x_n) \text{ is a formula of first} \\ \text{order set theory;} \\ 0 \quad \text{otherwise.} \end{cases}$$

Then $\exists f(f`1 = a_1 \wedge \ldots \wedge f`n = a_n \wedge [R(\alpha), f \models m])$ if and only if $R(\alpha) \overset{2}{\models} m^*(\underline{a}_1, \ldots, \underline{a}_n)$, or $R(\alpha) \overset{1}{\models} \ulcorner\varphi(a_1, \ldots, a_n)\urcorner$ if and only if $R(\alpha) \overset{2}{\models} \ulcorner\varphi(\underline{a}_1, \ldots, \underline{a}_n)\urcorner^*$ for any formula $\varphi(x_1, \ldots, x_n)$ of ZF.

In particular, $R(\alpha) \overset{1}{\models} \ulcorner\varphi\urcorner$ if and only if $R(\alpha) \overset{2}{\models} \ulcorner\varphi\urcorner^*$, where $\varphi$ is an arbitrary closed formula of ZF.

We have given the above argument since in order to apply infinite induction in the succeeding argument it is useful to have the set of Gödelizations of formal expressions included in $\omega$, while it is more convenient to use $\ulcorner\varphi(\underline{a}_1, \ldots, \underline{a}_n)\urcorner^*$ to define the truth definition.

$\overset{1}{\models}$ is sometimes abbreviated by $\models$.

*Proof of the theorem.* Let $\varphi_0$ be provable in TT and $P$ be a proof-figure to $\varphi_0$. Since our inferences are those of the predicate calculus together with infinite induction, we can assign an ordinal number ($< \omega_1$) to $P$ and can prove the theorem by transfinite induction on the ordinal of $P$. Since the induction steps are easy to carry out, we shall show that $\mathscr{N}(\{\alpha_1 \mid \varphi_0(\alpha_1)\})$ is provable in NTT[1] for every axiom $\varphi_0(\alpha_0)$ of TT.

Case 1) $\varphi_0(\alpha_0)$ is a logical axiom or an axiom of ZF. Then

$$\{\alpha_1 | \varphi_0(\alpha_1)\} = \mathrm{On}$$

and the theorem is evident.

Case 2) $\varphi_0(\alpha_0)$ is $\mathrm{In}(\alpha_0)$. In this case, what must be proved is

$$\mathcal{N}(\{\alpha_1 | \mathrm{In}(\alpha_1)\}),$$

which is Proposition 11 in § 3.

Case 3) $\varphi_0(\alpha_0)$ is $\forall x \in R(\alpha_0)(A(x) \leftrightarrow A^{R(\alpha_0)}(x))$. This case is directly proved by using a special case of Axiom 2.5.

Case 4) $\varphi_0(\alpha_0)$ is

$$C_A \to \forall \gamma < \alpha_0 \, \exists \alpha < \alpha_0 \, \exists \beta \, \exists \beta_0 (\mathrm{In}(\alpha) \wedge \gamma < \alpha \wedge A(\alpha, \beta) \\ \wedge A(\alpha_0, \beta_0) \wedge \forall \ulcorner \varphi \urcorner \, \forall x \in R(\alpha) (\varphi^{R(\beta_0)}(\alpha_0, x) \leftrightarrow \varphi^{R(\beta)}(\alpha, x))),$$

where $\varphi^{R(\beta)}(\alpha, x)$ is to be understood as

$$\exists f (f` 0 = \alpha \wedge f` 1 = x \wedge [R(\beta), f \models \ulcorner \varphi(x_0, x_1) \urcorner]$$

(cf. § 1, Chapter I of [28]). Given any formula $A(\alpha, \beta)$ of ZF whose only free variables are $\alpha$ and $\beta$, we shall define the $A$-operation applied to the formulas of TT. The result of $A$-operation applied to a formula $\varphi$ shall be denoted by $\varphi^A$.

(1) $(b \in c)^A \leftrightarrow b \in c$.

(2) $(\alpha_0 \in b)^A \leftrightarrow \mathrm{On} \in b$.

(3) $(\alpha_0 \in \alpha_0)^A \leftrightarrow \mathrm{On} \in \mathrm{On}$.

(4) $(b \in a_0)^A \leftrightarrow \mathrm{Ord}(b)$.

(5) $(\forall x \psi(x))^A \leftrightarrow \forall X^A \psi^A(X^A)$, where $\psi^A(X^A)$ is obtained from $\psi^A(x)$ by replacing $x$ by $X^A$ at all its occurrences.

(6) $(\varphi_1 \wedge \varphi_2)^A \leftrightarrow \varphi_1^A \wedge \varphi_2^A$.

(7) $(\neg \psi)^A \leftrightarrow \neg \psi^A$.

*Remark.* The resulting formulas are those of $\mathrm{NTT}^1$ having only free set variables and bound set- or type $A$-variables.

We can define a formula $T_A(b)$ of $\mathrm{NTT}^1$ (the truth definition of type $A$) such that

(8) $C_A \wedge A(\alpha, \beta) \wedge a \in R(\alpha) \to ((T_A(\ulcorner(\varphi(\alpha_0, \underline{a}))^A \urcorner *))^{R(\alpha)} \leftrightarrow \varphi^{R(\beta)}(\alpha, a))$,

and

(9) $C_A \wedge A(\alpha, \beta) \wedge \beta + \omega + 1 < \alpha_2$
$\to (((T_A(\ulcorner(\varphi(a_0, \underline{a}))^A \urcorner *))^{R(\alpha)})^{R(\alpha_2)} \leftrightarrow (T_A(\ulcorner(\varphi(\alpha_0, \underline{\alpha}))^A \urcorner *))^{R(\alpha)})$,

where $\varphi(\alpha_0, a)$ is any formula of TT having only one free variable $a$. A more detailed argument shall be provided in I of supplement to this section.

In order to prove the theorem for this case, we need to show that $C_A$ imples $\mathcal{N}(B)$, where $B$ is

$$\{\alpha_1 \mid \forall\gamma < \alpha_1 \exists\alpha < \alpha_1 \exists\beta \exists\beta_1 (\mathrm{In}(\alpha) \wedge \gamma < \alpha \wedge A(\alpha,\beta) \wedge A(\alpha_1,\beta_1) \wedge \forall \ulcorner\varphi\urcorner \forall x \in R(\alpha)(\varphi^{R(\beta_1)}(\alpha_1, x) \leftrightarrow \varphi^{R(\beta)}(\alpha, x)))\}.$$

First we shall show $\mathcal{N}(B)$ follows from the following (10) under the assumption $C_A$:

(10) $\quad \mathcal{N}(B_1)$,

where

$$B_1 = \{\alpha_1 \mid (\forall_\gamma \exists\alpha(\gamma < \alpha \wedge \mathrm{In}(\alpha) \wedge \forall \ulcorner\varphi(x_i)\urcorner \forall x \in R(\alpha)(T_A(\ulcorner(\varphi(\underline{x}))^A\urcorner *) \leftrightarrow T_A(\ulcorner(\varphi(\underline{x}))^A\urcorner *)))^{R(\alpha_1)}\}.$$

By 2.5, $\mathcal{N}(\{\alpha_1 \mid C_A \leftrightarrow C_A{}^{R(\alpha_1)}\}$. Thus from (10) using 2.5, Propositions 5 and 11

(11) $\quad \mathcal{N}(B_2)$,

where $B_2 = \{\alpha_1 \mid (C_A \leftrightarrow C_A^{R(\alpha_1)}) \wedge \mathrm{In}(\alpha_1) \wedge \alpha_1 \in B_1\}$.

Let $\alpha_1 \in B_2$. Then, from $C_A$, (8) and (9),

$$\{\forall_\gamma < \alpha_1 \exists\alpha < \alpha_1 \exists\beta < \alpha_1 \exists\beta_1(\gamma < \alpha \wedge \mathrm{In}(\alpha) \wedge A(\alpha,\beta) \wedge A(\alpha_1,\beta_1) \wedge \forall \ulcorner\varphi(x_i)\urcorner \forall x \in R(\alpha)(\varphi^{R(\beta_1)}(\alpha_1, x) \leftrightarrow \varphi^{R(\beta)}(\alpha, x))\},$$

i.e., $\alpha_1 \in B$. Thus $\mathcal{N}(B)$ follows from (10) and 2.1 under the assumption $C_A$. Assume $C_A$. Then, by Proposition 5′ (10) follows from

$$\forall\gamma \exists\alpha(\gamma < \alpha \wedge \mathrm{In}(\alpha) \wedge \forall \ulcorner\varphi(x_i)\urcorner \forall x \in R(\alpha)(T_A(\ulcorner(\varphi(\underline{x}))^A\urcorner *) \leftrightarrow (T_A(\ulcorner(\varphi(\underline{x}))^A\urcorner *))^{R(\alpha)}).$$

Therefore what must be proved is

$$\mathcal{N}(\{\alpha / \forall \ulcorner\varphi(x_i)\urcorner \forall x \in R(\alpha)(T_A(\ulcorner\varphi(\underline{x}))^A\urcorner *) \leftrightarrow (T_A(\ulcorner(\varphi(\underline{x}))^A\urcorner *))^{R(\alpha)})\}),$$

since we have $\mathcal{N}(\{\alpha \mid \mathrm{In}(\alpha)\})$ by Proposition 11 in § 3. Using Proposition 4 in § 3, we have only to prove

$$\mathcal{N}(\{\alpha \mid \forall x \in R(\alpha)(T_A(\ulcorner(\varphi(\underline{x}))^A\urcorner *) \leftrightarrow (T_A(\ulcorner(\varphi(\underline{x}))^A\urcorner *))^{R(\alpha)})\})$$

for each integer $\ulcorner\varphi(x)\urcorner$, which is a special case of 2.5.

Case 5). Let $\Gamma$ be a subset of provable closed formulas of TT, which is strongly definable, and let $E(\alpha_0, a)$ represents $\Gamma$: i.e. $\Gamma$ and $E(\alpha_0, a)$ satisfy the following conditions.

1) The only free variable in $E(\alpha_0, a)$ is $a$.

2) For every numberal $n$, one of $E(\alpha_0, n)$ and $\neg E(\alpha_0, n)$ is TT-provable.

3) $E(\alpha_0, n)$ is TT-provable if and only if there exists a formula $\varphi$ in $\Gamma$ such that $n = \ulcorner\varphi\urcorner$.

$\varphi(\alpha_0)$ is

$$\forall\gamma\,\exists\alpha(\gamma<\alpha\wedge\mathrm{In}(\alpha)\wedge\forall\ulcorner\varphi(\alpha_0)\urcorner(E(\alpha_0,\ulcorner\varphi(\alpha_0)\urcorner)\to(R(\alpha)\models\ulcorner\varphi(\alpha_0)\urcorner))).$$

We can define a formula $T(\ulcorner\varphi(\underline{\alpha})\urcorner{*})$ of NTT[1] for which the following are NTT[1]-provable.

(1) $$\alpha_1<\alpha\wedge\alpha\in K_{\mathrm{II}}\to((T(\ulcorner\varphi(\underline{\alpha_1})\urcorner{*})^{R(\alpha)}\leftrightarrow R(\alpha)\models^{\!1}\ulcorner\varphi(\alpha_1)\urcorner)$$

for any closed formula $\ulcorner\varphi(\alpha_0)\urcorner$ of TT where $R(\alpha)\models^{\!1}\ulcorner\varphi(\alpha_1)\urcorner$ is the abbreviation of $\exists f(f`0=\alpha_1\wedge[R(\alpha),f\models^{\!1}\ulcorner\varphi(x_0)\urcorner])$.

(2) For each closed formula $\ulcorner\varphi(\alpha_0)\urcorner$ of TT,

$$T(\ulcorner\varphi(\underline{\alpha_1})\urcorner{*})\leftrightarrow\varphi(\alpha_1).$$

A more detailed argument shall be provided in II of supplement to this section. Now if $\varphi(\alpha_0)$ is in $\Gamma$, then $\mathscr{N}(\{\alpha_1|\varphi(\alpha_1)\})$ by the inductive hypothesis, and hence $\mathscr{N}(\{\alpha_1|T(\ulcorner\varphi(\underline{\alpha_1})\urcorner{*})\})$ from (2). On the other hand, if $\varphi(\alpha_0)$ is not in $\Gamma$, then $\neg E(\alpha_0,\ulcorner\varphi(\alpha_0)\urcorner)$ is provable in TT, hence $\mathscr{N}(\{\alpha_1|\neg E(\alpha_1,\ulcorner\varphi(\alpha_1)\urcorner)\})$. Therefore we have $\mathscr{N}(\{\alpha_1|E(\alpha_1,\ulcorner\varphi(\alpha_1)\urcorner)\to T(\ulcorner\varphi(\alpha_1)\urcorner{*}))$ for each integer $\ulcorner\varphi(\alpha_0)\urcorner$, from which $\mathscr{N}(B)$ is derived, where $B=\{\alpha_1|\forall\ulcorner\varphi(x_0)\urcorner(E(\alpha_1,\ulcorner\varphi(\alpha_1)\urcorner)\to T(\ulcorner\varphi(\underline{\alpha_1})\urcorner{*}))\}$. We also have

$$\mathscr{N}(\{\alpha|\forall\ulcorner\varphi(x_0)\urcorner((T(\ulcorner\varphi(\underline{\alpha_1})\urcorner{*}))^{R(\alpha)}\leftrightarrow T(\ulcorner\varphi(\underline{\alpha_1})\urcorner{*}))\}),$$

whence follows

$$\alpha_1\in B\to\mathscr{N}(\{\alpha|\forall\ulcorner\varphi(x_0)\urcorner(E(\alpha_1,\ulcorner\varphi(\alpha_1)\urcorner)\to R(\alpha)\models^{\!1}\ulcorner\varphi(\alpha_1)\urcorner)\})$$

from (1). Therefore $\alpha_1\in B\to\varphi_0(\alpha_1)$ by Proposition 12 in § 3, which implies $\{\alpha_1|\varphi_0(\alpha_1)\}\supseteqq B$. Hence the theorem holds.

In [28], we also presented several generalizations of TT. We shall prove that our theorem holds for all these generalizations as well

Case 6) $\varphi_0(\alpha_0)$ is

$$\forall x\in R(\alpha_0+1)\,\exists\beta<\alpha_0(\mathrm{In}(\beta)\wedge\forall y\in R(\beta)(x`y\in R(\beta)).$$

From Proposition 3 and Proposition 11 in § 3, follows

$$\forall F\,\exists\beta(\mathrm{In}(\beta)\wedge\forall y\in R(\beta)(F`y\in R(\beta))).$$

Using 2.5, we have

$$\mathscr{N}(\{\alpha_1|\forall x\in R(\alpha_1+1)\,\exists\beta<\alpha_1(\mathrm{In}(\beta)\wedge\forall y\in R(\beta)(x`y\in R(\beta)))\}),$$

whence follows the theorem.

Case 7) $\varphi_0(\alpha_0)$ is

$$C_A\to\forall\delta<\alpha_0\,\forall\gamma<\alpha_0\,\exists\alpha<\alpha_0\,\exists\beta\,\exists\beta_0(\mathrm{In}(\delta,\alpha)\wedge A(\alpha,\beta)\wedge A(\alpha_0,\beta_0)$$
$$\wedge\,\gamma<\alpha\wedge\forall\ulcorner\varphi(\alpha_0,x_1)\urcorner\,\forall x\in R(\alpha)(\varphi^{R(\beta_0)}(\alpha_0,x)\leftrightarrow\varphi^{R(\beta)}(\alpha,x))),$$

where $\mathrm{In}(\delta,\alpha)$ means "$\alpha$ is a hyper-inaccessible number of type $\delta$" and is defined by the following.

$$\mathrm{In}(0, \alpha) \leftrightarrow \mathrm{In}(\alpha).$$
$$\mathrm{In}(\delta+1, \alpha) \leftrightarrow \mathrm{In}(\delta, \alpha) \wedge \forall x \in R(\alpha+1)\, \exists \beta < \alpha (\mathrm{In}(\delta, \beta) \wedge \forall y \in R(\beta)(x' y \in R(\beta)).$$
$$\delta \in K_{\mathrm{II}} \rightarrow (\mathrm{In}(\delta, \alpha) \leftrightarrow \forall \mu < \delta\, \mathrm{In}(\mu, \alpha)).$$

By transfinite induction on $\delta$, we can prove $\forall \delta\, \mathscr{N}(\{\alpha | \mathrm{In}(\delta, \alpha)\})$ in the same way as in Case 6). Then, similarly as in Case 4), we see that what must be proved is

$$C_A \rightarrow \forall \delta\, \forall \gamma\, \exists \alpha (\mathrm{In}(\delta, \alpha) \wedge \gamma < \alpha \wedge \forall \ulcorner \varphi(x_1) \urcorner\, \forall x \in R(\alpha) (T_A(\ulcorner (\varphi(\underline{x}))^{A} \urcorner *) \leftrightarrow (T_A(\ulcorner (\varphi(\underline{x}))^{A} \urcorner *))^{R(\alpha)})).$$

Therefore it suffices to show

$$C_A \rightarrow \mathscr{N}(\{\alpha | \forall \ulcorner \varphi(x_1) \urcorner\, \forall x \in R(\alpha) (T_A(\ulcorner (\varphi(\underline{x}))^{A} \urcorner *) \leftrightarrow (T_A(\ulcorner (\varphi(\underline{x}))^{A} \urcorner *))^{R(\alpha)})\}),$$

which has already been proved in Case 4).

Case 8) $\varphi_0(\alpha_0)$ is

$$\forall \beta\, \forall \gamma\, \exists \alpha (\beta < \alpha \wedge \mathrm{In}(\gamma, \alpha) \wedge \forall \ulcorner \varphi(\alpha_0) \urcorner (E(\alpha_0, \ulcorner \varphi(\alpha_0) \urcorner) \rightarrow (R(\alpha) \overset{1}{\models} \ulcorner \varphi(\alpha_0) \urcorner))),$$

where $E(\alpha_0, a)$ satisfies the same properties as in Case 5).

By the same proof as in Case 5), we have

$$\alpha_1 \in B \rightarrow \mathscr{N}(\{\alpha | \forall \ulcorner \varphi(x_0) \urcorner (E(\alpha_1, \ulcorner \varphi(\alpha_1) \urcorner) \rightarrow R(\alpha) \overset{1}{\models} \ulcorner \varphi(\alpha_1) \urcorner)\}),$$

where

$$B = \{\alpha_1 | \forall \ulcorner \varphi(x_0) \urcorner (E(\alpha_1, \ulcorner \varphi(\alpha_1) \urcorner) \rightarrow T(\ulcorner \varphi(\alpha_1) \urcorner *))\}.$$

We also have $\mathscr{N}(B)$. Therefore the theorem holds in this case.

Case 9) $\varphi_0(\alpha_0)$ is obtained from one of $\varphi_0(\alpha_0)$ in Cases 2), 4) or 5) by replacing $\mathrm{In}(\alpha)$ by $\tilde{E}(\alpha)$, where $\tilde{E}(\alpha)$ is the abbreviation of the formula

$$\exists \beta (\mathrm{In}(\alpha) \wedge A(\alpha, \beta) \wedge \forall \ulcorner \varphi(\alpha_0) \urcorner (B(\ulcorner \varphi(\alpha_0) \urcorner) \rightarrow R(\beta) \overset{1}{\models} \ulcorner \varphi(\alpha) \urcorner)),$$

where $\alpha_0$ does not occur in $A(\alpha, \beta)$ and $B(\ulcorner \varphi(\alpha) \urcorner)$ satisfies the following conditions. ($\tilde{E}$ depends on $A$.)

1) The only free variable in $B(a)$ is $a$.

2) For every natural number $n$, one of $B(n)$ or $\neg B(n)$ is provable.

3) There exists a subsystem $\Gamma$ of provable closed formulas and $B(n)$ is provable if and only if $n = \ulcorner \varphi(\alpha_0) \urcorner$ for some $\varphi(\alpha_0)$ in $\Gamma$.

4) $C_A$ is provable.

5) $\forall \beta (A(\alpha_0, \beta) \rightarrow \varphi^{R(\beta)}(\alpha_0))$ is provable for any $\varphi(\alpha_0)$ in $\Gamma$.

As we see in the proof of Case 2), 4) and 5), what we must prove is $\mathcal{N}(\{\alpha_1|\tilde{E}(\alpha_1)\})$. Since in virtue of 4), $\tilde{E}(\alpha)$ is equivalent to

$$\mathrm{In}(\alpha) \wedge \forall \ulcorner\varphi(\alpha_0)\urcorner \exists\beta(A(\alpha,\beta) \wedge (B(\ulcorner\varphi(\alpha_0)\urcorner) \to R(\beta) \overset{1}{\models} \ulcorner\varphi(\alpha)\urcorner)),$$

what must be proved is

$$\mathcal{N}(\{\alpha_1|\exists\beta(A(\alpha,\beta) \wedge (B(\ulcorner\varphi(\alpha_1)\urcorner) \to R(\beta) \overset{1}{\models} \ulcorner\varphi(\alpha)\urcorner)\})$$

for each integer $\ulcorner\varphi(x_0)\urcorner$. If $\varphi(\alpha_0)$ is in $\Gamma$, then this is equivalent to

$$\mathcal{N}(\{\alpha|\exists\beta(A(\alpha,\beta) \to \varphi^{R(\beta)}(\alpha)\}),$$

which follows from 5) and the inductive hypothesis.

If $\varphi(\alpha_0)$ is not in $\Gamma$, then $\neg B(\ulcorner\varphi(\alpha_0)\urcorner)$, whence follows

$$\mathcal{N}(\{\alpha_1|\neg B(\ulcorner\varphi(\alpha_1)\urcorner)\})$$

and so

$$\mathcal{N}(\{\alpha|\exists\beta(A(\alpha,\beta) \wedge (B(\ulcorner\varphi(\alpha_1)\urcorner) \to R(\beta) \overset{1}{\models} \ulcorner\varphi(\alpha)\urcorner)\}).$$

**Supplement to § 5**

I. Definition of the truth definition for Case 4). We shall outline the definition of the truth definition $T_A$ for the formulas of type $A$ in the language with typed variables $X^{A_i}$ where $A_i(\alpha,\beta)$ is

$$C_A \wedge \exists\gamma(A(\alpha,\gamma) \wedge \beta = \gamma + i),$$

and the proof of the properties used in the proof of the theorem for Case 4). Note that $C_A \leftrightarrow C_{A_i}$ and $C_A \to (A(\alpha,\beta) \to A_i(\alpha,\beta+i))$ are provable. (This shall be made use of through the following argument.) $A$ shall be fixed throughout I and the notation $A_i$ is temporarily used here.

I.1. Gödelization. For simplicity's sake we shall write $X$ instead of $X^A$. This abbreviation shall be effective throughout I.

I.1.1. $\ulcorner\underline{a}\urcorner^* = \langle 0\, a\rangle$.
I.1.2. $\ulcorner x_i\urcorner^* = \langle 1\, i\rangle$.
I.1.3. $\ulcorner\underline{X}\urcorner^* = \langle 2\, X\rangle$.
I.1.4. $\ulcorner X_i\urcorner^* = \langle 3\, i\rangle$.
I.1.5. $\ulcorner X \in Y\urcorner^* = \langle 4\, \ulcorner X\urcorner^* \ulcorner Y\urcorner^*\rangle$.
I.1.6. $\ulcorner\varphi \wedge \psi\urcorner^* = \langle 5\, \ulcorner\varphi\urcorner^* \ulcorner\psi\urcorner^*\rangle$.
I.1.7. $\ulcorner\neg\varphi\urcorner^* = \langle 6\, \ulcorner\varphi\urcorner^*\rangle$.
I.1.8. $\ulcorner\forall x_i\,\varphi\urcorner^* = \langle 7\, \ulcorner x_i\urcorner^* \ulcorner\varphi\urcorner^*\rangle$.
I.1.9. $\ulcorner\forall X_i\,\varphi\urcorner^* = \langle 8\, \ulcorner X_i\urcorner^* \ulcorner\varphi\urcorner^*\rangle$.

I.2. We can define the following predicates.
I.2.1. $Cf_A(Y^{A\omega})$: "$Y^{A\omega}$ is a closed formula".
I.2.2. $Nl‘\ulcorner\varphi\urcorner^*$ is defined by $Nl‘\ulcorner X \in Y\urcorner^* = 0$;
$Nl‘\ulcorner\varphi \wedge \psi\urcorner^* = \max(Nl‘\ulcorner\varphi\urcorner^*, Nl‘\ulcorner\psi\urcorner^*) + 1$ ect.
I.2.3. $Cf_A(n, \ulcorner\varphi\urcorner^*) \leftrightarrow Cf_A(\ulcorner\varphi\urcorner^*) \wedge Nl‘\ulcorner\varphi\urcorner^* \leqq n$.

*Remark.* All formulas which are necessary in order to define the above predicates ("$\ulcorner X_i \urcorner *$ is a variable"; $\ulcorner Y^{A\omega} \urcorner *$ is a formula""; $sb(\ulcorner \varphi \urcorner *, \ulcorner \psi \urcorner *, X, d)$, which means "$\ulcorner \varphi \urcorner *$ is the result of a substitution of $X$ for $d$ in $\ulcorner \psi \urcorner *$", etc.), including those predicates themselves, are defined like the case of first order formulas. When we examine the definitions of the latter case we know that, if the variables are restricted to range over $R(\beta)$ for some $\beta$, then all quantifiers which appear in the definitions can be restricted to $R(\beta + i)$ for some $i \leqq \omega$. (Cf. the following I.3.) Concerning those definitions some properties which are similar to the case of first order formulas hold ("Substitution" is commutative with logical symbols, etc.). They are proved in NTT[1] by the following reason: Let us call one of those predicates $\varphi$. The types of variables are suitably chosen so that $\varphi^{R(\alpha)}$ is ZF-provable under the assumption $C_A$; by Axiom 2.5 $\exists \alpha (\varphi^{R(\alpha)} \leftrightarrow \varphi)$; for such an $\alpha$, $\varphi^{R(\alpha)}$ and so $\varphi$.

I.3. Definition of $T_A$. Let $\tilde{A}$ denote $A_\omega$ and $\tilde{\tilde{A}}$ denote $A_{\omega+1}$.

I.3.1. $F(X^{\tilde{\tilde{A}}}, n)$:

$$
\begin{aligned}
&\forall x \, \forall y (X^{\tilde{\tilde{A}}}[\ulcorner \underline{x} \in \underline{y} \urcorner *] \leftrightarrow x \in y) \\
&\qquad \wedge \forall x \, \forall X ((X^{\tilde{\tilde{A}}}[\ulcorner \underline{x} \in \underline{X} \urcorner *] \leftrightarrow x \in X) \wedge [X^{\tilde{\tilde{A}}}[\ulcorner \underline{X} \in \underline{x} \urcorner *] \leftrightarrow X \in x)) \\
&\qquad \wedge \forall X \, \forall Y (X^{\tilde{\tilde{A}}}[\ulcorner \underline{X} \in \underline{Y} \urcorner *] \leftrightarrow X \in Y) \\
&\qquad \wedge \forall \ulcorner \varphi_1 \urcorner * \, \forall \ulcorner \varphi_2 \urcorner * (Cf_A(n, \ulcorner \varphi_1 \wedge \varphi_2 \urcorner *) \\
&\qquad (X^{\tilde{\tilde{A}}}[\ulcorner \varphi_1 \wedge \varphi_2 \urcorner *] \leftrightarrow X^{\tilde{\tilde{A}}}[\ulcorner \varphi_1 \urcorner *] \wedge X^{\tilde{\tilde{A}}}[\ulcorner \varphi_2 \urcorner *])) \\
&\qquad \wedge \forall \ulcorner \varphi \urcorner * (Cf_A(n, \ulcorner \neg \varphi \urcorner *) \to (X^{\tilde{\tilde{A}}}[\ulcorner \neg \varphi \urcorner *] \leftrightarrow \neg X^{\tilde{\tilde{A}}}[\ulcorner \varphi \urcorner *])) \\
&\qquad \wedge \forall \ulcorner \forall x_i \, \varphi(x_i) \urcorner * (Cf_A(n, \ulcorner \forall x_i \, \varphi(x_i) \urcorner *) \\
&\qquad \to (X^{\tilde{\tilde{A}}}[\forall x_i \, \varphi(x_i) \urcorner *] \leftrightarrow \forall x \, X^{\tilde{\tilde{A}}}[\ulcorner \varphi(\underline{x}) \urcorner *])) \\
&\qquad \wedge \forall \ulcorner \forall X_i \, \varphi(X_i) \urcorner * (Cf_A(n, \ulcorner \forall X_i \, \varphi(X_i) \urcorner *) \\
&\qquad \to (X^{\tilde{\tilde{A}}}[\forall X_i^A \, \varphi(X_i) \urcorner *] \leftrightarrow \forall X \, X^{\tilde{\tilde{A}}}[\ulcorner \varphi(\underline{X}) \urcorner *])),
\end{aligned}
$$

where $X^{\tilde{\tilde{A}}}[t]$ denotes $t \in X^{\tilde{\tilde{A}}}$.

$F(X^{\tilde{\tilde{A}}}, n)$ has the following meaning: $X^{\tilde{\tilde{A}}}$ is the truth definition of the closed formulas $\varphi$ of type $A$ which satisfy $Nl` \ulcorner \varphi \urcorner * \leqq n$.

I.3.2. $T_A(X^{\tilde{A}}) \underset{\mathrm{Df}}{\leftrightarrow} \exists X^{\tilde{\tilde{A}}} (Cf_A(X^{\tilde{A}}) \wedge F(X^{\tilde{\tilde{A}}}, Nl` X^{\tilde{A}}))$.

I.4. For the $T_A$ which is thus defined, the following properties hold, which shall be made use of later.

**Theorem.** Let $\Phi(b_1, \ldots, b_m, X_1, \ldots, X_n)$ be any formula with only free variables $b_1, \ldots, b_m, X_1, \ldots, X_n$. Then

$$T_A(\ulcorner \Phi(\underline{c}_1, \ldots, \underline{c}_m, \underline{Y}_1, \ldots, \underline{Y}_n) \urcorner *) \leftrightarrow \Phi(c_1, \ldots, c_m, Y_1, \ldots, Y_n),$$

where $c_1, \ldots, c_m$ and $Y_1, \ldots, Y_n$ are any free variables of type 0 and $A$ respectively.

*Proof.* This is proved by using the comprehension axiom of type $\tilde{A}$. (Cf. Proposition 8 of § 3), since types of the variables which appear in the definition of $T_A$ are suitably chosen.

**Lemma.** *Assume* $C_A\ A(\alpha, \beta)$. *Then the following are provable.*

1) $a \in R(\beta) \wedge b \in R(\beta) \rightarrow (T_A^{R(\alpha)}(\ulcorner \underline{a} \in \underline{b} \urcorner *) \leftrightarrow a \in b)$.

*Assume that in the following* $\varphi(a_1, \ldots, a_n)$, $\psi(a_1, \ldots, a_n)$, *etc. are the formulas of type A with at most* $a_1, \ldots, a_n$ *as free variables. Assume also* $a_1 \in R(\beta) \wedge \cdots \wedge a_n \in R(\beta)$.

2. $T_A^{R(\alpha)}(\ulcorner \varphi(\underline{a}_1, \ldots, \underline{a}_n) \wedge \psi(\underline{a}_1, \ldots, \underline{a}_n)\urcorner *)$
$$\leftrightarrow (T_A^{R(\alpha)}(\ulcorner \varphi(\underline{a}_1, \ldots, \underline{a}_n)\urcorner *) \wedge R_A^{R(\alpha)}(\ulcorner \psi(\underline{a}_1, \ldots, \underline{a}_n)\urcorner *)),$$

3) $T_A^{R(\alpha)}(\ulcorner \neg \varphi(\underline{a}_1, \ldots, \underline{a}_n)\urcorner *) \leftrightarrow \neg T_A^{R(\alpha)}(\ulcorner \varphi(\underline{a}_1, \ldots, \underline{a}_n)\urcorner *)$.

4) $T_A^{R(\alpha)}(\ulcorner \forall x_i \varphi(x_i, \underline{a}_1, \ldots, \underline{a}_n)\urcorner *)$
$$\leftrightarrow \forall x \in R(\alpha)\, T_A^{R(\alpha)}(\ulcorner \varphi(\underline{x}, \underline{a}_1, \ldots, \underline{a}_n)\urcorner *).$$

5) $T_A^{R(\alpha)}(\ulcorner \forall X_i \varphi(X_i, \underline{x}_1, \ldots, \underline{x}_n)\urcorner *)$
$$\leftrightarrow \forall x \in R(\beta)\, T_A^{R(\alpha)}(\ulcorner \varphi(\underline{x}, \underline{x}_1, \ldots, \underline{x}_n)\urcorner *).$$

Now we shall give the proof of (8) and (9), which have been used in the proof of the theorem for case 4).

I.5. Proof of (8), i.e.,
$$C_A \wedge A(\alpha, \beta) \wedge a \in R(\alpha) \rightarrow ((T_A(\ulcorner (\varphi(\alpha_0, \underline{a}))^{A} \urcorner))^{R(\alpha)} \leftrightarrow \varphi^{R(\beta)}(\alpha, a))$$
for each formula $\varphi(\alpha_0, a)$ of TT with only free variable $a$.

*Proof.* Assume $C_A \wedge A(\alpha, \beta) \wedge a \in R(\alpha)$. Let $\Phi(a)$ denote $(\varphi(\alpha_0, a))^A$.
$$(T_A(\ulcorner (\varphi(\alpha_0, \underline{a}))^{A} \urcorner *))^{R(\alpha)} \leftrightarrow (T_A(\ulcorner \Phi(\underline{a})\urcorner *))^{R(\alpha)} \leftrightarrow \Phi^{R(\alpha)}(a)$$
by Lemma, under the assumption. Therefore it suffices to prove
$$C_A \wedge A(\alpha, \beta) \wedge a \in R(\alpha) \rightarrow ((\Phi(a))^{R(\alpha)} \leftrightarrow \varphi^{R(\beta)}(\alpha, a)).$$
We shall prove this in the general form, i.e.,
$$\varphi^{R(\beta)}(\alpha, a_1, \ldots, a_n) \leftrightarrow (\Phi(a_1, \ldots, a_n))^{R(\alpha)}$$
is NTT[1] provable under the assumption $C_A \wedge A(\alpha, \beta) \wedge a_1 \in R(\alpha) \wedge \cdots \wedge a_n \in R(\alpha)$, where $\Phi(a_1, \ldots, a_n)$ is $(\varphi(\alpha_0, a_1, \ldots, a_n))^A$.

1) Basis. $\varphi(\alpha_0, a_1, \ldots, a_n)$ is a prime formula.

Note that, if $\alpha \leqq \beta$, then $(a_i \in \alpha)^{R(\beta)} \leftrightarrow a_i \in \alpha$, $(\alpha \in a_i)^{R(\beta)} \leftrightarrow \alpha \in a_i$ and $(\alpha \in \alpha)^{R(\beta)} \leftrightarrow \alpha \in \alpha$, since $\mathrm{Ord}(\alpha)$ is absolute with respect to $R(\beta)$.

Case 1.1) $\varphi(\alpha_0, a_1, \ldots, a_n)$ is $a_i \in \alpha_0$.
$$((a_i \in \alpha_0)^A)^{R(\alpha)} \leftrightarrow \mathrm{Ord}^{R(\alpha)}(a_i) \leftrightarrow \mathrm{Ord}(a_i).$$
$(a_i \in \alpha)^{R(\beta)} \leftrightarrow a_i \in \alpha \leftrightarrow \mathrm{Ord}(a_i)$, (since $a_i \in R(\alpha) \rightarrow (\mathrm{Ord}(a_i) \leftrightarrow a_i \in \alpha)$).

Case 1.2) $\varphi(\alpha_0, a_1, \ldots, a_n)$ is $\alpha_0 \in a_i$.

$$((\alpha_0 \in a)^A)^{R(\alpha)} \leftrightarrow (\mathrm{On} \in a)^{R(\alpha)} \leftrightarrow \alpha \in a \leftrightarrow (\alpha \in a)^{R(\beta)}.$$

Case 1.3) $\varphi(\alpha_0, a_1, \ldots, a_n)$ is $\alpha_0 \in \alpha_0$.

$$((\alpha_0 \in \alpha_0)^A)^{R(\alpha)} \leftrightarrow (\mathrm{On} \in \mathrm{On})^{R(\alpha)} \leftrightarrow \alpha \in \alpha \leftrightarrow (\alpha \in \alpha)^{R(\beta)}.$$

Case 1.4) $\varphi(\alpha_0, a_1, \ldots, a_n)$ is $a_i \in a_j$.

$$((a_i \in a_j)^A)^{R(\alpha)} \leftrightarrow (a_i \in a_j)^{R(\alpha)} \leftrightarrow a_i \in a_j \leftrightarrow (a_i \in a_j)^{R(\beta)}.$$

2) Induction step.

Case 2.1) $\varphi(\alpha_0, a_1, \ldots, a_n)$ is of the form $\varphi_1 \wedge \varphi_2$ or $\neg\varphi_1$.

This case is easily proved from the inductive hypothesis.

Case 2.2) $\varphi(\alpha_0, a_1, \ldots, a_n)$ is $\forall y \psi(y, \alpha_0, a_1, \ldots, a_n)$.

Let $\Psi(y, a_1, \ldots, a_n)$ denote $(\psi(y, \alpha_0, a_1, \ldots, a_n))^A$. Then

$$(\varphi(\alpha_0, a_1, \ldots, a_n))^A \quad \text{or} \quad \Phi(a_1, \ldots, a_n) \quad \text{is} \quad \forall Y^A \Psi(Y^A, a_1, \ldots, a_n).$$

$$\begin{aligned}(\Phi(a_1, \ldots, a_n))^{R(\alpha)} & \\ &\leftrightarrow (C_A \wedge \exists\beta(A(\alpha, \beta) \wedge \forall y \in R(\beta)\, \psi^{R(\beta)}(y, \alpha, a_1, \ldots, a_n))) \\ &\quad \vee (\neg C_A \wedge \forall y \in R(\alpha)\, \psi^{R(\beta)}(y, \alpha, a_1, \ldots, a_n))\end{aligned}$$

by the inductive hypothesis. Therefore under our assumption,

$$\begin{aligned}(\Phi(a_1, \ldots, a_n))^{R(\alpha)} &\leftrightarrow \forall y \in R(\beta)\, \psi^{R(\beta)}(y, \alpha, a_1, \ldots, a_n) \\ &\leftrightarrow \varphi^{R(\beta)}(\alpha, a_1, \ldots, a_n).\end{aligned}$$

I.6. Proof of (9), i.e.,

$$\begin{aligned}&C_A \wedge A(\alpha, \beta) \wedge a \in R(\alpha) \wedge \beta + \omega + 1 < \alpha_2 \\ &\quad \to ((T_A(\ulcorner(\varphi(\alpha_0, \underline{a}))^A\urcorner *))^{R(\alpha)})^{R(\alpha_2)} \leftrightarrow (T_A(\ulcorner(\varphi(\alpha_0, \underline{a}))^A\urcorner *))^{R(\alpha)})\end{aligned}$$

for any formula $\varphi(a_0, a)$ of TT with only free variable $a$.

*Proof.* Assume $C_A \wedge A(\alpha, \beta) \wedge \beta + \omega + 1 < \alpha_2$. Then we can prove the following:

1′) $a \in R(\beta) \wedge b \in R(\beta) \to ((T_A^{R(\alpha)}(\ulcorner \underline{a} \in \underline{b} \urcorner *))^{R(\alpha_2)} \leftrightarrow a \in b)$.

Let $\varphi(a_1, \ldots, a_n)$ and $\psi(a_1, \ldots, a_n)$ etc. in the following formulas of type $A$, and assume $a_1 \in R(\beta) \wedge \cdots \wedge a_n \in R(\beta)$.

2′) $T_A^{R(\alpha)}(\ulcorner\varphi(\underline{a}_1, \ldots, \underline{a}_n) \wedge \psi(\underline{a}_1, \ldots, \underline{a}_n)\urcorner *)^{R(\alpha_2)}$

$$\leftrightarrow (T_A^{R(\alpha)}(\ulcorner\varphi(\underline{a}_1, \ldots, \underline{a}_n)\urcorner *))^{R(\alpha_2)} \wedge (T_A^{R(\alpha)}(\ulcorner\varphi(\underline{a}_1, \ldots, \underline{a}_n)\urcorner *))^{R(\alpha_2)}.$$

3′) $(T_A^{R(\alpha)}(\ulcorner\neg\varphi(\underline{a}_1, \ldots, \underline{a}_n)\urcorner *))^{R(\alpha_2)}$

$$\leftrightarrow \neg(T_A^{R(\alpha)}(\ulcorner\varphi(\underline{a}_1, \ldots, \underline{a}_n)\urcorner *))^{R(\alpha_2)}.$$

4′) $(T_A^{R(\alpha)}(\ulcorner\forall x_i \varphi(x_i, \underline{a}_1, \ldots, \underline{a}_n)\urcorner *))^{R(\alpha_2)}$

$$\leftrightarrow \forall x \in R(\alpha)\, (T_A^{R(\alpha)}(\ulcorner\varphi(\underline{x}, \underline{a}_i, \ldots, \underline{a}_n)\urcorner *))^{R(\alpha_2)}.$$

5′) $T_A^{R(\alpha)}(\ulcorner\forall X_i \varphi(X_i, \underline{a}_1, \ldots, \underline{a}_n)\urcorner *)^{R(\alpha_2)}$

$$\leftrightarrow \forall x \in R(\beta)\, (T_A^{R(\alpha)}(\ulcorner\varphi(\underline{x}, \underline{a}_1, \ldots, \underline{a}_n)\urcorner *))^{R(\alpha_2)}.$$

Now we can prove the proposition by using 1) — 5) in Lemma and 1')—5').

II. Definition of $T$ and the proof of (1) and (2) in Case 5).

II.1 Gödelization of the formulas of first order set theory.

II.1.1. $\ulcorner \underline{a} \urcorner^* = \langle 0\, a \rangle$.

II.1.2. $\ulcorner x_i \urcorner^* = \langle 1\, i \rangle$.

II.1.3. $\ulcorner x \in y \urcorner^* = \langle 2 \ulcorner x \urcorner^* \ulcorner y \urcorner^* \rangle$.

II.1.4. $\ulcorner \neg \varphi \urcorner^* = \langle 3 \ulcorner \varphi \urcorner^* \rangle$.

II.1.5. $\ulcorner \varphi \wedge \psi \urcorner^* = \langle 4 \ulcorner \varphi \urcorner^* \ulcorner \psi \urcorner^* \rangle$.

II.1.6. $\ulcorner \exists x_i \varphi \urcorner^* = \langle 5\, i \ulcorner \varphi \urcorner^* \rangle$.

II.1.7. $Cf(a)$, which means that a is a closed formula of first order set theory is defined as usual, regarding that $\ulcorner \underline{a} \in \underline{b} \urcorner^*$ is a closed formula.

II.1.8. $Nl'a$ is defined as in I so that $Nl' \ulcorner \underline{a} \in \underline{b} \urcorner^* = 0$ and $Nl' \ulcorner \varphi \wedge \psi \urcorner^* = \max(Nl' \ulcorner \varphi \urcorner^*, Nl' \ulcorner \psi \urcorner^*) + 1$ etc. hold.

II.1.9. $\ulcorner \varphi(\#) \urcorner^* \underset{\mathrm{Df}}{=} Sb(\ulcorner \varphi(x_i) \urcorner^*, \#, \ulcorner x_i \urcorner^*)$, which means that "$\varphi(\#)$ is the result of the substitution of $\#$ for $x_i$ in $\varphi(x_i)$", is defined as usual.

II.2. Truth definition of first order set theory in NTT[1]. We shall abbreviate $X^{A_0}$ by $X$, where $A_0(\alpha, \beta) \underset{\mathrm{Df}}{\leftrightarrow} \beta = \alpha + 1$. $X[t]$ denotes $t \in X$.

II.2.1.
$$\begin{aligned} F(X, n) \leftrightarrow\ & \forall x \forall y (X[\ulcorner \underline{x} \in \underline{y} \urcorner^*] \leftrightarrow x \in y) \\ & \wedge \forall \ulcorner \varphi_1 \urcorner^* \forall \ulcorner \varphi_2 \urcorner^* (Cf(\ulcorner \varphi_1 \wedge \varphi_2 \urcorner^*) \wedge Nl' \ulcorner \varphi_1 \wedge \varphi_2 \urcorner^* \leqq n \\ & \qquad \to (X[\ulcorner \varphi_1 \wedge \varphi_2 \urcorner^*] \leftrightarrow X[\ulcorner \varphi_1 \urcorner^*] \wedge X[\ulcorner \varphi_2 \urcorner^*])) \\ & \wedge \forall \ulcorner \varphi \urcorner^* (Cf(\ulcorner \varphi \urcorner)^* \wedge Nl' \ulcorner \neg \varphi \urcorner^* \leqq n \\ & \qquad \to (X[\ulcorner \neg \varphi \urcorner^*] \leftrightarrow \neg X[\ulcorner \varphi \urcorner^*])) \\ & \wedge \forall \ulcorner \forall x_i \varphi(x_i) \urcorner^* (Cf(\ulcorner \forall x_i \varphi(x_i) \urcorner^*) \wedge Nl' \ulcorner \forall x_i \varphi(x_i) \urcorner^*) \leqq n \\ & \qquad \to (X[\ulcorner \forall x_i \varphi(x_i) \urcorner^*] \leftrightarrow \forall x X[\ulcorner \varphi(\underline{x}) \urcorner^*])). \end{aligned}$$

II.2.2. $T(a) \underset{\mathrm{Df}}{\leftrightarrow} Cf(a) \wedge \exists X (F(X, Nl'a) \wedge X[a])$. The following lemmas are provable in NTT[1] by using the comprehension axiom of type $A_0$.

**Lemma 1.** *Let $\varphi(\alpha_0, a_1, \ldots, a_n)$ be a formula of $TT$ with only free variables $a_1, \ldots, a_n$. Then*

$$T(\ulcorner \varphi(\underline{\alpha}, \underline{b}_1, \ldots, \underline{b}_n) \urcorner^*) \leftrightarrow \varphi(\alpha, b_1, \ldots, b_n)$$

*for any free variables $\alpha, b_1, \ldots, b_n$.*

**Lemma 2.** *Assume $\alpha \in K_{\mathrm{II}}$. Then the following are provable.*

1) $x \in R(\alpha) \wedge y \in R(\alpha) \to (T^{R(\alpha)}(\ulcorner \underline{x} \in \underline{y} \urcorner^*) \leftrightarrow x \in y)$.

*In 2)—4) $\varphi$, $\varphi_1$, $\varphi_2$ are formulas of first order.*

2) $\ulcorner \varphi_1 \urcorner^* \in R(\alpha) \wedge \ulcorner \varphi_2 \urcorner^* \in R(\alpha) \to (T^{R(\alpha)}(\ulcorner \varphi_1 \wedge \varphi_2 \urcorner^*)$
$\leftrightarrow T^{R(\alpha)}(\ulcorner \varphi_1 \urcorner^*) \wedge T^{R(\alpha)}(\ulcorner \varphi_2 \urcorner^*))$.

3) $\ulcorner \neg \varphi \urcorner^* \in R(\alpha) \to (T^{R(\alpha)}(\ulcorner \neg \varphi \urcorner^*) \leftrightarrow \neg T^{R(\alpha)}(\ulcorner \varphi \urcorner^*))$.

4) $\ulcorner \forall x_i \varphi(x_i)\urcorner * \in R(\alpha)$

$$\rightarrow (T^{R(\alpha)}(\ulcorner \forall x_i \varphi(x_i)\urcorner *) \leftrightarrow \forall x \in R(\alpha)\, T^{R(\alpha)}(\ulcorner \varphi(\underline{x})\urcorner *)).$$

II.3. Now we shall prove (1) and (2).

II.3.1. Proof of (1):

$$\alpha_1 < \alpha \wedge \alpha \in K_{\mathrm{II}} \rightarrow ((T(\ulcorner \varphi(\underline{a}_1)\urcorner *))^{R(\alpha)} \leftrightarrow R(\alpha) \overset{1}{\models} \ulcorner \varphi(\alpha_1)\urcorner).$$

Assume $\alpha_1 < \alpha \wedge \alpha \in K_{\mathrm{II}}$. It is evident from Lemma 2 that

$$(T(\ulcorner \varphi(\underline{\alpha}_1)\urcorner *))^{R(\alpha)} \leftrightarrow R(\alpha) \overset{2}{\models} \ulcorner \varphi(\underline{\alpha}_1)\urcorner *$$

is provable. Then by 3) of the explanation of satisfaction relations,

$$(T(\ulcorner \varphi(\underline{\alpha}_1)\urcorner *))^{R(\alpha)} \leftrightarrow R(\alpha) \overset{1}{\models} \ulcorner \varphi(\alpha_1)\urcorner.$$

II.3.2. Proof of (2): $T(\ulcorner \varphi(\underline{\alpha}_1)\urcorner *) \leftrightarrow \varphi(\alpha_1)$. This is evident from Lemma 1.

## § 6. Several Generalizations of NTT

6.1. When we defined typed variables in our language, we took only formulas $A(\alpha, \beta)$ having $\alpha$ and $\beta$ as the only free variables to determine types. One of the natural generalizations of NTT is to extend this so that any formula $A(\alpha, \beta; b)$ (without $\mathscr{N}$) having $\alpha$, $\beta$ and $b$ as its sole free variables determines a type. The counterpart of $C_A$ is now

$$C_A(b){:}\ \forall \alpha \exists! \beta A(\alpha, \beta; b) \wedge \forall \alpha \forall \beta (A(\alpha, \beta; b) \rightarrow \alpha \leqq \beta),$$

where $b$ is regarded as a parameter. We think this system is very plausible.

6.2. In NTT we introduced the notion $\mathscr{N}$ as an auxiliary one in order to study set theory developed in a language with typed variables, so that we stated only few axioms concerning $\mathscr{N}$ and left the notion as vague as possible. However when we wish to consider the notion of $\mathscr{N}$ itself, we certainly should introduce the axiom schema of specification (Aussonderungsaxiom) for all formulas (possibly containing $\mathscr{N}$), which is evidently equivalent to the axiom schema of replacement for all formulas (cf. Proposition 7, § 1). Moreover, we would like to extend Axiom 2.5 in such a way that it applies also to formulas containing $\mathscr{N}$. For this, the following axiom seems plausible:

$$\forall X \mathscr{N}(\{\alpha \mid \exists y \forall x \in R(\alpha)(\mathscr{A}(x, \mathscr{N}, X) \leftrightarrow \mathscr{A}^{R(\alpha)}(x, y, X \cap R(\alpha))\}),$$

where $\mathscr{A}(x, y, X)$ denotes the formula obtained from $\mathscr{A}(x, \mathscr{N}, X)$ by replacing every occurrence of the form $\mathscr{N}(T)$ in $\mathscr{A}(x, \mathscr{N}, X)$ by $T \in y$.

*Remark.* It would be better if we could define the relativization $\mathscr{A}^{R(\alpha)}$ directly for any formula $\mathscr{A}$ containing $\mathscr{N}$ and introduce

(1) $$\mathscr{N}(\{\alpha \mid \mathscr{A} \leftrightarrow \mathscr{A}^{R(\alpha)}\}),$$

where $\mathscr{A}$ is any closed formula, as an axiom schema. However many definitions of this kind usually lead to a contradiction, and we have to weaken (1).

The two axioms,

$$\mathscr{A} \to \mathscr{N}(\{\alpha \mid \mathscr{A}^{R(\alpha)}\}) \tag{2}$$

and

$$\exists X \, \forall x (x \in X \leftrightarrow \mathscr{A}(x)) \tag{3}$$

are much weaker than (1). We can define a truth definition of $\mathscr{A}$ by $\mathscr{N}(\{\alpha \mid \mathscr{A}^{R(\alpha)}\})$ in the system having (2) and (3). On the other hand, (3) is derivable from $\mathscr{N}(\{\alpha \mid \mathscr{A} \leftrightarrow \mathscr{A}^{R(\alpha)}\})$ in most cases. If $\mathscr{A}$ is an arbitrary formula, this means $\mathscr{N}(\{\alpha \mid \mathscr{A} \leftrightarrow \mathscr{A}^{R(\alpha)}\})$ yields a contradiction.

For this reason, we have to give up the schema of the form $\mathscr{N}(\{\alpha \mid \mathscr{A}(X) \leftrightarrow \mathscr{A}^{R(\alpha)}(R(\alpha) \cap X)\})$, if we wish to extend the notion $\mathscr{A}^{R(\alpha)}$ to the case where $\mathscr{A}$ contains $\mathscr{N}$. The only way to extend $\mathscr{A}^{R(\alpha)}$ is to return to the method presented in [24].

First let us take the second system presented in § 4 of [24], with the following modification. We introduce $\mathscr{N}(\ )$ and $\mathscr{N}(\ ,\ )$ instead of $\mathscr{I}(\ )$ and $\mathscr{I}(\ ,\ )$ respectively. The relativization can be defined as in [24]. Axioms 6 and 7 are eliminated from Group $H$ and an $\mathscr{N}$-inference

$$\frac{\mathscr{A}(X)}{\mathscr{N}(\{\alpha \mid \mathscr{A}^{R(\alpha)}(R(\alpha) \cap X)\})}$$

should be added instead.

Second, consider the first system presented in § 1 of [24]. Here we take Axiom F5′ instead of F5 and $\mathscr{G}$-inference should have the form

$$\frac{\mathscr{A}(\mathscr{G}, X)}{\mathscr{G}(\{< b, R(\alpha) > \mid \mathscr{A}^{R(\alpha)}(b, R(\alpha) \cap X)\}}\,.$$

However, we believe that NTT and its generalizations are more plausible than these.

At this opportunity, we would like to give the definition of relativization to a on page 231 of [24], lines 17 and 18, which is missing there. In order to state $\mathscr{I}$-inference, we must define the notation $U^a$ for a formula or a term $U$ and a set variable $a$ not contained in $U$. $U^a$ is obtained from $U$ by transforming all the quantifiers: $\mathscr{M}(\ )$, $\mathscr{I}(\ )$ and $\mathscr{I}(\ ,\ )$ in $U$ as follows: $\forall A$, $\exists B$, $(Kx)$, $\mathscr{M}(T)$, $\mathscr{I}(T)$, $\mathscr{I}(A, B)$ to $\forall A (A \subseteqq a \vdash\ )$, $\exists B (B \subseteqq a \wedge)$, $(Kx)\, a\, (X \in a)$, $T^a \in a$, $\mathscr{I}(a, T^a)$ and $\mathscr{I}(A^a, B^a)$ respectively.

6.3. At first we considered only $\mathrm{NTT}^1$ for which Solovay devised the following elegant and strengthened version[3]: Solovay's system ist developed in a first order language having one predicate $\varepsilon$ and constants $N$ and $\varkappa$. We shall use capital italics $X$, $Y$, ... to denote variables.

---

[3] The author is grateful to Professor R. Solovay for his valuable comments.

Axioms

I. 1) all the axioms of ZF including the axiom of choice. We define ordinals as usual and use small Greek letters $\alpha, \beta, \gamma, \dots$ as variables ranging over ordinals.

2) $\mathrm{Ord}(\varkappa)$.

$R(\alpha)$ is defined as usual and small italics $x, y, \dots$ denote variables ranging over $R(\varkappa)$.

3) $\forall x \cdots \forall z (\varphi(x, \dots, z) \leftrightarrow \varphi^{R(\varkappa)}(x, \dots, z))$, where $\varphi$ is a formula containing no $N$ or $\varkappa$ and having only free variables ranging over $R(\varkappa)$.

II. 1) $N \subseteqq P(\varkappa)$,

2) $0 \notin N$,

3) $\forall X \forall Y (X \subseteqq Y \subseteqq \varkappa \wedge X \in N \to Y \in N)$,

4) $\forall X \forall \beta (\beta < \varkappa \wedge \forall \gamma < \beta (X_\gamma \in N) \to \bigcap_{\gamma < \beta} X_\gamma \in N))$,

5) $\forall x (\bar{\bar{x}} < \varkappa \to \varkappa - x \in N)$,

6) $\forall X (\forall x (X_x \in N) \to \{\alpha \mid \alpha < \varkappa \wedge \forall x \in R(\alpha) (\alpha \in X_x)\} \in N)$,

III. $\forall X \subseteqq R(\varkappa) (\{\alpha \mid \alpha < \varkappa \wedge \forall x \in R(\alpha) (\varphi(X, x)$

$$\leftrightarrow \varphi^{R(\alpha)}(X \cap R(\alpha), x))\} \in N,$$

where $\varphi(X, x)$ is a formula containing no $N$ and having $X$ and $x$ as its only free variables and where $\varphi^{R(\alpha)}(X, x)$ is obtained from $\varphi(X, x)$ by restricting all variables ranging over $R(\varkappa)$ to range over $R(\alpha)$ and replacing $\varkappa$ by $\alpha$.

IV. Infinite induction.

NTT[1] is interpreted in this system if we regard $\varkappa$ as On and $X \in N$ as $\mathscr{N}(X)$. The formulas $\mathscr{A}$ of NTT[1] translate into $\mathscr{A}^*$ where $\mathscr{A}^*$ are defined as follows. If $\mathscr{A}$ is a prime formula, $\mathscr{A}^*$ is $\mathscr{A}$. If $\mathscr{A}$ is $\mathscr{B} \wedge \mathscr{C}$ or $\neg \mathscr{B}$, then $\mathscr{A}^*$ is $\mathscr{B}^* \wedge \mathscr{C}^*$ or $\neg \mathscr{B}^*$ respectively. If $\mathscr{A}$ is $\forall x \mathscr{B}(x)$, then $\mathscr{A}^*$ is $\forall x \in R(\varkappa) \mathscr{B}^*(x)$. If $\mathscr{A}$ is $\forall X^A \mathscr{B}(X^A)$, then $\mathscr{A}^*$ is

$$[C_A \wedge \exists \beta (A(\varkappa, \beta) \wedge \forall x \in R(\beta) \mathscr{B}^*(x))] \vee [\neg C_A \wedge \forall x \in R(\varkappa) \mathscr{B}^*(x)].$$

Now if $\mathscr{A}$ is a theorem of NTT[1], then $\mathscr{A}^*$ is provable in SOLOVAY's system.

SOLOVAY has proved that his system can be interpreted in a system with a measurable cardinal.

6.4. Although the author has not yet found any reasonable justification of 2-valued measurable cardinals, we anticipate being able to prove the following: if an axiom of strong infinity $\exists \alpha A(\alpha)$ is well-justified (where $A(\alpha)$ is intentionally understood as "$\alpha$ is a measurable cardinal"), then there exists a formula $B(\alpha)$ of one argument for which we have

$$\mathscr{N}(\{\alpha \mid B(\alpha)\}) \wedge (\exists \alpha B(\alpha) \to \exists \alpha A(\alpha)).$$

For this reason, it is significant to consider a system which is an extreme generalization of both NTT and the axiom of the existence of measurable cardinals and to examine whether it is inconsistent or not. To carry out this program, we present the system MTT; MTT is obtained from NTT by adding the following axioms

$$(1) \qquad \forall A\, \mathscr{N}(\{\alpha \mid \exists x\, \forall y \in R(\alpha)\, (\mathscr{A}(\mathscr{N}, A, y) \leftrightarrow \mathscr{A}^{R(\alpha)}(x, A \cap R(\alpha), y))\})$$

and

$$(2) \qquad \forall X \subseteqq \mathrm{On}\, (\mathscr{N}(X) \vee \mathscr{N}(\mathrm{On} - X)).$$

# Chapter II

# A Classification of the Width of the Universe

## § 1. Properties of Recursive, Arithmetical and Hyperarithmetical Functions

As we have explained in the introduction, the concepts concerning only the length of the universe but not the width of the universe are useful for the classification of the width.

In [23], [25], and [26], we defined recursive functions and arithmetical functions of ordinal numbers. We can also define the functions which are "arithmetical in $f$" for any given function $f$ of ordinal numbers. Those are obtained by adding the following equation to the defining equations of arithmetical functions:

$$h(a_1, \ldots, a_n) = f(a_1, \ldots, a_n).$$

Using this extension of the notion of arithmetical functions, we can modify some theorems in [25]. In order to carry this out, we shall define some related notions, i.e., "almost universal" and "relative constructibility", following Hajnal.

**Definition.** A class $M$ is called *almost universal* if $\forall x \subseteqq M\, \exists y \in M (x \subseteqq y)$ holds

**Definition.** A class is said to be *constructible with respect to M* (denoted $\mathscr{L}_M(A)$) if $A \subseteqq M \wedge \forall x \in M\, ((x \cap A) \in M)$.

The following theorem is useful for the definition of relative constructibility.

**Theorem.** The following two conditions are equivalent for a class $M$:

1) $M$ is complete, almost universal and closed with respect to Gödel's eight fundamental operations.

2) $M$ is a complete model of ZF such that $M \supseteqq \mathrm{On}$.

*Proof.* The proof that 1) implies 2) is given in Hajnal [10] following Gödel's construction by using the axiom of choice. We shall show that the use of the axiom of choice can be eliminated, i.e. we shall prove $\mathscr{L}_M(Q_i``\bar{A})\,(i = 5, \ldots, 8)$ and $\mathscr{L}_M(M \cap Q_4``\bar{A})$, where $\bar{A}$ denotes a class constructible with respect to $M$. Since $Q_i``\bar{A} \subseteqq M$ $(i = 5, \ldots, 8)$ and $M \cap Q_4``\bar{A} \subseteqq M$, it suffices to show $\bar{a} \cap Q_i``\bar{A} \in M$ $(i = 4, \ldots, 8)$ for any element $\bar{a}$ of $M$. Let $G(i, \bar{A})`\bar{b}$ be

$$\{x \mid x \in \bar{A} \wedge \langle x\bar{b}\rangle \in Q_i \wedge \forall y (y \in \bar{A} \wedge \langle y\bar{b}\rangle \in Q_i \rightarrow \operatorname{rank}(x) \leqq \operatorname{rank}(y))\}$$

and let $c(i, \bar{A}, \bar{a})$ be $\bigcup_{b \in a} G(i, \bar{A})`\bar{b}$. Then

$$G(i, \bar{A})`\bar{b} \subseteqq \bar{A},$$
$$\bar{x} \in \bar{A} \wedge \langle \bar{x}\bar{b}\rangle \in Q_i \quad \text{for every} \quad \bar{x} \in G(i, \bar{A})`\bar{b},$$
$$c(i, \bar{A}, \bar{a}) \subseteqq \bar{A}$$

and, since $M$ is almost universal, there exist a set $\bar{d}$ such that

$$c(i, \bar{A}, \bar{a}) \subseteqq \bar{d}.$$

Let $\bar{d}(i, \bar{A}, \bar{a})$ be $\bar{d} \cap \bar{A}$. Then $c(i, \bar{A}, \bar{a}) \subseteqq \bar{d}(i, \bar{A}, \bar{a}) \subseteqq \bar{A}$ and $G(i, \bar{A})`\bar{b} \subseteqq \bar{d}(i, \bar{A}, \bar{a})$ for every element $\bar{b}$ of $\bar{a}$.

We claim $\bar{a} \cap Q_i``\bar{A} = \bar{a} \cap Q_i``\,\bar{d}(i, \bar{A}, \bar{a})$. $\bar{a} \cap Q_i``\bar{A} \supseteqq \bar{a} \cap Q_i``\bar{d}(i, \bar{A}, \bar{a})$ is clear. $\bar{a} \cap Q_i``\,\bar{A} \subseteqq \bar{a} \cap Q_i``\,\bar{d}(i, \bar{A}, \bar{a})$ is proved as follows:

$$\begin{aligned} \bar{b} \in \bar{a} \cap Q_i``\bar{A} &\rightarrow \bar{b} \in \bar{a} \wedge \exists \bar{x} \in G(i, \bar{A})`\bar{b}\,(\langle \bar{b}\bar{x}\rangle \in Q_i) \\ &\rightarrow \bar{b} \in \bar{a} \wedge \bar{b} \in Q_i``(G(i, \bar{A})`\bar{b}) \\ &\rightarrow \bar{b} \in \bar{a} \cap Q_i``\,\bar{d}(i, \bar{A}, \bar{a}). \end{aligned}$$

Therefore $\bar{a} \cap Q_i``\,\bar{A} = \bar{a} \cap Q_i``\,\bar{d}(i, \bar{A}, \bar{a}) = \mathfrak{F}_i(\bar{a}, \bar{d}(i, \bar{A}, \bar{a})) \in M$. This completes the proof.

*Proof* that 2) implies 1). Let $M$ be a complete model of ZF such that $M \supseteqq \mathrm{On}$. Since $\exists x (x = \mathfrak{F}_i(\bar{a}, \bar{b}))$ is absolute with respect to $M$, it suffices to prove $\forall x \subseteqq M\; \exists \bar{y} (x \subseteqq \bar{y})$. It is proved by transfinite induction on $\alpha$ that $\alpha = \operatorname{rank}(\bar{x})$ and $\operatorname{rank}(\bar{x}) < \alpha$ are absolute with respect to $M$. Moreover,

$$R^M(\alpha) = \{\bar{y} \mid (\operatorname{rank}(\bar{y}) < \alpha)^M\} = \{\bar{y} \mid \operatorname{rank}(\bar{y}) < \alpha\} = R(\alpha) \cap M,$$

$R^M(\alpha) \in M$ i.e. $R(\alpha) \cap M \in M$, since $M$ is a model of ZF. We have

$$\exists \alpha (x \subseteqq R(\alpha) \cap M) \quad \text{for every } x \text{ such that} \quad x \subseteqq M.$$

**Definition.** Given a class $A$, $L(A)$ is the smallest class among the classes $M$ which satisfy the following.

i) $M$ is a complete model of ZF such that $M \supseteqq \mathrm{On}$.

ii) $\mathscr{L}_M(A)$.

The existence of $L(A)$ can be proved by an extended method of Gödel's construction and the above theorem.

*Remarks.* 1) If $A \subseteqq L$, then $L(A) = L_A$, where $L_A$ is the class defined by LÉVY [13].

2) If $A$ is a set $a$, then $L(A)$ is the smallest class among the classes $M$ satisfying the following conditions.

i) $M$ is a complete model of ZF such that $M \supseteqq \mathrm{On}$.

ii) $a \in M$.

**Definition.** Let $f$ be a function from On to On.

i) We define $L(f)$ by $L(f) = L(G_f)$, where $G_f$ denotes the graph of $f$.

ii) A function $g$ from On to On is called *constructible in* $f$ if $G_g$ is constructible in the model $L(f)$.

Using the definitions given above, we shall state generalized results of Theorems 1 and 2 of § 2 in [25]. In this paper we shall use functions $g_0, g_1, g_2$ (definied in [27]) and also $K_1, K_2$ (defined in [7]), although $g_i(\alpha)$ is equal to $K_i{}^{\backprime}\alpha$ $(i = 0, 1, 2)$.

**Theorem 1.** A function $g$ is constructible in $f$ if $g$ is numerically arithmetical in $f$, where "numerically arithmetical" means that any ordinal number is regarded as a constant.

*Proof.* If $g$ is numerically arithmetical in $f$, then

$$g(a) = b \leftrightarrow A(f, a, b, \alpha_1, \ldots, a_n)$$

for some $A$ arithmetical in $f$, where $\alpha_1, \ldots, \alpha_n$ are ordinal numbers regarded as constants. Since $\mathrm{On} \subseteqq L(f)$, this theorem is proved in the same way as in the proof of Theorem 1′ of § 2 in [25].

**Theorem 2.** Let $A(\alpha_1, \ldots, \alpha_n)$ be a formula with only free variables $\alpha_1, \ldots, \alpha_n$ on ordinal numbers. Then $A(\alpha_1, \ldots, \alpha_n)$ is $L(f)$-expressible if and only if it is arithmetical in $f$, where $A(\alpha_1, \ldots, \alpha_n)$ is called $L(f)$-expressible if it is constructed by using only $\in, \alpha_1, \ldots, \alpha_n, f, \wedge, \neg$ and $\forall x \in L(f)$.

*Proof.* The first half of the theorem is obvious. In order to prove the second half we shall first show that $L(f)$ can be defined by following GÖDEL's (or LÉVY's) construction ([7] and [13]).

Let $m(\alpha)$ be defined by:

$$m(\alpha) = \mu\gamma(\alpha < \gamma \wedge g_0(\gamma) = 0 \\ \wedge \forall\beta < \gamma\, \forall\delta < \gamma\, \forall i < 9(j(i, \beta, \delta) < \gamma) \wedge \forall\beta < \alpha(m^\alpha(\beta) < \gamma)),$$

where $m^\alpha(x) = m(x)$ if $x < \alpha$; 0 otherwise. Evidently $m(\alpha)$ is recursive and

(1) $$m(\alpha) = j(0, m(\alpha), 0) \in K_{\mathrm{II}}.$$

Let $\mathfrak{F}_i$ $(i = 1, \ldots, 8)$ be GÖDEL's fundamental operations. Given a function $f$, we define the function $F_f$ on ordinal numbers as follows:

$$F_f`\alpha = \begin{cases} F_f``\alpha & \text{if } g_0(\alpha) = 0 \\ \mathfrak{F}_i(F_f`K_1`\alpha, F_f`K_2`\alpha) & \text{if } \forall\gamma < \alpha(\alpha \neq m(\gamma) + 1) \text{ and} \\ & \qquad 1 \leqq g_0(\alpha) = i \leqq 8 \\ f \cap F_f`K_1`\alpha & \text{if } \alpha = m(\gamma) + 1 \text{ for some } \gamma. \end{cases}$$

Then $L^*(f)$ is defined to be $F_f``$ On. We shall use a notational convention $\tilde{x}$ to denote a member of $L^*(f)$. We define $\mathrm{Od}_f`\tilde{x}$ to be $\mu\gamma(F_f`\gamma = \tilde{x})$.

We claim $L^*(f) = L(f)$, where $L(f)$ is the smallest complete almost universal model of ZF such that $f$ is constructible with respect to $L(f)$.

In order to show $L(f) \subseteqq L^*(f)$, it suffices to prove: (i) $L^*(f)$ is complete. (ii) $L^*(f)$ is almost universal. (iii) $L^*(f)$ is closed with respect to $\mathfrak{F}_i$ $(i = 1, \ldots, 8)$. (iv) $f \subseteqq L^*(f)$ and (v) $\forall \tilde{x}(\tilde{x} \cap f \in L^*(f))$.

For (i), it is sufficient to show $F_f`\alpha \subseteqq F_f``\alpha$ by transfinite induction on $\alpha$. Since other cases are treated as in GÖDEL [7], we shall only consider the case where $\alpha = m(\beta) + 1$ for some $\beta$ and show that $f \cap F_f`m(\beta) \subseteqq F_f``\alpha$ (where $m(\beta) = j(0, m(\beta), 0)$ by$_f$` (1)). $F\, m(\beta) \subseteqq F_f``m(\beta)$ by the inductive hypothesis, from which follows $f \cap F_f`m(\beta) \subseteqq F_f``\alpha$.

For (ii), suppose $x \subseteqq L^*(f)$ and show $\exists\beta(x \subseteqq F_f`\beta)$. By the axiom of replacement, $x \subseteqq F_f``\alpha$ for some $\alpha$. We can take such an $\alpha$ with $g_0(\alpha) = 0$: $x \subseteqq F_f``\alpha = F_f`\alpha$.

For (iii), we shall show $\{a\, \tilde{b}\} \in L^*(f)$: all other cases are seen in [7]. Let $\alpha$ be $j(1, \mathrm{Od}_f`\tilde{a}, \mathrm{Od}_f`\tilde{b})$ if $j(1, \mathrm{Od}_f`\tilde{a}, \mathrm{Od}_f`\tilde{b})$ is not of the form $m(\gamma) + 1$ for any $\gamma$, otherwise let $\alpha$ be $j(1, \mathrm{Od}_f`\tilde{b}, \mathrm{Od}_f`\tilde{a})$. Then $\{\tilde{a}\, \tilde{b}\} = F_f`\alpha$.

For (iv), since $L$ is known to be the smallest class satisfying (i)—(iii), it suffices to show $f \subseteqq L$, which is obvious.

For (v), it is ZF-provable that $\tilde{x} \cap f = \tilde{x} \cap F_f``\alpha$ for some $\alpha$ by (i) and the axiom of replacement. We may assume $\alpha = m(\gamma)$ for some $\gamma$ so that $F_f``\alpha = F_f`\alpha$ (by (1)) and $f \cap F_f`\alpha = F_f`(\alpha + 1)$. Therefore

$$\tilde{x} \cap f = \tilde{x} \cap F_f`(\alpha + 1) \in L^*(f).$$

Next we shall show $L^*(f) \subseteqq L(f)$. It suffices to prove that $L^*(f)$ is absolute with respect to a complete almost universal model $M$ of ZF such that $M \supseteqq$ On and $f$ is constructible with respect to $M$. Let $M$ be such a model. We shall show

$$\bar{x} = F_f`\alpha \leftrightarrow (\bar{x} = F_f`\alpha)^M,$$

where $\bar{x}$ denotes an element of $M$, by transfinite induction on $\alpha$ separating cases according to $K_0`\alpha$. Since all other cases are treated as usual, we show only the case where $\alpha = m(\gamma) + 1$ for some $\gamma$:

$$(\bar{x} = F_f`\alpha)^M \leftrightarrow (\bar{x} = f \cap F_f`K_1`\alpha)^M$$
$$\leftrightarrow \forall \bar{y} (\bar{y} \in \bar{x} \leftrightarrow \exists \delta (\bar{y} = \langle f(\delta)\, \delta \rangle \wedge \bar{y} \in F_f`K_1`\alpha))$$
$$\leftrightarrow \bar{x} = F_f`\alpha,$$

by the inductive hypothesis and the fact $\langle f(\delta)\,\delta\rangle \in M$.

In the following we shall identify $L^*(f)$ with $L(f)$. Now we shall show that an $L(f)$-expressible formula is arithmetical in $f$. For this, it is sufficient to show that the formulas of the forms $F_f`\alpha \in F_f`\beta$, $F_f`\alpha = F_f`\beta$ and $F_f`\gamma \in f$ are recursive in $f$. We shall prove this by transfinite induction on $\max(\alpha, \beta)$ with $\gamma < \max(\alpha, \beta)$. We follow the proof of 12.6 of [7]. As the inductive hypothesis let us assume for an arbitrary pair $\alpha, \beta$ of ordinal numbers with $\zeta = \max(\alpha, \beta)$:

H 1) "$F_f`\xi \in F_f`\eta$" is recursive in $f$ for every $\xi, \eta < \zeta$;

H 2) "$F_f`\xi = F_f`\eta$" is recursive in $f$ for every $\xi, \eta < \zeta$,

and show non-trivial cases, i.e.

0) "$F_f`\alpha \in f$" is recursive in $f$ where $\alpha < \zeta$;

1) "$F_f`\alpha \in F_f`\zeta$" is recursive in $f$ where $\alpha < \zeta$;

2) "$F_f`\alpha = F_f`\zeta$" is recursive in $f$ where $\alpha < \zeta$;

3) "$F_f`\zeta \in F_f`\beta$" is recursive in $f$ where $\beta < \zeta$.

*Proof* of 0).

$$F_f`\alpha \in f \leftrightarrow \exists \gamma < \alpha\, \exists \delta < \alpha (F_f`\alpha = \langle F_f`\gamma\, F_f`\delta\rangle \wedge F_f`\gamma = f(F_f`\delta)),$$

the right side of which is recursive in $f$ by H 1).

New we show 2) and 3) are reduced to 1):

*Proof* that 1) implies 2).

$$F_f`\alpha = F_f`\zeta \leftrightarrow \forall \gamma < \alpha (F_f`\gamma \in F_f`\alpha \rightarrow F_f`\gamma \in F_f`\zeta)$$
$$\wedge\ \forall \gamma < \zeta (F_f`\gamma \in F_f`\zeta \rightarrow F_f`\gamma \in F_f`\alpha),$$

the right side of which is recursive in $f$ by H 1) and 1).

*Proof* that 1) and 2) implies 3).

$$F_f`\zeta \in F_f`\beta \leftrightarrow \exists \gamma < \beta (F_f`\gamma = F_f`\zeta \wedge F_f`\gamma \in F_f`\beta),$$

the right side of which is recursive in $f$ by 2) and H 1).

From the above consideration we can easily verify that the predicates $F_f`\alpha \in F_f`\beta$, $F_f`\alpha = F_f`\beta$ and $F_f`\gamma \in f$ can actually be defined by recursion as follows: Let $D(f, \alpha, \beta, \zeta)$ be defined by

$$D(f, \alpha, \beta, \zeta) = \begin{cases} j(g_0(D(f, \alpha, \beta, \max(\alpha, \beta))), g_1(D(f, \alpha, \beta, \max(\alpha, \beta))), c) \\ \text{if} \quad \max(\alpha, \beta) < \zeta \\ j(1, 1, 1) \quad \text{if} \quad \zeta \leqq \max(\alpha, \beta) \\ j(1, 0, 1) \quad \text{if} \quad \alpha = \beta = \zeta \\ j(a_1, b, 1) \quad \text{if} \quad \alpha < \beta = \zeta \\ j(a_2, b, 1) \quad \text{if} \quad \beta < \alpha = \zeta, \end{cases}$$

where $c = 0$ or 1 according as $F_f`\alpha \in f$ or not; the condition can be expressed by using $D(f, \xi, \eta, \lambda)$ with $\lambda < \zeta$ and recursive predicates (cf. the proof of 0)): $a_1 = 0$ or 1, according as $F_f`\alpha \in F_f`\beta$ or not; the condition can be expressed by using $D(f, \xi, \eta, \lambda)$ with $\lambda < \zeta$ and recursive predicates (cf. the proof of 1) from H 1) and H 2)): $b = 0$ or 1, according as $F_f`\alpha = F_f`\beta$ or not; the condition can be expressed as above, cf. the proof of 2) from 1): $a_2 = 0$ or 1, according as $F_f`\alpha \in F_f`\beta$ or not; the condition can be expressed as above (cf. the proof of 3) from 1) and 2)). Then $D(f, \alpha, \beta, \zeta)$ is recursive in $f$: $g_0(D(f, \alpha, \beta, \max(\alpha, \beta)))$, $g_1(D(f, \alpha, \beta, \max(\alpha, \beta)))$ and $g_2(D(f, \alpha, \beta, \alpha + 1))$ become the characteristic functions of $F_f`\alpha \in F_f`\beta$. $F_f`\alpha = F_f`\beta$ and $F_f`\alpha \in f$ respectively.

Thus we only have to show 1) assuming H 1) and H 2). We separate cases according to $K_f`\zeta$. Since all cases are treated similarly, we give some examples: The case where $K_f`\zeta = 0$:

$$F_f`\alpha \in F_f`\zeta \leftrightarrow \exists \gamma < \zeta (F_f`\alpha = F_f`\gamma),$$

the right side of which is reduced to H 2). The case where $K_0`\zeta = 1$ and $\zeta = m(\gamma) + 1$:

$$F_f`\alpha \in F_f`\zeta \leftrightarrow F_f`\alpha \in f \wedge F\alpha_f` \in F_f`m(\gamma),$$

the right side of which is reduced to 0) and H 1).

Thus we can complete the proof of the second half of Theorem 2.

Now we shall consider the hyperarithmetical functions defined in § 5, Chapter II, of [28]. In this article we call those functions hyperarithmetical "in the narrow sense". We can define the hyperarithmetical functions (in the wider sense). The definition of hyperarithmetical functions and the detailed discussion of those functions are seen in [20].

**Definition.** We say a formula $\varphi(a)$ is *provably* $\Delta_1^{\text{Ord}}$ *in* $h$ if there exist formulas $A(f, a, h)$ and $B(f, a, h)$ which are arithmetical in $f$ and $h$ and

$$\varphi(a) \leftrightarrow \forall f\, A(f, a, h) \leftrightarrow \exists f\, B(f, a, h)$$

is provable in a simple type theory, assuming the general comprehension axiom and the axiom of choice.

In the following $\mathscr{L}_h(a)$ denotes $\mathscr{L}_{L(h)}(a)$.

**Lemma.** 1) $\mathscr{N}(\{a \mid L^{R(\alpha)}(h \cap R(a)) = L(h \cap R(a)) \cap R(a)\})$.

2) If $y \in R(\alpha + 1) \cap L(h \cap R(\alpha))$ and $L^{R(\alpha)}(h \cap R(\alpha)) = L(h \cap R(\alpha)) \cap R(\alpha)$, then $\mathscr{L}^{R(\alpha)}_{h \cap R(\alpha)}(y)$.

3) $R^{L(h \cap R(\alpha))}(\alpha) = L(h \cap R(\alpha)) \cap R(\alpha)$ for all $\alpha$.

*Proof.* 1) Since $L^{R(\alpha)}(h \cap R(\alpha)) \subseteqq L(h \cap R(\alpha)) \cap R(\alpha)$ for every inaccessible $\alpha$ and $\mathscr{N}(\{\alpha \mid \text{In}(\alpha)\})$ by Proposition 11 in § 3 of Chapter 1, we have

$$\mathscr{N}(\{\alpha \mid L^{R(\alpha)}(h \cap R(\alpha)) \subseteqq L(h \cap R(\alpha)) \cap R(\alpha)\})$$

by 2.2. To prove

$$\mathcal{N}(\{\alpha \mid L^{R(\alpha)}(h \cap R(\alpha)) \supseteqq L(h \cap R(\alpha)) \cap R(\alpha)\}),$$

we define $f$ by

$$f(\alpha) = \mu\beta(L^{R(\beta)}(h \cap R(\beta)) \supseteqq L(h) \cap R(\alpha) \wedge \mathrm{In}(\beta)).$$

From the construction of $L(h)$, $f(\alpha) > 0$ for every $\alpha > 0$. Since $\mathcal{N}(\{\alpha \mid \mathrm{In}(\alpha) \wedge \forall \gamma < \alpha(f(\gamma) < \alpha)\})$ by Proposition 3, 5 and 11 of § 3, Chapter I, it suffices to show

$$\text{(1)} \qquad L^{R(\alpha)}(h \cap R(\alpha)) \supseteqq L(h \cap R(\alpha)) \cap R(\alpha)$$

for each $\alpha$ such that $\mathrm{In}(\alpha)$ and $\forall \gamma < \alpha(f(\gamma) < \alpha)$. Let $\alpha$ be such an ordinal number and $x$ an arbitrary element of $L(h \cap R(\alpha)) \cap R(\alpha)$. Then there exists an ordinal number $\beta_x < \alpha$ such that $x \in L(h) \cap R(\beta_x)$. By our assumption on $\alpha$ $f(\beta_x) < \alpha$ and therefore

$$L(h) \cap R(\beta_x) \subseteqq L^{R(f(\beta_x))}(h \cap R(f(\beta_x))) \subseteqq L^{R(\alpha)}(h \cap R(\alpha)),$$

i.e. $x \in L^{R(\alpha)}(h \cap R(\alpha))$. Thus we have (1).

2) Suppose

$$\text{(1)} \qquad y \in R(\alpha + 1) \cap L(h \cap R(\alpha))$$

and

$$\text{(2)} \qquad L^{R(\alpha)}(h \cap R(\alpha)) = L(h \cap R(\alpha)) \cap R(\alpha).$$

We shall show $\mathcal{L}^{R(\alpha)}_{h \cap R(\alpha)}(y)$, i.e.

$$\text{(3)} \qquad y \subseteqq L^{R(\alpha)}(h \cap R(\alpha))$$

and

$$\text{(4)} \qquad \forall x \in L^{R(\alpha)}(h \cap R(\alpha))(x \cap y \in L^{R(\alpha)}(h \cap R(\alpha)).$$

We have (3) by (1), (2) and completeness of $L(h \cap R(\alpha))$. For (4) let $x$ be any element of $L^{R(\alpha)}(h \cap R(\alpha))$. Then

$$x \in L(h \cap R(\alpha)) \cap R(\alpha) \qquad \text{(by (2))}$$

and therefore

$$x \cap y \in L(h \cap R(\alpha)) \cap R(\alpha) \qquad \text{(by (1))}.$$

3) is proved by transfinite induction on $\alpha$.

**Theorem 3.** Let $h$ be a function of ordinal numbers and let $A(f, a, h)$ and $B(f, a, h)$ be arithmetical in $h$ such that

$$\text{(1)} \qquad \forall f\, A(f, a, h) \leftrightarrow \exists f\, B(f, a, h)$$

is provable possibly by using the comprehension axiom of a simple type theory and the axiom of choice. Then

$$\forall f\, A(f, a, h) \leftrightarrow \forall f(\mathscr{L}_h(f) \to A(f, a, h)) \leftrightarrow \exists f(\mathscr{L}_h(f) \wedge B(f, a, h)) \tag{2}$$

is provable.

*Proof.* Let $C(a, h)$ be short for the formula (2) and let $\alpha$ be an ordinal number satisfying the following conditions (3)—(5):

$$\mathrm{In}(\alpha) \tag{3}$$

$$\forall x\, C(x, h) \leftrightarrow (\forall x\, C(x, h))^{R(\alpha)} \tag{4}$$

where $(C(h))^{R(\alpha)}$ denotes $C^{R(\alpha)}(h \cap R(\alpha))$.

$$L^{R(\alpha)}(h \cap R(\alpha)) = L(h \cap R(\alpha)) \cap R(\alpha). \tag{5}$$

(The existence of such an $\alpha$ is assured by Propositions 5 and 11 of § 3, Chapter I, Axiom 2.5, and Lemma 1).) Then $R(\alpha + n)$ is a model of the comprehension axiom for the $n$-th simple type theory and the axiom of choice for $R(\alpha)$, and the second order variables are interpreted to range over $R(\alpha + 1)$. Therefore from the assumption (1) follows

$$\forall x \in R(\alpha)(\forall y \in R(\alpha + 1)\, A^{R(\alpha)}(y, x, h \cap R(\alpha)) \leftrightarrow \exists y \in R(\alpha + 1)\, B^{R(\alpha)}(y, x, h \cap R(\alpha)). \tag{6}$$

By (1), (3) and Lemma 3),

$$\forall y \in R(\alpha + 1) \cap L(h \cap R(\alpha))\, A^{R(\alpha) \cap L(h \cap R(\alpha))}(y, a, h \cap R(\alpha)) \leftrightarrow \exists y \in R(\alpha + 1) \cap L(h \cap R(\alpha))\, B^{R(\alpha) \cap L(h \cap R(\alpha))}(y, a, h \cap R(\alpha)). \tag{7}$$

Let $a \in R(\alpha)$. Then we have successively the following implications and equivalences:

$\forall y \in R(\alpha + 1)\, A^{R(\alpha)}(y, a, h \cap R(\alpha))$

$\to \forall y \in R(\alpha + 1)(\mathscr{L}^{R(\alpha)}_{h \cap R(\alpha)}(y) \to A^{R(\alpha)}(y, a, h \cap R(\alpha))$

$\to \forall y \in R(\alpha + 1) \cap L(h \cap R(\alpha))\, A^{R(\alpha)}(y, a, h \cap R(\alpha))$ (by Lemma 2) and (5))

$\leftrightarrow \forall y \in R(\alpha + 1) \cap L(h \cap R(\alpha))\, A^{R(\alpha) \cap L(h \cap R(\alpha))}(y, a, h \cap R(\alpha))$ (by Lemma 3))

$\leftrightarrow \exists y \in R(\alpha + 1) \cap L(h \cap R(\alpha))\, B^{R(\alpha) \cap L(h \cap R(\alpha))}(y, a, h \cap R(\alpha))$ (by (7))

$\leftrightarrow \exists y \in R(\alpha + 1) \cap L(h \cap R(\alpha))\, B^{R(\alpha)}(y, a, h \cap R(\alpha))$ (by Lemma 3))

$\to \exists y \in R(\alpha + 1)(\mathscr{L}^{R(\alpha)}_{h \cap R(\alpha)}(y) \wedge B^{R(\alpha)}(y, a, h \cap R(\alpha))$ (by (5) and Lemma 2))

$\to \exists y \in R(\alpha + 1)\, B^{R(\alpha)}(y, a, h \cap R(\alpha))$

$\to \forall y \in R(\alpha + 1)\, A^{R(\alpha)}(y, a, h \cap R(\alpha))$ (by (6)).

That is, we have $(\forall x\, C(x, h))^{R(\alpha)}$, which implies (2) by (4).

*Remark.* For the proof of the above theorem, the second order reflection principle is sufficient, but, to make the argument simpler, we proved the theorem assuming NTT[1].

**Lemma.** The predicate $\mathscr{W}_e(f)$, which expresses "$f$ is a well-ordering", is provably $\Delta_1^{\mathrm{Ord}}$ in $f$. Moreover, it can be expressed as:

$$\mathscr{W}_e(f) \leftrightarrow \forall g\, A(g \restriction \omega, f) \leftrightarrow \exists h\, B(h, f),$$

where $A(g, f)$ and $B(g, f)$ are arithmetical in $g$ and $f$.

*Proof.* Let $\mathscr{L}_e(f)$ be the predicate which expresses that $f$ is a linear ordering. $\mathscr{L}_e(f)$ is arithmetical in $f$. Let $a \overset{f}{<} b$ be short for $\langle ab\rangle \in f$. $\mathscr{W}_e(f)$ is expressible by

$$(1) \qquad \mathscr{L}_e(f) \wedge \forall g \neg \forall i (g(i+1) \overset{f}{<} g(i)),$$

or

$$\forall g (\mathscr{L}_e(f) \wedge \neg \forall i (g \restriction \omega (i+1) \overset{f}{<} g \restriction \omega (i))).$$

Let $W(B)$ be $\exists h\, \exists \beta (h$ is an isomorphism from $\beta$ onto $B$ where $B$ is well-ordered by $f)$. Then (1) is equivalent to

$$\mathscr{L}_e(f) \wedge \forall \alpha\, W(f \restriction \alpha \times \alpha),$$

or, by using the axiom of choice,

$$\exists h (\mathscr{L}_e(f) \wedge \forall \alpha\, \exists \beta (h(\alpha) \text{ is an isomorphism from } \beta \text{ onto } f \restriction \alpha \times \alpha)).$$

**Theorem 4.** The predicate $\mathscr{W}_e(f)$ is arithmetical in $f$.

*Proof.* By Lemma

$$\mathscr{W}_e(f) \leftrightarrow \forall g\, A(g \restriction \omega, f) \leftrightarrow \exists h\, B(h, f).$$

By Theorem 3,

$$\mathscr{W}_e(f) \leftrightarrow (\mathscr{W}_e(f))^{L(h)} \leftrightarrow \forall g (\mathscr{L}_f(g) \rightarrow A(g \restriction \omega, f)).$$

The last formula is equivalent to $\forall g (\mathscr{L}_f(g \restriction \omega) \rightarrow A(g \restriction \omega, f))$ or

$$\forall x \in L(f) (x\, \mathscr{F}_n\, \omega \wedge \mathscr{W}(x) \subseteqq \mathrm{On} \rightarrow A(x, f)).$$

Evidently this is arithmetical in $f$.

**Theorem 5.** If $g$ is numerically hyperarithmetical in $f$, then $\mathscr{L}_f(g)$, where "numerically hyperarithmetical in $f$" means hyperarithmetical in $f$ when any ordinal number is regarded as a constant.

*Proof.* By Theorems 1 and 4 and the definition of "numerically hyperarithmetical".

**Theorem 6.** If $\mathscr{L}_e(f) \wedge \neg \mathscr{W}_e(f)$, then we can define a function $g$ such that $\forall i (g(i+1) \overset{f}{<} g(i))$ and $g$ is arithmetical in $f$.

*Proof*. Assume $\mathscr{L}_e(f)$. Then $\neg \mathscr{W}_e(f) \leftrightarrow \exists g\, \forall i (g(i+1) \overset{f}{<} g(i))$. Since $\mathscr{W}_e(f)$ is arithmetical in $f$ by Theorem 4, it is $L(f)$-expressible; that is, $\neg \mathscr{W}_e(f)$ is absolute with respect to $f$:

$$(\neg \mathscr{W}_e(f))^{L(f)} \leftrightarrow \neg \mathscr{W}_e(f).$$

From this we have successively:

$$\exists g (\mathscr{L}_f(g) \wedge \forall i (g(i+1) \overset{f}{<} g(i));$$

$$\exists g (\mathscr{L}_f(g) \wedge \forall i (g \upharpoonright \omega` i + 1 \overset{f}{<} g \upharpoonright \omega` i);$$

$$\exists x \in L(f)\, \forall i (x` i + 1 \overset{f}{<} x` i);$$

$$\exists \alpha\, \forall i ((F_f` \alpha)` i + 1 \overset{f}{<} (F_f` \alpha)` i).$$

Let $\alpha_0$ be $\mu\alpha\, \forall i ((F_f`\alpha)` i + 1 \overset{f}{<} (F_f`\alpha)` i)$ and let $g$ be defined by: $g \upharpoonright \omega = F_f`\alpha_0$ and $g`a = 0$ for every $a \geqq \omega$. Then $g$ is arithmetical in $f$ and $\forall i (g(i+1) < g(i))$.

Our definition of "hyperarithmetical" does not depend only on the length of the universe, since $\mathscr{W}_e(f)$ appears in this notion, while it is fairly constructible. If we accept the second order reflection principle, however, then we may consider that the notion of "hyperarithmetical" completely depends on the length. Though, in [28], the author anticipated that the class of $\Delta_1^{\mathrm{Ord}}$ functions is identical with the class of hyperarithmetical function in the narrow sense, MAYBERRY has given the negative answer [20]. However, it is still quite possible that $\Delta\, TC$[4] holds, i.e. if $f$ is a $\Delta_1^{\mathrm{Ord}}$ function and $\alpha \geqq \omega$, then $\overline{\overline{f(\alpha)}} \leqq \overline{\overline{\alpha}}$ (cf. 5.7, Chapter II, of [28]).

For the classification of $\aleph$ we recall the definitions of TC and HTC.

TC: If $f$ is arithmetical and $\alpha \geqq \omega$, then $\overline{\overline{f(\alpha)}} \leqq \overline{\overline{\alpha}}$.

HTC: If $f$ is hyperarithmetical and $\alpha \geqq \omega$, then $\overline{\overline{f(\alpha)}} \leqq \overline{\overline{\alpha}}$.

Further detailed classifications will be considered in a later paper.

*Remark.* TC holds if and only if, for any formula $\varphi$ of first order,

$$\forall \alpha (\forall x \in (R(\aleph_\alpha) \cap L) (\varphi^{R(\aleph_\alpha) \cap L}(x) \leftrightarrow \varphi^L(x)).$$

*Proof.* This is easily proved by applying Theorem 2, § 2, of [25].

## § 2. Valuation of $2^{\aleph_\alpha}$

Throughout this section, $m^n$ with cardinals $m$ and $n$ denotes cardinal exponentiation.

After we fix $\aleph$ in a certain way, for instance assuming TC, HTC or $\Delta$TC in § 1, $\aleph$ is very useful for a rough valuation of $2^{\aleph_\alpha}$. $2^{\aleph_\alpha}$ is measured

[4] NAMBA recently proved that the existence of a measurable cardinal implies $\Delta TC$.

by the function $f$ such that $2^{\aleph_\alpha} = \aleph_{f(\alpha)}$. Therefore what matters is to characterize $f$, that is the classification of $f$. Concerning this, EASTON [5] has given a solution, namely if $\aleph_\alpha$ is regular, then the classification of $f$ is reduced to that of the functions which satisfy certain conditions (cf. [5]). However, the final solution has not been given yet. Various problems remain unsolved concerning this question.

1) In [5] the conditions on the function $f$ for the case $\aleph_\alpha$ is singular have not been found yet, or we cannot evaluate $2^{\aleph_\alpha}$ at present when $\aleph_\alpha$ is singular.

2) In [5], it is assumed that GCH holds in the original model.

3) In [5], some conditions on $\aleph$ and cofinality of cardials are assumed, i.e., $\aleph_{f(\alpha)}$ is not cofinal with any cardinal less than or equal to $\aleph_\alpha$.

Because of 3), in order to classify $2^{\aleph_\alpha}$, we must first evaluate $\aleph$ and cofinality of cardinals. Therefore more general question on the evaluation of $2^{\aleph_\alpha}$ is expressed as follows.

Problem 1. Suppose $\alpha$ is a countable ordinal number and $m = F``\alpha$ is a model of ZF. Consider a function $g$ from $\alpha$ into $\alpha$ and $L^m(g)$. Let $m_1$ be $L^m(g)$. Assume the following conditions:

(1) $m_1$ is a model of ZF.

(2) $g$ is compatible with $cf^{m_1}$, i.e., for all $\beta < \alpha$; $g(\beta) \leqq \beta$, $g(\beta) = g(cf^{m_1}(\beta))$, and $cf^{m_1}(g(\beta)) = g(\beta)$. Then the question is: Does there exist a model $n$ of ZF such that $\mathrm{On}^n = \alpha$ and $cf^n(\beta) = g(\beta)$ for $\beta < \alpha$?

We should remark that this problem arises, because $cf^n(\beta)$ might be different from $cf^m(\beta)$ (or in the weaker form, $\aleph_\alpha^m \neq \aleph_\alpha^n$ might be the case), while in Easton model $cf^m(\beta) = cf^n(\beta)$ is proved. Since $cf^n(\beta)$ is a function of $\beta$, we must consider various possibilities of the functions which define $cf^n(\beta)$.

Problem 2. Assume that Problem 1 is affirmatively answered. Then the second question is: Does there exist a model among the models satisfying the conditions in Problem 1 in which GCH holds?

If Problem 2 is affirmatively answered, then we may conclude that the classification of $2^{\aleph_\alpha}$ is completed for the case $\aleph_\alpha$ is regular by EASTON [5].

Now we generalize the problem further and consider the classification of $\aleph_\alpha^{\aleph_\beta}$ for all $\alpha$ and $\beta$. Using ZERMELO-KÖNIG's theorem, TARSKI proved in [30] that the valuation problem of $\aleph_\alpha^{\aleph_\beta}$ can be reduced to the valuation problem of $\aleph_\alpha^{cf(\aleph_\alpha)}$. We can express this fact in terms of recursive functions of ordinal numbers in the following way.

**Lemma 1.** *$\aleph$ is recursive in $cf$.*

*Proof.* $\aleph_\alpha$ is expressed as follows:

$$\begin{aligned} \aleph_0 &= \omega, \\ \aleph_{\alpha+1} &= \mu\gamma(\gamma \in K_{\mathrm{II}} \wedge \aleph_\alpha < \gamma \wedge cf(\gamma) = \gamma), \\ \aleph_\alpha &= \sup_{\beta<\alpha} \aleph_\beta \quad \text{for} \quad \alpha \in K_{\mathrm{II}}. \end{aligned}$$

**Definition.** We define the functions $F(\alpha, \beta)$, $cf^*(\alpha)$ and $f(\alpha)$ as follows:

$$F(\alpha, \beta) = \aleph_\alpha^{\aleph_\beta},$$
$$\aleph_{cf^*(\alpha)} = cf(\aleph_\alpha),$$
$$\aleph_{f(\alpha)} = F(\alpha, cf^*(\alpha)).$$

**Lemma 2.** 1) *$\overline{\overline{\gamma}}$ is recursive in $\aleph$ (i.e. recursive in $cf$).*

2) *$cf^*$ is recursive in $cf$.*

3) *If $\aleph_\gamma$ is regular, then $2^{\aleph_\gamma} = \aleph_{f(\gamma)}$.*

*Proof.* 1) $\overline{\overline{\gamma}} = \mu x \leqq \gamma\, \exists \delta (\aleph_\delta = x \wedge \aleph_{\delta+1} > \gamma)$.

2) $cf^*(\alpha) = \mu x \leqq \alpha (\aleph_x = cf(\aleph_\alpha))$.

3) If $\aleph_\gamma$ is regular, then

$$\aleph_{f(\gamma)} = \aleph_\gamma^{\aleph_\gamma} = 2^{\aleph_\gamma}.$$

**Lemma 3.** *Let $G(\alpha)$ be $2^{\aleph_\alpha}$. Then $G$ is recursive in $cf$ and $f$.*

*Proof.* We prove that

$$G(\alpha) = \begin{cases} \aleph_{f(\alpha)} & \text{if } cf(\aleph_\alpha) = \aleph_\alpha \text{ (Case 1)}, \\ \aleph_{f(h(\alpha))} & \text{if } cf(\aleph_\alpha) < \aleph_\alpha \wedge \forall \lambda < \alpha\, \exists \nu < \alpha (G^\alpha(\lambda) < G^\alpha(\nu)), \\ & \text{where } \aleph_{h(\beta)} = \sup_{\xi<\beta} \aleph_{f(\xi)} \text{ (Case 2)}. \\ \aleph_{f(\gamma)} & \text{otherwise, where } \aleph_\gamma = \max(\aleph_\delta, cf(\aleph_\alpha)) \text{ and} \\ & \delta = \mu\lambda (\forall \nu < \alpha (\lambda \leqq \nu \rightarrow G^\alpha(\lambda) = G^\alpha(\nu)) \\ & \quad \wedge \lambda \in K_{\mathrm{I}}) \text{ (Case 3)}; \end{cases}$$

where $G^\alpha(\beta) = G(\beta)$ if $\beta < \alpha$; 0 otherwise.

Case 1. By Lemma 2.3).

Case 2. In this case $\alpha \in K_{\mathrm{II}}$. Let $\varkappa$ be $cf(\aleph_\alpha)$, $\aleph_\alpha = \sum_{\xi<\varkappa} \chi_{\alpha_\xi}$, where $\{\alpha_\xi\}$ is an increasing sequence of successor ordinal numbers whose limit is $\alpha$.

$$G(\alpha) = 2^{\sum_{\xi<\varkappa} \aleph_{\alpha_\xi}} = \prod_{\xi<\varkappa} 2^{\aleph_{\alpha_\xi}} = \prod_{\xi<\varkappa} \aleph_{f(\alpha_\xi)} \qquad \text{(by Lemma 2.3)}$$
$$\leqq \prod_{\xi<\varkappa} \aleph_{h(\alpha)} = \aleph_{h(\alpha)}^{\varkappa} \leqq (2^{\aleph_\alpha})^{\aleph} = 2^{\aleph_\alpha \cdot \varkappa} = 2^{\aleph_\alpha}.$$

Therefore it suffices to show $\varkappa = cf(\aleph_{h(\alpha)})$. Since

$$\aleph_{h(\alpha)} = \lim_{\xi<\varkappa} 2^{\aleph_{\alpha_\xi}}$$

$\varkappa$ is cofinal with $\aleph_{h(\alpha)}$ (consider the subsequence $\{\alpha_{\xi_\nu}\}$ of $\{\alpha_\xi\}$ such that $2^{\aleph_{\alpha_{\xi_\eta}}} < 2^{\aleph_{\alpha_{\xi_\zeta}}}$ if $\eta < \zeta$) and $cf(\aleph_{h(\alpha)}) = \varkappa$.

Case 3. By the definition $\gamma < \alpha$ and $\aleph_\gamma$ is regular.

$$2^{\aleph_\gamma} \leqq G(\alpha) \leqq 2^{\sum\limits_{\xi<\varkappa} \aleph_{\alpha_\xi}} \leqq 2^{\aleph_\gamma \cdot \varkappa} = 2^{\aleph_\gamma} = \aleph_{f(\gamma)} \leqq 2^{\aleph_\alpha},$$

where $\varkappa = cf(\aleph_\alpha)$.

**Theorem.** $F(\alpha, \beta)$ is recursive in $cf$ and $f$.

*Proof.* We prove the theorem by transfinite induction on $j(\alpha, \beta)$, separating three cases:

Case 1. The case where $\aleph_\alpha \leqq \aleph_\beta$, so that $F(\alpha, \beta) = 2^{\aleph_\beta}$. This case is reduced to Lemma 3.

Case 2. The case where $cf(\aleph_\alpha) > \aleph_\beta$.

$$F(\alpha, \beta) = \sum_{\gamma<\aleph_\alpha} \gamma^{\aleph_\beta} \leqq (\lim_{\gamma<\aleph_\alpha} \gamma^{\aleph_\beta}) \times \aleph_\alpha.$$

Therefore $F(\alpha, \beta) = \mu x \, \forall \delta < \alpha (F(\delta, \beta) \leqq x) \times \aleph_\alpha$.

Case 3. The case where $\aleph_\alpha > \aleph_\beta \geqq cf(\aleph_\alpha)$.

3.1. The subcase where $\exists \gamma < \alpha (\aleph_\alpha \leqq F(\gamma, \beta))$. Let $\gamma_0$ be the least such $\gamma$. Then

$$F(\alpha, \beta) \leqq \aleph_{\gamma_0}^{\aleph_\beta \times \aleph_\beta} = \aleph_{\gamma_0}^{\aleph_\beta} \leqq F(\alpha, \beta),$$

i.e. $F(\alpha, \beta) = F(\gamma_0, \beta)$.

3.2. The subcase where $\forall \gamma < \alpha (F(\gamma, \beta) < \aleph_\alpha)$.

$$F(\alpha, cf^*(\alpha)) \leqq F(\alpha, \beta) \leqq \prod_{\xi<cf(\aleph_\alpha)} \aleph_{\alpha_\xi}^{\aleph_\beta} \leqq \prod_{\xi<cf(\aleph_\alpha)} \aleph_\alpha$$
$$\leqq F(\alpha, cf^*(\alpha)),$$

i.e. $F(\alpha, \beta) = \aleph_{f(\alpha)}$.

*Remark.* The following properties of $f$ are ZF-provable. Let $\alpha^*$ be short for $cf^*(\alpha)$.

1) $cf(\aleph_{f(\alpha)}) > cf(\aleph_\alpha)$.

2) $f(\alpha) > \alpha$

3) $\aleph_\alpha < \aleph_\beta \wedge cf(\aleph_\alpha) \leqq cf(\aleph_\beta) \to f(\alpha) \leqq f(\beta)$

4) $\alpha \leqq f(\alpha^*) \to f(\alpha) = f(\alpha^*)$.

5) Assume $f(\alpha^*) < \alpha \wedge \exists \beta < \alpha (\aleph_\alpha \leqq \aleph_{f(\beta)} \wedge cf(\aleph_\beta) < cf(\aleph_\alpha))$, and let $\lambda$ be the ordinal number $\gamma$ such that $cf(\aleph_\gamma) \leqq cf(\aleph_\alpha)$ and $\alpha \leqq f(\gamma)$. Then $f(\alpha) = f(\lambda)$.

6) $\alpha \leqq f(\beta) \wedge cf(\aleph_\alpha) \leqq cf(\aleph_\beta) \to f(\alpha) \leqq f(\beta)$.

*Proof.* 1) and 2) are proved from KÖNIG's lemma, and 3) is obvious.

4) Since $\alpha^* \leqq \alpha$, $\aleph_{f(\alpha)} \geqq \aleph_{f(\alpha^*)}$, i.e. $f(\alpha) \geqq f(\alpha^*)$. From $\alpha \leqq f(\alpha^*)$, we have $\aleph_\alpha \leqq \aleph_{f(\alpha^*)} = F(\alpha^*, \alpha^*)$. Then $F(\alpha, \alpha^*) \leqq F(\alpha^*, \alpha^*)$, i.e. $f(\alpha) \leqq f(\alpha^*)$.

5) Under the assumption of 5), let $\nu$ be the least $\gamma$ such that

$$\aleph_\alpha \leqq F(\gamma, \alpha^*).$$

Then from the assumption $\nu < \alpha$ and $F(\gamma, \alpha^*) < \aleph_\alpha$ for every $\gamma < \nu$. We shall show $f(\alpha) = f(\nu)$ and $\lambda = \nu$. Firstly $cf(\aleph_\nu) \leqq cf(\aleph_\alpha)$; otherwise

$$\aleph_\alpha \leqq F(\nu, \alpha^*) = \aleph_\nu \times \sup_{\gamma<\nu} \aleph_\gamma^{\aleph_{\alpha^*}} = \aleph_\nu < \aleph_\alpha,$$

yielding a contradiction. Secondly $\nu \in K_{\mathrm{II}}$, for if $\nu \in K_{\mathrm{I}}$,

$$\aleph_\nu = cf(\aleph_\nu) \leqq cf(\aleph_\alpha),$$

and

$$\aleph_\alpha \leqq F(\nu, \alpha^*) \leqq F(\alpha^*, \alpha^*) = \aleph_{f(\alpha^*)},$$

contradicting $\alpha > f(\alpha^*)$. Finally

$$F(\alpha, \alpha^*) \leqq F(\nu, \alpha^*) \leqq \prod_{\gamma<\nu} F(\gamma, \alpha^*) \leqq F(\nu, \nu^*) \leqq F(\alpha, \alpha^*),$$

i.e. $f(\alpha) = f(\nu)$, (and $\alpha \leqq f(\nu)$). Since $\lambda \leqq \nu$ from the above, and

$$\aleph_\alpha \leqq F(\lambda, \lambda^*) \leqq F(\lambda, \alpha^*)$$

by the assumption, yielding $\nu \leqq \lambda$, $\lambda = \nu$.

6) Assume $\alpha \leqq f(\beta)$ and $cf(\aleph_\alpha) \leqq cf(\aleph_\beta)$. Then

$$F(\alpha, \alpha^*) \leqq F(f(\beta), \alpha^*) = \aleph_\beta^{\aleph_{\beta*} \times \aleph_{\alpha*}} = F(\beta, \beta^*),$$

i.e. $f(\alpha) \leqq f(\beta)$.

We are inclined to think that the cardinality of a power set is either *very* small or *very* large. The extreme of one way is the generalized continuum hypothesis from which we can obtain many interesting consequences. As the extreme of the other way we have already proposed two hypotheses at the end of [26]:

1. $\overline{\overline{P(a)}}$ is not arithmetical in all cardinals less than $\overline{\overline{P(a)}}$.

2. $\overline{\overline{P(a)}}$ is not arithmetical in $\chi$, where $\chi$ is the function such that $\chi(a)$ is the least cardinal greater than $a$.

Now we would like to propose several stronger hypotheses with the same motivation as the hypothesis $\Delta$TC presented in [28] concerning the limit of the expression of cardinality.

The predicate "$b = \overline{\overline{P(a)}}$" is expressible in the $\Pi_1^{\mathrm{Ord}\,1} \wedge \Sigma_1^{\mathrm{Ord}\,1}$-form in the function $\aleph$, or equivalently, $\Pi_1^{\mathrm{Ord}\,1} \wedge \Sigma_1^{\mathrm{Ord}\,1}$-form in the class Card of all cardinal numbers; but it seems not to be expressible in any lower form in the hierarchies of this kind, i.e. the following similar hypotheses seem to be plausible:

P 1. If $a \geqq \omega$, then "$b = \overline{\overline{P(a)}}$" is not expressible in $\Delta_1^{\mathrm{Ord}\,1}$ in the function $\aleph$ (or equivalently in the class Card).

P 2. If $a \geqq \omega$, then "$b = \overline{\overline{P(a)}}$" is not expressible in $\Delta_1^{\mathrm{Ord}\,1}$ in $\aleph$ (or Card) and $f$, where $f$ is any function from $a_0 < \overline{\overline{P(a)}}$ into $\overline{\overline{P(a)}}$.

P 3. Let $a \geqq \omega$ and $g$ be expressible in $\Delta_1^{\mathrm{Ord}\,1}$ in Reg and $f$, where Reg is the class of all regular cardinals and $f$ is any function from an ordinal $a_0 < \overline{\overline{P(a)}}$ into $\overline{\overline{P(a)}}$. Then $g(b) < \overline{\overline{P(a)}}$ for every $b < \overline{\overline{P(a)}}$.

Let $\forall^\circ f$ mean that $f$ ranges over all bounded functions (i.e., functions $f$ such that $\{x \mid f(x) \neq 0\}$ is bounded). We can consider a hierarchy concerning $\forall^\circ f$ and $\exists^\circ f$ (instead of that concerning $\forall f$ and $\exists f$). In this hierarchy we defined $\Pi_1^{\mathrm{Ord}\,1}$ or $\Sigma_1^{\mathrm{Ord}\,1^\circ}$ similarly.

By P 1°, P 2° and P 3° we understand the hypotheses obtained from P 1, P 2 and P 3 respectively by replacing $\Delta_1^{\mathrm{Ord}\,1}$ by $\Delta_1^{\mathrm{Ord}\,1^\circ}$. Obviously P 3 $\rightarrow$ P 2 $\rightarrow$ P 1; P 3° $\rightarrow$ P 2° $\rightarrow$ P 1° and P$n$ $\rightarrow$ P$n$° $(n = 1, 2, 3)$. P$n$° can easily be expressed in set theory of a first order language. Therefore it is interesting that the relative consistency of P$n$° with ZF and the axiom of choice can be considered to be the consistency problem concerning ZF.

Now we state simple implications from P 3°: Since $a^+$ (the least cardinal $> a$) can be expressed in $\Pi_1^{\mathrm{Ord}\,1^\circ} \wedge \Sigma_1^{\mathrm{Ord}\,1^\circ}$, Tarski's hypothesis:

(1) $a \geqq \omega \rightarrow \overline{\overline{P(a)}}$ is weakly inaccessible, follows immediately from P 3°. Since the first weakly inaccessible cardinal greater than $b$ is in $\Delta_1^{\mathrm{Ord}\,1^\circ}$ in Reg, the $b$-th weakly inaccessible cardinal is in $\Delta_1^{\mathrm{Ord}\,1^\circ}$ in Reg. Hence

(2) $a \geqq \omega \wedge b < \overline{\overline{P(a)}} \rightarrow$ the $b$-th weakly inaccessible cardinal $< \overline{\overline{P(a)}}$. Since "$f(a) = a$" is in $\Delta_1^{\mathrm{Ord}\,1^\circ}$ in $f$, the $b$-th fixed point of $f$ for $f$ in $\Delta_1^{\mathrm{Ord}\,1^\circ}$ in Reg is again in $\Delta_1^{\mathrm{Ord}\,1}$ in Reg. Therefore starting with (2) and following this method we can obtain much stronger statement.

## § 3. Degree of Undefinability of $2^{\aleph_\alpha}$

Throughout this section the expression $m^n$ denotes the set of functions from $n$ to $m$. Cardinal exponentiation is denoted by $\overline{\overline{m^n}}$.

Assume that a scale $\aleph$ and the function $f$ satisfying $\overline{\overline{2^{\aleph_\alpha}}} = \aleph_{f(\alpha)}$ have been determined. It is true that under this assumption we can roughly measure how large $\overline{\overline{2^{\aleph_\alpha}}}$ is, but this still does not give us an adequate information about $2^{\aleph_\alpha}$, i.e., "how complex $2^{\aleph_\alpha}$ is". For example, it does not give any key to solve the most basic and important question: What sort of family of functions does the one-to-one and onto map from $\aleph_{f(\alpha)}$ to $2^{\aleph_\alpha}$ belong to?

Let us start with a simple case. If the above function is defined depending only on the length of the universe, then we can conclude that the status of $2^{\aleph_\alpha}$ is well-known. Recursive functions, arithmetical functions and hyperarithmetical functions are the examples of the functions depending only on the length of the universe. If this is not possible, the best we can do is to pick one element of $2^{\aleph_\alpha}$ and find out the status of $2^{\aleph_\alpha}$ assuming certain properties of this designated element. That is, if there exists an element $g$ of $2^{\aleph_\alpha}$ such that the one-to-one and onto map from $\aleph_{f(\alpha)}$ to $2^{\aleph_\alpha}$ is constructible in $g$, then we may say that the nature of $2^{\aleph_\alpha}$ is rather simple. More generally, suppose we have $\beta$ functions of $2^{\aleph_\alpha}$ and know that the function in question is constructible in those functions. Then, assuming that there exists a function $g$ from $\beta$ into $2^{\aleph_\alpha}$, we can express the function in question constructively in $g$. It seems that the $\beta$ defined as above well characterizes the undefinability of $2^{\aleph_\alpha}$. The possible supremum of such $\beta$ is evidently $\aleph_{f(\alpha)}$ itself.

We have seen in § 1 that $L(f)$, i.e. the smallest complete model $M$ of ZF such that $M \supseteqq \mathrm{On}$ and $f$ is constructible in $M$, can be defined by the Gödel-Levy construction. If we restrict $f$ to be bounded, then $L(f)$ can be defined as in § 2 of [25] where $f$ is taken to be a function from $\omega$ into $\omega$.

**Definition.** Let $f$ be a function of ordinal numbers and $\alpha$ an ordinal number, and let $\tilde{\alpha}$ be $j(3, \gamma, j(3, \gamma, \delta))$, where $\gamma = j(4, \gamma_1, \mathrm{Od}`\alpha)$, $\gamma_1 = j(0, j(0, \mathrm{Od}`\alpha, \mathrm{Od}`\alpha), 0)$ and $\delta = j(6, j(0, j(0, \gamma, 0), 0), \gamma)$, so that $F`\tilde{\alpha} = \alpha \times \alpha$. We define the function $F_{f,\alpha}$ as follows:

$$F_{f,\alpha}`\xi = \begin{cases} F_{f,\alpha}``\xi & \text{if } g_0(\xi) = 0 \\ \mathfrak{F}_i(F_{f,\alpha}`K_1`\xi, F_{f,\alpha}`K_2`\xi) & \text{if } 1 \leqq g_0(\xi) = i \leqq 8 \wedge \xi \neq j(1, \tilde{\alpha}, 0) \\ f \cap (\alpha \times \alpha) & \text{if } \xi = j(1, \tilde{\alpha}, 0). \end{cases}$$

$L(f, \alpha)$ is defined to be $F_{f,\alpha}``\mathrm{On}$. $\mathrm{Od}_{f,\alpha}$ is defined by

$$(\langle y x\rangle \in \mathrm{Od}_{f,\alpha} \leftrightarrow \langle x y\rangle \in F_{f,\alpha} \wedge \forall z(z \in y \rightarrow \neg \langle x z\rangle \in F_{f,\alpha})) \\ \wedge\ \mathrm{Od}_{f,\alpha} \subseteqq V \times \mathrm{On}.$$

**Lemma.** *Let $f$ be a function of ordinal numbers and $\alpha$ an ordinal number. Then we have:*

1) $F \upharpoonright j(1, \tilde{\alpha}, 0) = F_{f,\alpha} \upharpoonright j(1, \tilde{\alpha}, 0)$,
2) $\mathrm{Od}`a < j(1, \tilde{\alpha}, 0) \rightarrow \mathrm{Od}_{f,\alpha}`a = \mathrm{Od}`a$,
3) $\mathfrak{F}_{f,\alpha}`\xi \subseteqq \mathfrak{F}_{f,\alpha}``\xi$ for every ordinal number $\xi$.

*Proof.* 1) and 2) are obvious from the definition of $F_{f,\alpha}$ and $\mathrm{Od}_{f,\alpha}$. 3) can be proved by transfinite induction on $\xi$. Since the other cases are done like 9.5, [7], we consider the case where $\xi = j(1, \tilde{\alpha}, 0)$. Since $\mathfrak{F}_{f,\alpha}`\xi \subseteqq \alpha \times \alpha = \mathfrak{F}_{f,\alpha}`\tilde{\alpha}$ and $\tilde{\alpha} < \xi$, we have $\mathfrak{F}_{f,\alpha}`\xi \subseteqq \mathfrak{F}_{f,\alpha}``\xi$ by using the inductive hypothesis $\mathfrak{F}_{f,\alpha}`\tilde{\alpha} \subseteqq \mathfrak{F}_{f,\alpha}``\tilde{\alpha}$.

**Theorem.** $L(f, \alpha)$ is complete, almost universal, and closed with respect to $\mathfrak{F}_1 - \mathfrak{F}_8$; $L(f, \alpha)$ is a model of ZF.

**Proposition.** For any function $f$ of ordinal numbers

$$L(f) = \bigcup_{\alpha} L(f, \alpha).$$

Given a function $f$ of ordinal numbers and an ordinal number $\alpha$, we shall give an arithmetical definition of the notions $\mathfrak{F}_{f,\alpha}`\xi \in \mathfrak{F}_{f,\alpha}`\eta$ and $\mathfrak{F}_{f,\alpha}`\xi = \mathfrak{F}_{f,\alpha}`\eta$, (the arithmetized notions of which will be seen to be recursive in $f$ and $\tilde{\alpha}$) following [25] and [27]. We shall use notions and notations defined in [27] concerning the model construction, e.g. $\in$, $\equiv$ Od, ..., mostly without mention. In [27], $u(a)$, $a^{\dagger}b$ and $n(a)$ are defined as follows:

$$u(0) = 0 \wedge \forall x(x > 0 \to u(x) = \mu z_{z<x'} \forall y(y < x \wedge y \in x \to u^x(y) < z)),$$

where $u^{\alpha}(b) = u(b)$ if $b < a$; 0 otherwise,

$$a^{\dagger}b = \mu y_{y<a}(\langle y, b\rangle \in a)$$

and

$$n(0) = 0 \wedge \forall x(x > 0 \to n(x) = \mu y(\forall z(z < x \to n(z) \in y) \\ \wedge \forall z(z < y \wedge z \in y \to \exists u(u < x \wedge n(u) \equiv z)))):$$

$u(b) = c$ means that $b$ is the $c$-th ordinal in the arithmetized model; $n(a)$ expresses the $a$-th ordinal number in the model.

Let $H_1(f, \alpha: h, a)$ be

$$(g_1(a) \neq j(1, \tilde{\alpha}, 0) \wedge H_1(h, a)) \vee (g_1(a) = j(1, \tilde{\alpha}, 0) \\ \wedge \exists x \exists y(x < g_2(a) \wedge y < g_2(a) \wedge x \in n(\alpha) \wedge y \in n(\alpha) \\ \wedge = (h: g^2(a); \langle x; y\rangle) \wedge x \equiv n(f(u(y))))).$$

**Proposition.** There exists a function $fn_{f,\alpha}$ recursive in $f$ and $\tilde{\alpha}$ such that the following conditions hold:

$$g^1(a) > g^2(a) \wedge g_0(g^1(a)) = 0 \to fn_{f,\alpha}(a) = 0;$$
$$g^1(a) > g^2(a) \wedge g_0(g^1(a)) = 1 \\ \to (fn_{f,\alpha}(a) = 0 \leftrightarrow H_1(f, \alpha: fn^a_{f,\alpha}, a));$$
$$g^1(a) > g^2(a) \wedge g_0(g^1(a)) = i \\ \to (fn_{f,\alpha}(a) = 0 \leftrightarrow H_i(fn^a_{f,\alpha}, a)) \quad (1 < i < 9);$$
$$g^1(a) \leqq g^2(a) \to (fn_{f,\alpha}(a) = 0 \leftrightarrow H_9(fn^a_{f,\alpha}, a)).$$

We shall use the following abbreviations: $b \in_{f,\alpha} a$ for $\leqslant(fn_{f,\alpha}, a, b)$; $a \equiv_{f,\alpha} b$ for $=(fn_{f,\alpha}, a, b)$; $\mathrm{Od}_{f,\alpha}(a)$, $C_{f,\alpha}(a)$, $n_{f,\alpha}(a)$, $u_{f,\alpha}(a)$ and $a^{\dagger f,\alpha}\, b$ are obtained from $\mathrm{Od}(a)$, $C(a)$, $n(a)$, $u(a)$ and $a^{\dagger}b$ respectively by replacing $\in$ and $\equiv$ by $\in_{f,\alpha}$ and $\equiv_{f,\alpha}$. $\langle a, b\rangle_{\alpha}$ is defined to be $\{\{a, a\}_{\alpha}, \{a, b\}_{\alpha}\}_{\alpha}$, where $\{a, b\}_{\alpha}$ is $j(1, b, a)$ if $a = \tilde{\alpha}$ and $b = 0$; $j(1, a, b)$ otherwise.

We can prove the following propositions (whose proofs are omitted here) like the corresponding proof in [25] or [27].

**Corollary.** (1) $fn_{f,\alpha}^{j(1,\tilde{\alpha},0)}(a) = fn^{j(1,\tilde{\alpha},0)}(a)$.

(2) $a < j(1,\tilde{\alpha},0) \wedge b < j(1,\tilde{\alpha},0) \to (a \equiv_{f,\alpha} b \leftrightarrow a \equiv b)$.

(3) $a < j(1,\tilde{\alpha},0) \wedge b < j(1,\tilde{\alpha},0) \to (a \in_{f,\alpha} b \leftrightarrow a \in b)$.

**Proposition.** 3.2.1—3.2.6 and 3.2.8—3.2.28 in [27] hold by reading $\in_{f,\alpha}$; $\equiv_{f,\alpha}$, $\mathrm{Od}_{f,\alpha}$, $C_{f,\alpha}$, $\subsetneqq_{f,\alpha}$ and $\subset_{f,\alpha}$ instead of $\in$, $\equiv$, Od, $C$, $\subsetneqq$ and $\subset$ respectively. Instead of 3.2.7 there we have

$$g_0(a) = 1 \wedge a \neq j(1,\tilde{\alpha},0) \to (b \in_{f,\alpha} a \leftrightarrow b \equiv_{f,\alpha} g_1(a) \vee b \equiv_{f,\alpha} g_2(a)).$$

$$b \in_{f,\alpha} j(1,\tilde{\alpha},0) \leftrightarrow \exists x \, \exists y (y \in_{f,\alpha} n(\alpha) \wedge b \equiv_{f,\alpha} \langle x, y\rangle_\alpha \wedge x \equiv_{f,\alpha} n(f(u(y))))\,.$$

**Proposition.** (1) $F_{f,\alpha}{}^{\backprime}\xi \in F_{f,\alpha}{}^{\backprime}\eta \leftrightarrow \xi \in_{f,\alpha} \eta$.

(2) $F_{f,\alpha}{}^{\backprime}\xi = F_{f,\alpha}{}^{\backprime}\eta \leftrightarrow \xi \equiv_{f,\alpha} \eta$.

(3) $\mathrm{Od}_{f,\alpha}{}^{\backprime} F_{f,\alpha}{}^{\backprime}\xi = \mathrm{Od}_{f,\alpha}(\xi)$.

(4) $F_{f,\alpha}{}^{\backprime}\xi \in \mathrm{On} \to n_{f,\alpha}(F_{f,\alpha}{}^{\backprime}\xi) = \mathrm{Od}_{f,\alpha}{}^{\backprime} F_{f,\alpha}{}^{\backprime}\xi$.

(5) $\xi = F_{f,\alpha}{}^{\backprime}\eta \leftrightarrow n_{f,\alpha}(\xi) =_{f,\alpha} \eta$.

For a bounded function $g$ of ordinal numbers $|g|$ is defined by

$$|g| = \mu\alpha(\forall\beta(\beta \geqq \alpha \to g(\beta) = 0) \wedge \forall\beta(\beta < \alpha \to g(\beta) < \alpha))\,.$$

If $f$ is a bounded function of ordinal numbers, $L(f, |f|)$ is the smallest complete almost universal model $M$ of ZF such that $f$ is constructible with respect to $M$.

**Theorem 1.** For a given function $f$ of ordinal numbers, the following conditions are equivalent for a bounded function $g$.

1) There exists an ordinal number $\alpha$ such that $g$ is recursive in $f$ and $\alpha$.

2) There exists an ordinal number $\alpha$ such that $g$ is arithmetical in $f$ and $\alpha$.

3) $g \upharpoonright |g| \in L(f)$.

*Proof.* 1) $\to$ 2). Evident. 2) $\to$ 3). By Theorem 1 in § 1. 3) $\to$ 1). Let $g \upharpoonright |g| \in L(f)$. Then $g \upharpoonright |g| \in L(f,\alpha)$ for some $\alpha$ so that $\mathrm{Od}_{f,\alpha}{}^{\backprime} g \upharpoonright |g| = \beta$ for some $\beta$. By the previous propositions we have:

$$\forall\gamma(\gamma < |g| \to (g(\gamma) = \delta \leftrightarrow \langle n(\delta), n(\delta)\rangle_\alpha \in_{f,\alpha} \beta)$$
$$\wedge \, \forall\gamma(\gamma \geqq |g| \to (g(\gamma) = \delta \leftrightarrow \delta = 0))\,,$$

so that $g$ is recursive in $f$, $|g|$, $\alpha$ and $\beta$, i.e. $g$ is recursive in $f$ and $j(|g|, j(\alpha,\beta))$.

**Theorem 2.** Under the assumption of the second order reflection principle, the following conditions are equivalent for a bounded function $g$:

1) $g \upharpoonright |g| \in L(f)$.

2) There exists an ordinal number $\alpha$ such that $g$ is hyperarithmetical in $f$ and $\alpha$.

*Proof.* 1) $\rightarrow$ 2): Obvious from Theorem 1. 2) $\rightarrow$ 1). If $g$ is hyperarithmetical in $f$ and $\alpha$, then $g$ is constructible with respect to $f$ by Theorem 5 in § 1, and so is $g \upharpoonright |g|$. Since $g \upharpoonright |g|$ is a set, this implies $g \upharpoonright |g| \in L(f)$.

*Remark.* If $f$ is bounded, e.g., if $f \in 2^{\aleph_\alpha}$, we can simplify the construction of $L(f)$ by taking $L(f) = F_f$" On and $F_f$ to be defined by

$$F_f{}^{\backprime}\xi = \begin{cases} F_f{}^{\backprime\backprime}\xi & \text{if } g_0(\xi) = 0 \\ \mathfrak{F}_i(F_f{}^{\backprime}K_1{}^{\backprime}\xi, F_f{}^{\backprime}K_2{}^{\backprime}\xi) & \text{if } 1 \leqq g_0(\xi) = 1 \leqq 8 \wedge \xi \neq j(1, \mathrm{Od}^{\backprime}\aleph_\alpha, 0) \\ f & \text{if } \xi = j(1, \mathrm{Od}^{\backprime}\aleph_\alpha, 0). \end{cases}$$

(In particular, if $f \in \omega^\omega$, then $L(f) = F_f$" On, where $F_f$ is obtained by replacing Od'$\aleph_\alpha$ by $\omega$ in the above definition.) We can arithmetize the notions concerning $L(f)$ with $f$ bounded in the same way as in the case for $L(f, \alpha)$. By $\in_f$, $\equiv_f$, $\mathrm{Od}_f$, $C_f$, $n_f$, $u_f$ and $\dagger_f$ we understand the notions corresponding to $\in_{f,\alpha}$, $\equiv_{f,\alpha}$, $\mathrm{Od}_{f,\alpha}$, $C_{f,\alpha}$, $n_{f,\alpha}$, $u_{f,\alpha}$ and $\dagger_{f,\alpha}$ respectively.

We shall refer to each one of the equivalent properties mentioned above as "$g$ is constructible in $f$".

**Proposition 1.** $\forall f \in \aleph_\alpha^{\aleph_\alpha} \exists g \in 2^{\aleph_\alpha}$ ($f$ is constructible in $g$).

*Proof.* We define $g$ by

$$g(a) = \begin{cases} 1 & \text{if } \; f(g^1(a)) = g^2(a) \wedge a < \aleph_\alpha. \\ 0 & \text{otherwise.} \end{cases}$$

Then $f$ is recursive in $g$ and $\aleph_\alpha$, i.e. $f \in L(g)$.

**Proposition 2.** $\forall f_1 \in \aleph_\alpha^{\aleph_\alpha} \forall f_2 \in \aleph_\alpha^{\aleph_\alpha} \exists g \in 2^{\aleph_\alpha}$ ($f_1$ and $f_2$ are constructible in $g$). Similarly, $\forall f_1 \in \aleph_\alpha^{\aleph_\alpha} \ldots \forall f_n \in \aleph_\alpha^{\aleph_\alpha} \exists g \in 2^{\aleph_\alpha}$ ($f_1, \ldots, f_n$ are constructible in $g$).

*Proof.* Let $f(a) = j(f_1(a), f_2(a))$. Then $f_1$ and $f_2$ are constructible in $f$ and $f \in \aleph_\alpha^{\aleph_\alpha}$. By Proposition 1, we have $g$ in which $f$ is constructible.

**Proposition 3.** Let $g$ be a well-ordering whose order-type is $b$. Then the order preserving one-to-one map $f$ of $b$ onto $\vartheta(g)$ is constructible in $g$.

*Proof.*

$$f(a) = \mu x (\forall y < a (f(y) \neq x) \wedge \forall z (\forall y < a (f(y) \neq z) \rightarrow \neg z \overset{g}{<} x)),$$

if $a < b$; $f(a) = 0$ otherwise.

**Proposition 4.** Let $g$ be a well-ordering whose order-type is $b$. Then the order preserving one-to-one map $h$ of $\vartheta(g)$ onto $b$ is constructible in $g$.

*Proof.* Let $f$ be the order preserving one-to-one map of $b$ onto $\vartheta(g)$. Then $f$ is constructible in $g$ by Proposition 3 and $h(a) = \mu x (a = f(x))$.

**Proposition 5.** Let $\overline{\overline{\beta}} = \overline{\overline{\gamma}} = \aleph_\alpha$ and let $\varkappa$ be a function from $\beta$ to $\gamma$. Then there exists a function $h \in 2^{\aleph_\alpha}$ such that $\varkappa$ is constructible in $h$.

*Proof.* Let $f_1$ and $f_2$ be the well-ordering of $\aleph_\alpha$ whose order-types are $\beta$ and $\gamma$ respectively, and let $g_1$ and $g_2$ be the order preserving (w.r.t. orders of $f_1$, $\beta$ and $f_2$, $\gamma$) one-to-one maps of $\aleph_\alpha$ onto $\beta$ and of $\gamma$ onto $\aleph_\alpha$ respectively. By Propositions 4 and 3, $g_1$ and $g_2$ are constructible in $f_1$ and $f_2$ respectively. Let $\varkappa^*$ be $g_2 \circ \varkappa \circ g_1$. Then $\varkappa^* \in \aleph_\alpha^{\aleph_\alpha}$ and $\varkappa$ is constructible in $f_1, f_2$ and $\varkappa^*$. By Proposition 2 this completes the proof.

**Proposition 6.** (Due to HAJNAL [10]). Let $\lambda$ and $\tilde{\lambda}$ be ordinal numbers such that $\lambda < \tilde{\lambda}$ and $\aleph_{\tilde{\lambda}}$ is regular, and let $h$ be a function such that $h \mathscr{F}_n \aleph_\lambda \wedge \mathscr{W}(h) \subseteqq P(\aleph_\lambda)$. Then there exists a complete and almost universal class $M$ such that $h \in M$, $M$ is a model of ZF,

$$\overline{\overline{(2^{\aleph_\varkappa^M})^M}}^M \leqq \aleph_{\tilde{\lambda}}^M \quad \text{for every} \quad \eta < \tilde{\lambda}$$

and $\aleph_\alpha^M = \aleph_\alpha$ for every $\alpha \leqq \tilde{\lambda}$.

Although HAJNAL does not prove this theorem in this form in [10], we can prove the theorem in this form by slight modifications of the proof there.

**Definition.** The *first degree of undefinability* of $2^{\aleph_\alpha}$ is the least $\beta$ which satisfies the condition

$$\exists h (h \mathscr{F}_n \beta \wedge \mathscr{W}(h) \subseteqq 2^{\aleph_\alpha} \wedge \exists g \in L(h) (2^{\aleph_\alpha} = g`` \aleph_{f(\alpha)})) .$$

(We know the existence of such a $\beta$, for if we set $\beta = \aleph_{f(\alpha)}$ and $g = h$, where $h$ is the one-to-one and onto map from $\beta$ to $2^{\aleph_\alpha}$, the condition is satisfied.)

**Theorem 3.** The first degree $\beta$ of undefinability of $2^{\aleph_\alpha}$ is one of the following values:

1) $\beta = 0$,

2) $\beta = 1$.

3) $\beta = \aleph_\gamma$, where $\aleph_{\gamma+1} = \aleph_{f(\alpha)}$ and $\aleph_\gamma$ is singular. We say that this degree $\beta$ is $\infty - 1$.

4) $\beta = \aleph_{f(\alpha)}$. We say that this degree $\beta$ is $\infty$[5].

*Proof.* First we prove that if $0 < \beta < \aleph_{\alpha+1}$, then $\beta = 1$. Let $0 < \beta < \aleph_{\alpha+1}$. Then there exists a function $h$ from $\aleph_\alpha \times \beta$ into 2 such that $2^{\aleph_\alpha} = g`` \aleph_{f(\alpha)}$ for some function $g \in L(h)$, and a one-to-one function $\varkappa$ from $\aleph_\alpha$ onto $\aleph_\alpha \times \beta$. By Proposition 5, there exists a function $f_1 \in 2^{\aleph_\alpha}$ such that $\varkappa$ is constructible in $f_1$. Let $f_2$ be $h \circ \varkappa$. Then $h$ is constructible in $f_1$ and $f_2$. Therefore there exists a function $h_0 \in 2^{\aleph_\alpha}$ such

[5] Professor R. SOLOVAY informed us that if there are arbitrarily large measurable cardinals, then only this case ($\infty$ case) is possible.

that $h$ is constructible in $h_0$ by Proposition 2. Define $\tilde{h}$ by

$$\tilde{h}(j(a,b)) = \begin{cases} h_0(a) & \text{if } b = 0, \\ 0 & \text{otherwise}. \end{cases}$$

Then $\tilde{h}$ is a function from $\aleph_\alpha \times 1$ into 2 and $g \in L(\tilde{h})$.

Next consider the case where $\aleph_{\alpha+1} \leqq \beta \leqq \aleph_{f(\alpha)}$. We shall first show $\beta = \overline{\overline{\beta}}$. Let $h$ be a function from $\aleph_\alpha \times \beta$ into 2 and such that $2^{\aleph_\alpha} = g``\aleph_{f(\alpha)}$ for some $g \in L(h)$. Let $\aleph_\gamma$ be $\overline{\overline{\beta}}$. Then there exists a one-to-one map $f_2$ from $\aleph_\gamma$ onto $\aleph_\alpha \times \beta$, and a function $f_1 \in 2^{\aleph_\gamma}$ such that $f_2$ is constructible in $f_1$. Let $f_3$ be $h \circ f_2$. Then $h$ is constructible in $f_1$ and $f_3$, i.e. $h$ is constructible in a function $h_0 \in 2^{\aleph_\gamma}$ by Proposition 2. Thus we have proved that $h_0 \mathscr{F}_n \aleph_\gamma \wedge \mathscr{W}(h_0) \subseteqq 2 \wedge \exists g \in L(h_0)(2^{\aleph_\alpha} = g``\aleph_{f(\alpha)})$, i.e. $\beta = \aleph_\gamma$.

Now let $\aleph_\delta$ be the least regular cardinal such that $\aleph_\gamma \leqq \aleph_\delta \leqq \aleph_{f(\alpha)}$. Take $\alpha$, $\delta$ and $h_0$ as $\lambda$, $\tilde{\lambda}$ and $h$ in Proposition 6 respectively. Then

$$\aleph_{f(\alpha)} = \overline{\overline{2^{\aleph_\alpha}}} = \overline{\overline{(2^{\aleph_\alpha}_M)}}^M \leqq \overline{\overline{(2^{\aleph_\alpha}_M)}}^{M\,M} \leqq \aleph_\delta^M \leqq \aleph_\delta .$$

Therefore $\delta = f(\alpha)$. This implies that the first regular cardinal $\aleph_\delta$ such that $\aleph_\gamma \leqq \aleph_\delta \leqq \aleph_{f(\alpha)}$ must be $\aleph_{f(\alpha)}$ if there exists one. Since $\aleph_{\gamma+1} > \aleph_\gamma$ and $\aleph_{\gamma+1}$ is regular, either $\aleph_\gamma = \aleph_{f(\alpha)}$ or $\aleph_{\gamma+1} = \aleph_{f(\alpha)}$, i.e. $\beta = \aleph_{f(\alpha)}$ or $\beta = \aleph_\gamma$, where $\aleph_\gamma$ is singular. This completes the proof of the theorem.

Now we consider the case where the first degree $\beta$ of undefinability of $2^{\aleph_\alpha}$ is $\infty - 1$. This case is really exceptional. The consistency of this case was proved in [4]. In every such case we know so far, $cf(\gamma) = \aleph_\alpha$, where $\beta = \aleph_\gamma$. The question whether there exists a case where the first degree $\beta$ $(= \aleph_\gamma)$ of undefinability of $2^{\aleph_\alpha}$ is $\infty - 1$ and such that $cf(\gamma) \neq \aleph_\alpha$ is open. The consistency of the case where the first degree of undefinability of $2^{\aleph_\alpha}$ is $\infty$ was proved by Lévy [14] and [15]; cf. also [4]. Our theorem shows that the first degree of undefinability of $2^{\aleph_\alpha}$ is either very small or very large.

Let us consider the case that the first degree $\beta$ of undefinability of $2^{\aleph_\alpha}$ is 0 or 1. In either case, we obviously have by [7] or [13] that $\overline{\overline{2^{\aleph_\alpha}}} = \aleph_{\alpha+1}$, i.e. $f(\alpha) = \alpha + 1$.

Let $h$ be a function such that $h \mathscr{F}_n \beta, \mathscr{W}(h) \subseteqq 2^{\aleph_\alpha}$ and

$$\exists g \in L(h)(2^{\aleph_\alpha} = g``\aleph_{f(\alpha)}),$$

where $\beta = 0$ or 1, i.e. a function required to exist in the definition of the first degree. We can identify $L(h)$ with $L$ or $L(h`0)$, where $h`0 \in 2^{\aleph_\alpha}$, according as $\beta$ is 0 or 1. For the sake of typographical simplicity we shall use the subscript * to express arithmetized notions concerning $L(h)$, i.e. * is the void expression or the subscript $h`0$, according as $\beta = 0$ or

1. For any $g \in 2^{\aleph_\alpha}$, we define

$$\mathrm{Od}(h; g) = \mu x (x \subseteqq^* \aleph_\alpha \wedge \forall y < \aleph_\alpha (g(y) = x^{\dagger^*} y))$$

and for any $g_1, g_2 \in 2^{\aleph_\alpha}$, $g_1 <_h g_2$ by $\mathrm{Od}(h; g_1) < \mathrm{Od}(h; g_2)$. Then obviously $g_1 <_h g_2$ is an $\aleph_{\alpha+1}$-type well ordering of $2^{\aleph_\alpha}$ (for $x \subseteqq^* \aleph_\gamma$ implies $\forall y (y \in^* x \to \mathrm{Od}^*(y) < \aleph_\gamma)$, which implies $\mathrm{Od}^*(x) < \aleph_{\gamma+1}$. (Cf. the remark after Theorem 2.)

Let $R(g_1, g_2)$ be recursive in $h$, where $g_1, g_2$ are in $2^{\aleph_\alpha}$. Then $\exists \beta < \mathrm{Od}(h; g)(R(f, \{x\} \beta^{\dagger^*} x))$ is recursive in $h$, since $R(f, \{x\} \beta^{\dagger^*} x)$ is recursive in $h$.

In the following we shall consider number-theoretic functions, i.e. functions in $\omega^\omega$. As a special case of $L(e, \alpha)$ for any $e \in \omega^\omega$, we can consider $L(e)$ to be defined as in § 2 of [25], i.e. $L(e) = \mathscr{W}(F_e)$, where $F_e$ is defined by

$$F_e{}^{\backprime}\alpha = \begin{cases} F_e{}^{\backprime\backprime}\alpha & \text{if } g_0(\alpha) = 0 \\ \mathfrak{F}_i(F_e{}^{\backprime} K_1{}^{\backprime}\alpha, F_e{}^{\backprime} K_2{}^{\backprime}\alpha) & \text{if } \alpha \neq \omega + 1 \text{ and } 1 \leqq i = g_0(\alpha) \leqq 8 \\ e & \text{if } \alpha = \omega + 1 . \end{cases}$$

We can arithmetize the notions concerning $L(e)$ as in the case for $L(f, \alpha)$. $\in_e$, $\equiv_e$, $\mathrm{Od}_e$, $C_e$, $n_e$, $u_e$ and $\dagger_e$ will denote the notions corresponding to $\in_{f,\alpha}$, $\equiv_{f,\alpha}$, $\mathrm{Od}_{f,\alpha}$, $C_{f,\alpha}$, $n_{f,\alpha}$, $u_{f,\alpha}$ and $\dagger_{f,\alpha}$ respectively. $\langle a, b\rangle^*$ is defined to be $\{\{a, a\}^*, \{a, b\}\}\}^*$, where $\{a, b\}^* = j(1, b, a)$ if $a = \omega$ and $b = 0$; $j(1, a, b)$ otherwise.

**Theorem 4.** Let $h \in \omega^\omega$, $\omega^\omega \in L(h)$, and let $P(f, a)$ be an analytic predicate, where $f$ ranges over $\omega^\omega$ and $a$ ranges over $\omega$. $P(f, a)$ is expressible in both two function quantifier forms with the scope recursive in $h$ in the Kleene hierarchy (denoted $\Delta_2$), if and only if it is expressible as a predicate recursive in $h$ in the sense of ordinal numbers. (This is an extension of Theorem 7 in § 2 of [25].)

*Proof.* (Only if): The assumption $\omega^\omega \in L(h)$ implies that every number theoretic function is expressed in $L(h)$. Therefore we can follow the argument developed in § 8 of [29].

In order to prove the first half of the theorem we introduce some notions and notations: Let $Fn'(h: b, c)$ be

$$\forall u \forall v \forall w (u < b \wedge v < b \wedge \langle v, u\rangle \in_h b \wedge \langle w, u\rangle^* \in_h b \to v \equiv_h w)$$
$$\wedge \forall x (x < c \to \exists y (y < b \wedge O_h(y) \wedge \langle y, n_h(x)\rangle^* \in_h b)) .$$

$$\mathrm{Od}(h: f) = \mu x (Fn'(h: w, \omega) \wedge \forall y (y < \omega \to f(y) = u_h(x \dagger_h n_h(y)))) ,$$

for every $f \in \omega^\omega$.

**Lemma.** *Every predicate recursive in $h$ with only function variables over $\omega^\omega$ is expressible in both 1-qe-form in $h$ (cf.* [29] *for the definition of qe-form).*

*Proof*. This is proved like Theorem 2 of [29] or Lemma on p. 186 of [25] with the following modifications:

a) We define $\omega$-quasi-elementary functions in $h$ by adding $f(a) = h(a)$ to the definition of $\Omega$-quasi-elementary functions.

b) Theorem 5 on p. 185 of [25] should be changed as follows:

Let $f$ be recursive in $h$ in the Kleene hierarchy. Then

$$\exists x (Fn'(h: x, \omega) \wedge \forall y (y < \omega \to f(y) = u_h(x^{\dagger_h} n_h(y))))$$

is provable for every $f \in \omega^\omega$, (which can be proved from the assumption $\omega^\omega \in L(h)$).

*Proof* of the first half of Theorem 4. The argument is like that of Theorem 6, § 2 of [25] and Proposition 12 of [29].

Let $\nu(f, g, a)$ be the following function recursive in $f$ and $g$:

$$\nu(f, g, 0) = g(0);$$
$$\nu(f, g, n+1) = f(\nu(f, g, n)).$$

We can follow the proof of Theorem 6, § 2 of [25] with the following modification. Let $M_9(h: \alpha_0; \psi_0, \psi_1, \psi_2, \psi_9)$ be

$$\begin{aligned}&\forall x \forall y (D(\alpha_0, x) \wedge D(\alpha_0, y) \to (y = \psi_9(x) \leftrightarrow (\alpha_0(x, \psi_2(0)) = 0\\ &\qquad \wedge x \neq \psi_2(0) \wedge \exists u \exists v (x = \nu(\psi_0, \psi_1, u) \wedge y = \nu(\psi_0, \psi_1, v) \wedge v = h(u)))\\ &\qquad \vee (\alpha_0(\psi_2(0), x) = 0 \wedge y = \psi_1(0))))\\ &\qquad \wedge \forall x (D(\alpha_0, x) \to D(\alpha_0; \psi_9(x))).\end{aligned}$$

This will be used only when $W(\alpha_0)$, $M_0(\alpha_0; \psi_0)$, $M_1(\alpha_0, \psi_1)$ and $M_2(\alpha; \psi_0, \psi_2)$, $\psi_9$ corresponds to $h$.

Let $M_{10}(f; \alpha_0, \psi_0, \psi_1, \psi_2, \beta)$ be obtained from $M_9(h, \alpha_0, \psi_0, \psi_1, \psi_2, \psi_9)$ by replacing $\psi_9$ and $h$ by $\beta$ and $f$, respectively. Let $C(f, a)$ be a function quasi-elementary in $h$ containing only the function variable $f$ ranging over $\omega^\omega$. Then the predicate $b = C(f, a)$ is expressible in the Kleene hierarchy as follows:

$$\begin{aligned}&\underline{\forall} \hat{\alpha}_0 \underline{\forall} \hat{a} \underline{\forall} \hat{b} \underline{\forall} \psi_0 \underline{\forall} \psi_1 \ldots \underline{\forall} \psi_9 \underline{\forall} \beta \underline{\forall} \hat{h}_1 \ldots \underline{\forall} \hat{h}_l \underline{\forall} \hat{k}_1 \ldots \underline{\forall} \hat{k}_j \underline{\forall} \hat{g}\\ &(W(\alpha_0) \wedge D(\alpha_0, \hat{a}) \wedge D(\alpha_0, \hat{b}) \wedge = (\varphi_1, \alpha_0 \ulcorner \hat{a}) \wedge = (\varphi_2, \alpha_0 \ulcorner \hat{b})\\ &\qquad \wedge M_0(\alpha_0, \psi_0) \wedge \ldots \wedge M_8(\alpha_0, \psi_0, \psi_1, \psi_4, \psi_5, \psi_6, \psi_8)\\ &\qquad \wedge M_9(h; \alpha_0, \psi_0, \psi_1, \psi_2, \psi_9) \wedge M_{10}(f; \alpha_0, \psi_0, \psi_1, \psi_2, \beta)\\ &\qquad \wedge Cl(\alpha_0; \psi_6, \hat{k}_1, \ldots, \hat{k}_j)\\ &\qquad \wedge [g(f, a) = C(f, a)]^{\hat{\alpha}_0}\\ &\qquad \underline{\to} \hat{b} = \hat{g}(f, \hat{a})),\end{aligned}$$

where $h_1, \ldots, h_l$ are auxiliary functions of $[g(f, a) = C(f, a)]$ and $k_1, \ldots, k_j$ are the functions introduced by (XII) on p. 181 of [25] applied to the construction of $C$, or the form obtained from the above by replacing the underlined $\forall$'s and $\rightarrow$ by $\exists$'s and $\wedge$ respectively.

**Corollary.** Let $h \in \omega^\omega$. There exists an $\aleph_1$-type $\Delta_2$-well ordering $<$ of $2^{\aleph_0}$ such that, for any predicate $R(f, g)$ $\Delta_2$ in $h$, the predicate $\exists g_1 < g\, R(f, g_1)$ is $\Delta_2$ in $h$. (Here $\Delta_2$ is in the sense of KLEENE's hierarchy.)

*Proof.* Define $g_1 < g$ by $\mathrm{Od}(h; g_1) < \mathrm{Od}(h; g)$. $<$ is recursive (in the sense of ordinal numbers) and $R(f, g)$ is recursive in $h$ from Theorem 4. Since $\exists g_1 < g\, R(f, g_1) \leftrightarrow R(f, \mu g_1 < g\, R(f, g_1))$ and $\mu g_1 < g\, R(f, g_1)$ is recursive in $h$, $\exists g_1 < g\, R(f, g_1)$ is recursive in $h$, or using Theorem 4 again, it is $\Delta_2$ in $h$.

This is an improvement of ADDISON's (C) in [2]. Therefore we can repeat ADDISON's work [1] and [2] and KURATOWSKI's work [11] and get many results of projective set theory in those cases.

*Remark 1.* Under the assumption of HTC- $\forall f \ulcorner \omega$ is not a hyperarithmetical operation, where $\forall f \ulcorner \omega(\ldots)$ means $\forall f(\vartheta(f) = \omega \rightarrow \ldots)$, and

$$\forall f \ulcorner \omega\, (\forall n (f(n) < a) \rightarrow \exists x < a \forall n (f(n) < x))$$

is not provable $\Sigma_1^{\mathrm{Ord}\,1}$.

*Proof.* Since $\aleph_1$ is expressed by

$$\mu a (\forall f \ulcorner \omega (\forall n (f(n) < a) \rightarrow \exists x < a \forall n (f(n) < x))),$$

where it should be noticed that the scope of $\forall f \ulcorner \omega$ is primitive recursive in $f$, $\forall f \ulcorner \omega$ being hyperarithmetical would imply $\aleph_1$ being hyperarithmetical, which contradicts HTC. The second half is obvious from the above.

*Remark 2.* An expression of $\aleph_1$. Let $f$ be a well-ordering, and let $|a|_f$ denote the order type of $\{b \mid b \overset{f}{<} a\}$ with respect to $f$, i.e. $|a|_f = \mu x \forall b (b \overset{f}{<} a \rightarrow (|b|_f < x))$. Then $|a|_f$ is hyperarithmetical in $f$ (for it is obtained by a hyper-recursion). If we assume the second order reflection principle, then $\aleph_1$ can be expressed by $\forall f \in \omega^\omega$ followed by a hyperarithmetical predicate, i.e. $\aleph_1 = \mu x (\forall f \in \omega^\omega$ ("$f$ is a well-ordering of $\omega$" $\rightarrow |\omega|_f < x))$, where $|\omega|_f = \mu x \forall b < \omega (|b|_f < x)$, and "$f$ is a well-ordering of $\omega$" is arithmetical in $f$ by our assumption of the second order reflection principle and Theorem 4 of § 1 in this chapter. This also implies that under the assumption of the second order reflection principle and HTC, this predicate

$$\forall f \in \omega^\omega \text{ ("}f\text{ is a well ordering of }\omega\text{"} \rightarrow |\omega|_f < a)$$

is not provable $\Sigma_1^{\mathrm{Ord}\,1}$.

## References

1. ADDISON, J. W.: Separation principles in the hierarchies of classical and effective descriptive set theory. Fundamenta Mathematicae **46**, 123—134 (1958).
2. — Some consequences of the axiom of constructibility. Fundamenta Mathematicae **46**, 337—357 (1959).
3. BERNAYS, P.: Zur Frage der Unendlichkeitsschemata in der axiomatischen Mengenlehre. Essays on the Foundations of Mathematics. Jerusalem 1961, pp. 3—49.
4. COHEN, P. J.: The independence of the axiom of choice, Mimeographed notes. Mathematics Department, Stanford University (May, 1963), 32 pp. (See also The independence of the continuum hypothesis. Proceedings of the National Academy of Sciences **50**, 1143—1148 (1963), and **51**, 105—110 (1964).
5. EASTON, W. B.: Powers of regular cardinals. Ph. D. Thesis, Princeton University (October 1964) pp. 1—66.
6. FEFERMAN, S.: Some applications of the notions of forcing and generic sets. Fundamenta Mathematicae **56**, 325—345 (1965).
7. GÖDEL, K.: The consistency of the axiom of choice and of the generalized continuum-hypothesis with the axioms of set theory. Princeton 1951.
8. — What is Cantor's continuum problem? The American Mathematical Monthly **54**, 515—525 (1947).
9. — Remarks before the Princeton Bicentennial Conference on Problems in Mathematics, pp. 84—88 (1964). The Undecidable.
10. HAJNAL, A.: On a consistency theorem connected with the generalized continuum problem. Acta Math. Acad. Sci. Hungaricae **12**, 321—376 (1961).
11. KURATOWSKI, C.: Ensembles projectifs et ensembles singuliers. Fundamenta Mathematicae **35**, 131—140 (1948).
12. LÉVY, A.: Axiom schemata of strong infinity in axiomatic set theory. Pacific J. Mathematics **10**, 223—238 (1960).
13. — A generalization of Gödel's notion of constructibility. J. Symbolic Logic **25**, 147—155 (1960).
14. — (No title.) Mimeographed notes, 1—32 (1963).
15. — Definability in set theory I. Proceedings of the 1964 International Congress for Logic, Methodology and Philosophy of Science held in Jerusalem, 1964, pp. 127—151.
16. —, and M. MACHOVER: Recursive functions of ordinal numbers, Amsterdam, to appear. [See abstract in Notices of American Mathematical Society **6**, 826 (1959).]
17. MAHLO, P.: Über lineare transfinite Mengen. Berichte über die Verhandlungen der Königlich Sächsischen Gesellschaft der Wissenschaften zu Leipzig. Mathematisch-Physikalische Klasse **63**, 187—225 (1911).
18. — Zur Theorie und Anwendung der $\varrho_0$-Zahlen. Berichte über die Verhandlungen der Königlich Sächsischen Gesellschaft der Wissenschaften zu Leipzig. Mathematisch-Physikalische Klasse **64**, 108—112 (1912).
19. — Zur Theorie und Anwendung der $\varrho_0$-Zahlen II. Berichte über die Verhandlungen der Königlich Sächsischen Gesellschaft der Wissenschaften zu Leipzig. Mathematisch-Physikalische Klasse **65**, 268—282 (1913).
20. MAYBERRY, J.: Hyperarithmetical functions of ordinal numbers. Thesis, The University of Illinois, 1966.
21. SHOENFIELD, J. R.: On the independence of the axiom of constructibility. Amer. J. Mathematics **81**, 537—540 (1959).

22. Shoenfield, J. R.: The problem of predicativity. Essays on the Foundations of Mathematics, 132–139. Jerusalem 1961.
23. Takeuti, G.: On the recursive functions of ordinal numbers. J. Mathematical Soc. Japan **12**, 119–128 (1960).
24. — Axioms of infinity of set theory. J. Mathematical Soc. Japan **13**, 220–233 (1961).
25. — Recursive functions and arithmetical functions of ordinal numbers. Proceedings of the 1964 International Congress for Logic, Methodology and Philosophy of Science held in Jerusalem, 1964, pp. 179–196.
26. — Transcendence of cardinals. J. Symbolic Logic **30**, 1–7 (1965).
27. — A formalization of the theory of ordinal numbers. J. Symbolic Logic **30**, 295–317 (1965).
28. — On the axiom of constructibility. Proceeding of Symposium of Logic, Computability and Automata at Rome (New York), (1965). To appear.
29. —, and A. Kino: On hierarchies of predicates of ordinal numbers. J. Mathematical Soc. Japan **14**, 199–232 (1962).
30. Tarski, A.: Quelques théorèmes sur les alephs. Fundamenta Mathematicae **7**, 1–14 (1925).

# Definition eines (relativ vollständigen) formalen Systems konstruktiver Arithmetik

EDUARD WETTE *

GÖDELS Resultate (1931) [5] — die Unvollständigkeit formaler Systeme, die eine Arithmetik enthalten, und die Unableitbarkeit einer Aussage, die die Widerspruchsfreiheit des betreffenden formalen Systems in arithmetischer Verschlüsselung ausdrückt — reizen wegen der konstruktiven Art ihres Beweises zu der Aufgabe, die konstruktive Mathematik mit der klassischen Mathematik in einen Wettbewerb treten zu lassen, bei dem es darum geht, welche Art des Mathematisierens unter dem unparteiischen Gesichtspunkt der finiten Metamathematik HILBERTS [9] zu einem formalistisch feinmaschigeren Netz von Postulaten führt. GÖDELS Umdeutung einer Arithmetik mit berechenbaren Funktionalen (1958) [7] schafft einen neuen Zugang zu Widerspruchsfreiheitsbeweisen; dort wird in dem früher von GENTZEN (1936) [4] behandelten Fall der reinen Zahlentheorie mit primitiv rekursiven Funktionen endlichen Typs über natürlichen Zahlen gearbeitet, dafür aber ohne transfinite Ordinalzahlen und ohne logische Partikeln — von der vollständigen Induktion reicht eine direkte (beim Schritt von $n$ auf $n + 1$ nicht auf eine Implikation reflektierende) Schlußweise aus, die von $\Phi(n) = \Phi(n + 1)$ zu $\Phi(1) = \Phi(m)$ übergeht.

Um auf konstruktiver Seite zu der eingangs genannten Aufgabe beizutragen, ist in erster Linie nach einer möglichst erschöpfenden Formalisierung jener Beweismethoden und Definitionsverfahren zu fahnden, die „über" Grundzahlen $0, 1, 2, \ldots$ verfügbar sind, ohne inhaltlich Zugeständnisse an die Vorstellung des Unendlichen als etwas wirklich Seienden zu machen (wie etwa in Theorien des Konstruierbaren, [8] 1.). Getreu einem umfassenderen, auch konstruktive Metatheorien einbeziehenden Plan, den ich 1957 in Amsterdam vorgetragen habe [21], soll die unter finite Kontrolle gebrachte Durchführung einer konstruktiven

* Professor BERNAYS bin ich zu Dank verpflichtet, weil er sich die Mühe gemacht hat, den größten Teil der detaillierten Ausführungen in der ersten Fassung dieser Arbeit zu überprüfen; seine Vorschläge zur Ergänzung und Abrundung, die er mir daraufhin mitteilte, habe ich bei der Überarbeitung zu berücksichtigen versucht. — Die Meinung des Referee hat mich bewogen, von einer mehr liberalen intuitionistischen Interpretation in der ersten Fassung zu einer strafferen dialogischen Deutung im Sinne von P. LORENZEN überzuwechseln.

Ordinalzahltheorie als Prüfstein für die Tragfähigkeit von formalen Systemen konstruktiver Arithmetik dienen. Das damals erwähnte Kodifikat einer normalen Sprache $\mathfrak{S}_1$ über Grundzahlen wird hier in verfeinerter Gestalt vorgelegt; es erscheint mir so weit vervollkommnet, daß man mit seiner Hilfe allein — ohne die gemäß [21] für diesen Zweck vorgesehenen höheren normalen Sprachschichten $\mathfrak{S}_\alpha$ — einen finit kontrollierten konstruktiven Wf.beweis für die nach BERNAYS und GÖDEL formalisierte axiomatische Mengenlehre [1, 2, 6] versuchen kann.[1] (Zur Einarbeitung in das Problem genügt, für jeden mengentheoretischen Deduktions-Stammbaum $d_\in$ nach Übersetzung in einen ordinalzahltheoretischen $d_<$, ähnlich der Konstruktion von TAKEUTI [19], einen passenden Bezirk $\mathrm{T}(d_<)$ von Ordinalzahlen, Typen und Funktionen aufzuzählen, der eine Umdeutung von $d_<$ löst, welche die erwähnte GÖDELsche Umdeutung in [7] auf Formeln mit transfinit rekursiv definierten Prädikaten erweitert.)

Die formalistische Aufarbeitung konstruktiver Intentionen ist im folgenden keine unnötige Zugabe (wie vom intuitionistischen Standpunkt aus, [8] 5.); vielmehr bildet sie in jedem Stadium der für eine Theorie über Ordinalzahlen erforderlichen Entwicklung konstruktiver Begriffsbildungen über Grundzahlen — Definition von Aussagen, Beweis von Implikationen — eine unentbehrliche Begleiterscheinung, die eben der Deutung selbst in wesentlichen Teilen überhaupt erst eine von imprädikativen Vorgriffen freie Konstruktivität verleiht.[2] Im Blickfeld eines so aus arithmetisch-kalkulatorischer Aktivität hervorgehenden Zugangs zur formalen Redeweise mit transfiniten Objekten (und über solche) ist es deshalb kaum noch als Opfer anzusehen, wenn wir uns metamathematisch ausdrücklich auf finite Schlußformen beschränken, die sich *direkt* — ohne Reflektieren auf Annahmen, die nicht bereits anschaulich zu verwirklichen sind — sicherstellen lassen.

Im Gegenteil: sollte der a priori ja nicht undenkbare Fall eintreten, daß uns eine Analyse des hier angesteuerten Wf.beweises unfreiwillig auch zur Herleitung „verkappter" Widersprüche im mengentheoretischen System anleitet — damit sind widersprechende Aussagen unter Verschlüsselungen gemeint, die nur nicht auf dem Wege systemgebundener Definitionen als intern falsch zu entschlüsseln sind — (oder gar zur Herleitung „unmittelbarer" Widersprüche), dann wäre dies ein Symptom für jenen leichten Zug ins Unwirkliche, der indirektem Schließen an-

---

1 Ein Zwiegespräch mit GÖDEL, das sich am 22. September 1960 im Institute for Advanced Study zu Princeton entspann, ließ einen solchen Fall als nicht direkt unmöglich erscheinen. Ich danke Professor GÖDEL für seine Bereitschaft zu dieser Begegnung, die von CLIFFORD SPECTOR im Anschluß an den Stanford-Kongreß 1960 vermittelt worden war. Seiner freundlichen und unvoreingenommenen Aufforderung, meine Ansicht zu publizieren und sie dem unpersönlichen Urteil formaler Systeme zu unterwerfen, möchte ich hiermit nachkommen.

2 Vergleiche [21], Etappe II (3. Absatz) und III (4. Absatz), mit [13] §§ 16, 17.

haftet; davon würde die klassische Denkart innerhalb des Systems gleichermaßen betroffen wie eine konstruktive außerhalb.

**0.1** Neuere Literatur zeigt, daß die Einstellung zum Unendlichen als etwas möglich Werdendem noch recht unterschiedliche Ansichten von einer mathematischen Disziplin wie der Analysis reeller Zahlen erlaubt [12, 15]. Streitigkeiten über die Konstruktivität einer Behauptung möge ein dialogisches Kriterium schlichten, das LORENZEN [14] instruktiv auseinandergesetzt hat. Dadurch treten gewisse Diskrepanzen gegenüber der intuitionistischen Mathematik auf: im dialogischen Sinn können weder Auswahlprinzip noch „bar induction" allgemein verteidigt werden, da die beiden Diskussionspartner, die sich unter fest vereinbarten Rechten und Pflichten angreifen, mit verdeckten Karten um Beendigung und Gewinn ihrer Partie spielen.

Nicht eingeweihte Leser mögen die Abschnitte **0.11** bis **0.14** kursorisch vornehmen und Einzelheiten überschlagen.

**0.11** Die Situation sei am Beispiel der bar-induktiv gefolgerten transfiniten Induktion $\bigwedge \varphi \bigvee x \neg \varphi x' \prec \varphi x \rightarrow \bigwedge \pi (\bigwedge x_0 (\bigwedge x (x \prec x_0 \rightarrow \pi x) \rightarrow \pi x_0) \rightarrow \bigwedge x \pi x)$ kurz vergegenwärtigt.[3] Der Opponent übernimmt für ein von ihm gewähltes, etwa primitiv rekursiv vorausgesetztes $R$ anstelle $\prec$ die Kettenbedingung $\bigwedge \varphi *$ und dann, nach seiner Wahl eines $P$ anstelle $\pi$, auch die $R$-Progressivität $\bigwedge x_0 *$ (für $P$); er bezweifelt $P m_0$ für ein von ihm ausgesuchtes $m_0$. Der Proponent wird nur im Glücksfall eine Strategie zum Gewinn von $P m_0$ wissen, und er beruft sich daher im allgemeinen auf die $R$-Progressivität in $m_0$ anstelle $x_0$; er hat nun die Verteidigung von $\bigwedge x (x R m_0 \rightarrow P x)$ zu übernehmen, während seinem Gegner die von $P m_0$ zufällt. Dieser wird, selbst wenn er eine Strategie kennt, die den Gewinn von $P m_0$ gewährleistet, $\bigwedge x (x R m_0 \rightarrow P x)$ für ein von ihm ausgesuchtes $m_1$ anstelle $x$ angreifen, indem er $m_1 R m_0$ übernimmt und $P m_1$ bezweifelt; würde er $P m_0$ verteidigen, hätte er die Partie verloren. Das Spiel wiederholt sich mit $m_1$ anstelle $m_0$ und $m_2$ anstelle $m_1$, ..[4]; der Opponent hat bei der Lenkung seiner $R$-Vorgängerkette $.. m_2 R m_1 R m_0$ lediglich darauf zu achten, daß er nicht auf ein $m_n$ stößt, für das er keinen $R$-Vorgänger mehr kennt. Sonst läuft er Gefahr, einen Bluff mit $m_{n+1} R m_n$ riskieren zu müssen, um seinen Angriff gegen $P m_n$ (und mit ihm die Partie) nicht aufzugeben — er hat aber keine Garantie, daß der Proponent jetzt oder später nicht $m_{n+1} R m_n$ angreift, statt nach neuen $R$-Progressivitäten zu fragen. Nach einem speziellen $f$ anstelle $\varphi$ wird der Proponent nur fragen, wenn er von einem groben Täuschungsversuch überzeugt ist; denn ihm fällt dann die Verteidigung von $f n_f' R f n_f$ zu, sobald der Opponent ein $n_f$ genannt hat, für das dieser $\neg f n_f' R f n_f$ behauptet — und um da zu gewinnen, muß der Proponent erst einmal selbst fähig sein, $f n_f$ und $f n_f'$ auszuwerten.

Der Opponent kann, obwohl er sich offiziell nicht auf solche Fragen einzulassen braucht, dem Proponenten gestatten, „dasjenige" $f$ anzugreifen, bei welchem der Wert von $f n$ seiner künftigen Wahl für $m_n$ gleich sein wird. Er sagt ein $N$ an und

---

[3] Von dieser Version der bar induction ging die Argumentation des Referee aus.

[4] Drei Punkte deuten unendlich viele, zwei Punkte endlich viele weitere Glieder an. Ein Punkt wird, wie $*$, †, .., als Platzhalter für die verschiedensten Objekte benutzt. Man unterscheide '.' von '**.**', den „dicken" Punkten, die in der inoffiziellen Notation (ab **1.61**) verwendet werden.

könnte entgegenkommenderweise sogar $m_N$ verraten, auch wenn er keinen $R$-Vorgänger von $m_N$ weiß. Greift nämlich der Proponent die $R$-Progressivität in $m_N$ auf, so vermag er seinem Partner zwar $Pm_N$ zuzuschieben — falls dieser die Induktionsannahme $\bigwedge x(x R m_N \rightarrow Px)$ nicht angreifen kann oder will —, aber damit ist der Proponent noch nicht in der Lage, seine zuletzt aufgestellte Behauptung $Pm_{n_0}$ gegen den Zweifel des Gegners pflichtgemäß zu verteidigen. Er mag von $m_N$ aus $Pm$ für manche $m$ gewinnen, die bezüglich $R$ „höher" als $m_N$ liegen; er mag dabei $Pm$ für einige $m$ durch vollständige Induktion gewinnen können; $Pm_{n_0}$ wird er im allgemeinen nicht zu erreichen vermögen — er müßte $R$ so beherrschen, daß er $\bigwedge \pi *$ nach einem Kalkül ableiten kann, dessen Regeln eine Herstellung von (rückwärts zu verfolgenden) Gewinnstrategien vermitteln, die gegen jeden Opponenten stichhaltig sind und die der Proponent überdies anzuwenden versteht.

Der Proponent wünscht eventuell, einen Dialog über Dialoge zu versuchen, und wendet ein, er *hätte* die Partie doch „an sich" nach $N - n_0$ Spielzyklen gewonnen, zumal wenn $m_N$ nachweislich keinen $R$-Vorgänger hat, wenn der Gegner also, sobald er $Pm_N$ anzweifelt, die Partie verliert. Der Opponent wird entgegnen, im Ernstfall *hätte* er $m_n$ anders gelenkt — oft hat bereits $m_{N-1}$ einen dazu geeigneten $R$-Vorgänger $\overline{m}_N$. Ein Proponent, der generell

$$\bigwedge \varphi (\bigvee x \neg \varphi x' R \varphi x \vee \bigwedge x \varphi x' R \varphi x)$$

behaupten will, verliert, da keine Strategie zum Gewinn von $A \vee \neg A$ vorhanden ist.

Vielleicht möchte sich der Proponent darauf berufen, es gäbe im Fall $\bigwedge \varphi \bigvee x \neg \varphi x' R \varphi x$ eine konstruktive Ordinalzahl $\varkappa$ mit der Eigenschaft, daß jedes $m$ durch eine Ordinalzahl $\nu_R(m) = \mu \nu \bigwedge x (x R m \rightarrow \nu_R(x) < \nu)$ zu bewerten sei, die $< \varkappa$; danach könnten die Glieder jeder Folge $\varphi$ gleichmäßig auf ihr Niveau $\nu_R(\varphi n)$ hin eingeschätzt werden. Der Opponent fragt dann — an Hand KLEENE l.c. [12] p. 192, 1955a (und l.c. [11] p. 527, 1944) — nach der Gödelnummer $k (\in O)$ eines solchen $\varkappa$ und nach der Bestimmung der Gödelnummern $k(m)$ von $\nu_R(m)$; ehe er sich jedoch auf einen endlosen Nebendialog über $k \in O$ einläßt, der ja beim Definitionsschritt $O$ 3. durch Stichproben auf $\bigwedge n \{\Phi(l, n_O)$ definiert und $= l_n\}$ und auf $\bigwedge n\, l_n \in O \wedge \bigwedge n\, l_n <_O l_{n+1}$ zu immer neuen Nummern der Form $3 \cdot 5^l$ und damit zum Regreß in $O$ 3. führen kann, wird er sich eine Herleitung von $k \in O$ vorlegen lassen und die benutzten Regeln auf ihre gewinnstrategische Relevanz prüfen.

Bei dieser Wendung wird folgendes einsichtig: eine induktive Definition ist im allgemeinen, auch wenn sie ein inklusions-isotones Definiens hat (wie die für $\in O$, $<_O$), dermaßen „offen", daß ihre formale Einführung in einen Kalkül dessen dialogische Relevanz aufhebt; denn die dialogische Definitheit einer solchen Definition ist nicht im voraus gesichert, etwa durch eine dialogisch schon gewonnene Vorgängerinduktion.

Nun kann der Proponent äußerlich nicht erkennen, wieviel Reflexionsstufen in einem Beweis der Vorgängerinduktion $\bigwedge \pi *$ bezüglich $R$ stecken mögen; er wird also die ordinale Höhe $\nu_R = \mu \nu \bigwedge x \nu_R(x) < \nu$ von $R$ im allgemeinen unterschätzen und der Opponent wird eine dialogisch relevante Herleitung von $\bigvee m\, \nu_R(m) = \varkappa$ anzugeben wissen. Ein Verteidiger der bar induction will überdies prinzipielle Gewißheit *vor* jeder Wahl von $R$ haben (: [8] 6.) und tritt, sofern er sich auf dialogische Sinngebung einläßt, hic et nunc gegen jeden Gegner an — unabhängig von Ort und Zeit.

**0.111** Man beachte, daß es wesentlich auf den Sinn von $\bigvee m$ ankommt, der in die Berechnungsschranke von $\Phi(l, n_O)$ eingeht. Bevor der reale Kern mengentheoretisch deduzierter Existenz nicht von einem Wf.beweis her geklärt und eingezäunt ist, halten wir uns hier an eine möglichst enge Existenzauslegung und fordern einen unmittelbaren Aufweis, der objektiver Entscheidung fähig ist, aber keiner Negation und auch keiner (indirekten) Reflexion.

Die konstruktiv deduzierte Existenz in der dialogischen Auslegung einer stets mit Gewinn zu verteidigenden Angabe wird zwar möglichst weit verfolgt, jedoch letztlich unter algorithmische Wörter-Rechnung gestellt (wie eine mengentheoretisch deduzierte Existenz). Und welches Wortspiel die „stärkeren" 'Existenzen' beinhaltet, läßt sich nicht ohne weiteres aus der ursprünglichen Deutung der Zeichen beurteilen. Die erste „nicht-konstruktive" Ordinalzahl kann unter deduktiver Relativierung (: $\mu\lambda \bigwedge k((\vdash_{\in} k \in O) \to |k| < \lambda)$) dialogisch-konstruktiv zu erreichen sein. (Zugespitzt: schon $\omega_0$ 'existiert' ohne direkt zu existieren.)

**0.12** Objektive Einigung erzielen die beiden Partner nur über „wahrheitsdefinite" Aussagen, für die ein direkt funktionierendes Entscheidungsverfahren festgelegt ist, dessen Ende durch Rückschritt vom Entscheid über ein Wort (d. i. eine normiert aufgebaute Zeichenreihe) auf eine feste Höchstzahl von Vorentscheidungen über seine Teilwörter zwangsläufig zuwege kommt — z.B. darüber, ob eine vorliegende Ableitung nach einem gegebenen Kalkül korrekt ist. Wie sich eine Partie über eine „beweisdefinite" Aussage entscheiden wird, z.B. über 'die Formel $A$ ist nach dem Kalkül $K$ ableitbar' (: $\vdash_K A$), hängt bereits von den Vorkenntnissen des Proponenten ab, z.B. von Erfahrungen beim Ableiten nach $K$.

**0.13** Vor Einführung einer induktiven Definition $Pm \leftrightarrow A(m, P)$ in einen Dialog ist deren dialogische Definitheit selbst zum Gegenstand des Dialogs zu machen. Um der Flucht in einen Regreß höherer Instanzen vorzubeugen, darf der Gegner objektive Auskunft über eine Fundierung verlangen, in bezug auf die $P$ durch Vorgängerinduktion nach $m$ mittels $A$ definiert sein soll.

Es ist also nicht damit getan, daß der Verteidiger der Definition ein bereits definiertes Vorgängerprädikat $R$ angibt, in bezug auf das er die Vorgängerinduktion $\bigwedge m_0(\bigwedge m(Rmm_0 \to D(m)) \to D(m_0)) \to \bigwedge m D(m)$ behauptet, um diese Behauptung ihrerseits dialogisch zu verteidigen. Denn die "Existenz" des einzuführenden Prädikats $P$ ist aufgrund der in [14] festgesetzten Vereinbarungen über die dialogische Verwendung logischer Partikeln offenbar nicht zu gewinnen. Vielmehr müssen die Dialogpartner neue Verabredungen treffen und gut verteilte Rechte und Pflichten festlegen, ehe sie Dialoge mit noch zu definierenden Prädikaten austragen.

**0.131** Einerseits darf der Angreifer verlangen, vorher zu sehen, wie die Vorgängerinduktion (bezüglich des vom Verteidiger angegebenen $R$) deduziert ist und nach welchen formalen Regeln; sonst kann er nicht beurteilen, ob er die Deduktion als Gewinnstrategie anerkennen muß oder ob er die Induktion bezweifeln soll. Andererseits ist zum Zeitpunkt der

Definition von $R$ die Vorgängerinduktion für solche Eigenschaften, die konkret erst später erfaßt werden können oder die dann gar die Deutung von Prädikaten betreffen, die gerade in Einführung begriffen sein werden, — also für Eigenschaften wie "$P$ ist in $m$ dialogisch definit" — offenkundig nur abstrakt vorwegzunehmen, wenn man imprädikative Vorgriffe vermeiden will.

Nicht bloß dialogische Interessen, sondern auch Gründe der technisch einwandfreien Wiedergabe konstruktiver Inhalte zwingen uns demnach, ein formales System aufzustellen, das den (dialogisch-)konstruktiven Definitions- und Beweismöglichkeiten zum Zeitpunkt der Definition von $R$ angepaßt ist und das abstrakte Prädikatmarken $\pi$, $\rho$, ...[4] mitführt, die selbst der Intention nach bedeutungsfrei sind — ähnlich den unbestimmten Rechenmarken in der Algebra — und die jederzeit für eine konkrete Metaüberlegung bereitstehen, soweit diese parallel zu einer fertig abgeleiteten formalen Deduktion abläuft.

Daß die Deutung nachher für ihre eigenen Definitheitseigenschaften auf solche Überlegungen unter Inanspruchnahme des zu deutenden formalen Systems angewiesen ist, zeugt von der innigen Verflechtung, die LORENZENS „operativer" Ansatz [13] zwischen den Fabrikaten eines Kalküls und dem mit ihnen gemeinten Sinn zustande gebracht hat. Steckt darin eine unentwirrbare Mixtur von Form und Inhalt? WEYL [27] S. 248 stimmt [13] zu. [13a] sucht den Begründungsanspruch abzustreiten. Um formalistisch unanfechtbar vorzugehen, fasse ich in **1.** alle Teilkodifikate zu einem Gesamtkodifikat zusammen und treibe dessen verbale Formalisierung bis zur gediegenen Kalkülisierung voran, die wohl keine Fehlanwendung mehr erlaubt.[5]

**0.132** Mit der Wahl eines $R$, der Aufstellung eines dialogisch relevanten formalen Systems und dem Aufweis einer Deduktion

$$(1) \qquad \vdash \bigwedge x_0 (\bigwedge x (R x x_0 \to \rho x) \to \rho x_0) \to \rho x$$

haben wir noch keine Gewißheit, ob $R$ die induktive Definition von $P$ tatsächlich zu einer Definition durch Induktion macht, ob $R$ also eine „passende" Fundierung ist, die sicherstellt, daß $A$ in $m_0$ nur in solchen $m$ vom zu definierenden $P$ abhängt, die $m_0$ bezüglich $R$ vorangehen. Wer die Definition von $P$ angreift, ist berechtigt, auch darüber objektiv informiert zu werden; und wer die Definition verteidigt, ist darum ver-

---

[5] Ohne vorherige Kalkülisierung könnte bereits die Prädikatquantifikation gemäß **0.2** strittig sein, etwa hinsichtlich des Umfangs, in dem Parameter zulässig sind (s. **3.21**), ebenso der offizielle Status einer Umschreibung variabler und transfiniter Indizes; die Prädikatsubstitution in Definitionsschemata unter Hypothesen und Bedingungen ist ebenfalls nicht trivial. Sonst wäre auch kaum an die Durchführbarkeit der hier lediglich verbal angedeuteten „koerziblen" Perfektion $\mathfrak{S}_1^{\vee}$ des Systems $\mathfrak{S}_1$ zu denken, die auf eine systemgebundene transfinite Iteration von Chiffrierungen reflektiert (s. **0.52**). Aber schon bei $\mathfrak{S}_1$ dürfte eine verbale Formalisierung für Betrachtungen wie in **3.24** nicht voll ausreichen.

pflichtet, außer (1) noch eine (dialogisch relevante) formale Deduktion

(2) $\vdash \bigwedge x(Rxx_0 \to (\pi_1 x \leftrightarrow \pi_2 x)) \to (A(x_0, \pi_1) \leftrightarrow A(x_0, \pi_2))$

aufzuweisen.

Nach Vorlage von (1) und (2) kann der Verteidiger behaupten, $P$ sei für alle $m$ dialogisch definit. Anhand (2) folgt: wenn das Definiendum $P$ schon für alle $m$, für die $Rmm_0$, dialogisch definit ist, dann ist auch $A(m_0, P)$ dialogisch definit; denn wie sich $P$ gegenüber seinem ($m_0$ vorangehenden) $R$-„Abschnitt" $P^{(R)}_{m_0}$ auch entwickeln mag, so steht $P$ jedenfalls für $m$, für die $Rmm_0$, mit $P^{(R)}_{m_0}$ in der (dialogisch-konstruktiven) Bisubjunktion $\leftrightarrow$. Wird nun in diesem Stadium der Verhandlungen zwischen den Partnern vereinbart, $Pm_0$ als Abkürzung für $A(m_0, P^{(R)}_{m_0})$ einzuführen, dann ist die $R$-Progressivität für "$P$ ist in $m$ dialogisch definit" bei $m_0$ gewonnen. Und anhand (1) folgt mithin die Behauptung.[6]

**0.133** Erst jetzt darf der Verteidiger der Definition von seinem Gegenspieler erwarten, daß dieser aufgrund der Erkenntnisse aus dem Metadialog anhand (1) und (2), den **0.132** latent enthält, folgenden zusätzlichen Dialogregelungen beistimmt: wird $Pm_0$ bezweifelt, so muß, wer $Pm_0$ behauptet hat, auch $A(m_0, P)$ behaupten; wer $Pm_0$ bezweifelt hat, muß später immer dann, wenn er $Rmm_0$ nicht angreift, seinerseits, je nach Fragerichtung des Partners, $Pm$ mit $A(m, P)$ beantworten und umgekehrt, und er verliert, sobald er sich gezwungen sieht, selbst $A(m_0, P)$ zu behaupten.

**0.134** Hegt der Opponent den Wunsch, den Dialog mittels Schrittzahlen zu verschärfen, so kommen als Schrittzahlen meines Ermessens allein natürliche Zahlen 1, 2, 3, ... in Betracht. Was hülfe es den Partnern, wenn ein transfinites Bewertungsverfahren des Proponenten aus endlichen Daten des Gegners errechnet, sie hätten beispielsweise von $\omega$ auf $10^{40}$ zurückzugreifen? — also, nach gegenwärtiger Lehre, auf die Größenordnung des Verhältnisses vom derzeitigen Weltalter zur (heutigen) Lichtzeit für den (heutigen) Elektronenradius-Weg.

Es sei betont, daß (1) kein allgemeines Verfahren liefert, das für „jede" Vorgängerkette von $m_0$ aus vorhersagt, nach wieviel Schritten sie abbricht; (1) gestattet im allgemeinen nicht einmal einen konstruktiven Schluß von $\bigvee m\, \underline{\varrho}\, m$ auf $\bigvee m_0(\underline{\varrho} m_0 \wedge \bigwedge m(\underline{\varrho} m \to \neg Rmm_0))$. Die Prozedur von **0.131** bis **0.133** legt eben nichts weiter fest als den dialogischen Sinn von $Pm$, nämlich den Ablauf gegenseitiger Widerlegungsversuche zweier Dialogpartner bei Behauptungen der Form $Pm$.

Ein auf derartige Weise (rekursiv, d.h. durch Induktion) definiertes Prädikat ist nicht mit einer Menge zu verwechseln, auf deren Elemente wie auf ein fertiges Ganzes reflektiert wird. Gewöhnlich steigern sich die Möglichkeiten, $Pk$ für konstante Argumente $k$ zu deduzieren, mit der Gewinnung oder der (dialogisch relevan-

[6] Das so eingeführte $P$ hängt zunächst von $R$ ab, ist aber bisubjunktiv unabhängig von der jeweiligen Fundierung $R$, wie Theorem 1 in **2.11** zeigt. Die korrekte Schreibweise für $P$, die **0.135** bringt, würde im Augenblick die Übersicht stören.

ten) Produktion neuer Definitions- und Beweismöglichkeiten, und ein nicht direkt zu entscheidendes Prädikat soll hier in seiner deduktiven Entfaltung begriffen werden.

**0.135** Es empfiehlt sich, die neuen Verfügungen unter **0.131** bis **0.133** an ein eigenes Operationssignal zu knüpfen und anstelle $P$ genauer $\xrightarrow{R} \pi x A(x, \pi)$ zu schreiben. Da, wo der „Rekursor" $\xrightarrow{*}$ die äußerste, nur vom Prädikatargument begleitete Operation bildet, setzt dann die metadialogische Offenlegung ein. Um die anschließende dialogische Verwendung von $\xrightarrow{R} \pi x A(x, \pi)$ ihrerseits wieder in formale Deduktionen einzubeziehen, gehen wir von (1) und (2) über zu

$$(3) \qquad \vdash \mathrm{A}(\mathrm{y}, \xrightarrow{\mathrm{R}} \pi\, \mathrm{x}\, \mathrm{A}) \leftrightarrow (\xrightarrow{\mathrm{R}} \pi\, \mathrm{x}\, \mathrm{A}\, \mathrm{y}).$$

Wohlgemerkt gehören (1), (2) formalistisch nicht ins Implikans eines Axioms, wo sie ins dialogische Wechselspiel geraten würden, sondern in die Prämissen eines Schlußschemas, weil sie dort finit-direkt abgeleitet sein wollen. Danach bleibt lediglich die Bestätigung übrig, daß die Konklusion dialogisch zu gewinnen ist.

**0.14** Die komplizierteren Definitionen dieser (aufweislich) konstruktiv deduzierbar rekursiven Art, die eine ausbaufähige konstruktive Ordinalzahltheorie erfordert, sind ebenfalls nach obigem Muster dialogisch deutbar. Da wir aus metadialogischen Beweggründen aber ohnehin bestrebt sind, eine aktiv erzielte Gewinnstrategie mit einem formalen System einzuholen und einzufangen, lassen wir weitere Einzelheiten der dialogischen Deutung vorerst beiseite und wenden uns lieber dem dabei entstehenden formalen Gesamtsystem $\mathfrak{S}_1$ zu.

**0.2** Verbal läßt sich $\mathfrak{S}_1$ im Rohbau so charakterisieren: es wird alles einkalkuliert, was der definitorische Übergang von (1), (2) zu (3) bei syntaktischer und deduktionstechnischer Abschließung in die intuitionistische Arithmetik hineinträgt. Der Vollständigkeitsgrad von $\mathfrak{S}_1$ erhöht sich beträchtlich, wenn der definitorische Übergang neben einer Argumentbedingung vor allem (beispielsweise im Hinblick auf das Deduktionstheorem) eine Definitionsvoraussetzung mitberücksichtigt und wenn Quantifikationen über denjenigen — aufzählbaren, aber unentscheidbaren — Prädikatbezirk einbezogen werden, welchen der Kalkül $\mathfrak{S}_1$ selbst produziert.

Es sei erneut hervorgehoben, daß metadialogische Anwendung einer formalen Deduktion mit freien Prädikatmarken den bedeutungsfreien Status der Marken nicht aufhebt und deswegen auch die Prädikatquantifikation nicht etwa imprädikativ beeinflußt. Der Unterschied zur dialogischen Verwendung der Grundzahlquantifikation [14] S. 27 ist bloß folgender: bei Wahlen von $P$ ist damit zu rechnen, daß der Partner sich erkundigt, wieso P ein $\mathfrak{S}_1$-Fabrikat ist (während bei Wahlen von $n$ kein Zweifel auftaucht, ob $n$ Grundzahl ist).

$\mathfrak{S}_1$ macht dann nicht allein die Eigenschaften der natürlichen Ordnung zwischen konstanten Ordinalzahlen zugänglich,[7] sondern setzt sich quasi selbst transfinit fort, solange in transfinit rekursiven Definitionen die auf Ordinalzahlen angewandten Operatoren einer natürlichen Beschränkung $< \underset{+}{t}$ durch einen (im allgemeinen unbeschränkten) Ordinalzahlterm $\underset{+}{t}$ unterliegen. $\mathfrak{S}_1$ setzt sich über diese Fortsetzungsrichtung hinaus sogar typentheoretisch transfinit fort, jedenfalls soweit es quantorenfreie Aussageformen anbelangt.

**0.21** Konstante Ordinalzahlen werden hier — ähnlich dem Ansatz in [21] (Etappe I, 1. Absatz, auch Fußnote 5) — durch Prädikatpaare PR dargestellt, für die

$$(0) \qquad \wedge x_0 (P x_0 \dot{\rightarrow} \wedge x (P x \dot{\rightarrow} R x x_0 \rightarrow \pi x) \rightarrow \pi x_0) \dot{\rightarrow} P x \rightarrow \pi x$$

in $\mathfrak{S}_1$ schon abgeleitet ist; als ständige Abkürzung für (0) dient $I(P, R; \pi, x)$. $I(P, R; \pi, x)$ drückt die Vorgängerinduktion *bei* der Bedingung P *bezüglich* R als Vorgängerprädikat (*für* $\pi$ *nach* x) aus.

Nicht (0) selbst soll also repräsentativ für eine Ordinalzahl sein, sondern erst die direkte Vorlage einer formalen Deduktion $\vdash_{\mathfrak{S}_1} I(P, R; \pi, x)$.

**0.22** Zur Darstellung eines beschränkten Ordinalzahlterms genügt ein Prädikat*form*paar $P(z_1, ..) R(z_1, ..)$, für das

$$\vdash I(P(z_1, ..), R(z_1, ..); \pi, x)$$

in $\mathfrak{S}_1$ gilt. Wesentlich ist dabei, daß die Deduktion für formale Variable $z_1, ..$ erfolgt und nicht etwa für „jedes" einzelne $\mathit{z}_1, ..$; halbformale Regeln, die auf inhaltliche Einsicht hin den Schluß von $\mathit{z}$ auf z ziehen wollen, sind hier nicht zugelassen, da man sie vom finiten Standpunkt aus nicht kontrollieren kann. Jedoch darf die Deduktion für $z_1, ..$ bereits abgeleitete Vorgängerinduktionen anwenden, die andere $P_0 R_0$ betreffen.

**0.23** Zur Darstellung eines unbeschränkten Ordinalzahlterms wird ein Prädikat*schema*paar $P(\pi_1, ..) R(\rho_1, ..)$ herangezogen, für das die Vorgängerinduktion bei $P(\pi_1, ..)$ bezüglich $R(\rho_1, ..)$ aus den hypothetisch angesetzten Vorgängerinduktionen bei $\pi_1$ bez. $\rho_1, ..$ in $\mathfrak{S}_1$ relativ deduziert ist.

**0.24** Bringt man auch noch die geschlossene Vorgängerinduktion

$$(\hat{0}) \qquad \wedge \pi \wedge x I(P, R; \pi, x)$$

bei P bezüglich R, mit Prädikatmarken $\pi \rho$ anstelle PR, in den hypothetischen Ansatz von Definitionen, so stellt das *unter* dieser Hypothese durch eine 0-stellige Prädikatmarke relativ explizit definierte Prädikat-

[7] Vgl. [21], S. 270.

schema ein (transfinites) Primaussageschema $\underset{+}{\pi}\alpha$ über Ordinalzahlen dar. Als ständige Abkürzung für $(\hat{0})$ wird $I(P, R)$ gebraucht.

**0.25** Zwar sieht $\mathfrak{S}_1$ allein Prädikatmarken erster Ordnung über Grundzahltupeln festen Typs vor (und keinerlei Prädikatmarken über Prädikatmarken), aber auf dem Wege formaler rekursiver Definitionen über Grundzahlen bietet das hier finit definierte Instrument genügend Handhaben, um nicht nur beschränkte transfinite Terme höheren Typs und entsprechende Termmarken einzuführen, sondern auch Prädikatformen und -schemata über Ordinalzahlen darzustellen, in denen unbeschränkte transfinite Terme und Termmarken vorkommen, deren Typen transfinite Ordnung erreichen, sei der Typ respektive die Ordnung nun konstant, beschränkt variabel oder markenmäßig unbestimmt, und mögen gar die Typen selbst erst mit beschränkt aufgebauten Typschemata transfinit rekursiv definiert sein.

**0.3** Was schließlich die, der Etappe I des Amsterdamer Plans entsprechend mittels Niveauvergleich innerhalb $\mathfrak{S}_1$ definierte natürliche Halbwohlordnung $\leq$ zwischen Ordinalzahlrepräsentanten anlangt, so ist konstruktiv eine „halb"stabilisierende formale Übersetzung $^{\cap}$ vorzunehmen, bei der die Prädikate zur Ordinalzahl- und Termrepräsentation keine Abschwächung erleiden dürfen. $\overset{\cap}{\leq}$ ist dann reflexiv, transitiv und schwach konnektiv.

Daß allein $\vdash I(P, R; \pi, x)$ ein Prädikatpaar PR zum Ordinalzahlrepräsentanten stempelt, also R „auf" P keinerlei Eigenschaften totaler oder partieller Ordnungen unterliegt, bewirkt eine gründliche Auffüllung der „Klasse" ordnungsisomorpher Wohlordnungen, die in der naiven Mengenlehre eine Ordinalzahl darstellt. Diese Auffüllung ist nützlich zur Ermittlung der ordinalen Höhe[8] verschiedenster Objekte: ob es sich nun um Definitionsfundierungen handelt, die ja i.a. weder transitiv noch unverzweigt sind, oder um fächerartige Typen (oder, für etwaige Reflexionen auf $\mathfrak{S}_1$ selbst, um chiffrierte Formen deduktiver Stammbäume).

Für die unbeschränkte transfinite Induktion

$$\underset{+}{(0)} \qquad \bigwedge\alpha_0(\bigwedge\alpha(\alpha < \alpha_0 \to \underset{+}{\pi}\alpha) \to \underset{+}{\pi}\alpha_0) \to \underset{+}{\pi}\alpha$$

ist in $\mathfrak{S}_1$ zwar allgemein keine Übersetzung mit $\pi\,\rho$ unter $I(\pi, \rho)$, $\pi_0\,\rho_0$ unter $I(\pi_0, \rho_0)$, .. als Vertreter der $\alpha, \alpha_0$, .. abzuleiten, wohl aber eine solche für die unbeschränkte schwache transfinite Induktion, die aus $\underset{+}{(0)}$ durch Einsetzung von $\neg\neg\underset{+}{\pi}$ anstelle von $\underset{+}{\pi}$ entsteht. Daraus kann für „stabile" Aussageformen $\underset{+}{A}(\alpha)$ über Ordinalzahlen, für die $\neg\neg\underset{+}{A}(\alpha) \to \underset{+}{A}(\alpha)$ abgeleitet ist — sehr viele lassen sich in $\mathfrak{S}_1$, von $\overset{\cap}{\leq}$ ausgehend, darstellen —, der Induktionssatz

$$\bigwedge\alpha_0(\bigwedge\alpha(\alpha < \alpha_0 \to \underset{+}{A}(\alpha)) \to \underset{+}{A}(\alpha_0)) \to \underset{+}{A}(\alpha)$$

---

[8] Vgl. dazu den letzten Absatz unter **0.11**.

kopiert werden, aus dem wieder Kopien mit der Induktionsfolgerung $\underset{+}{\mathrm{A}}(\underset{+}{\mathrm{t}})$ anstelle $\underset{+}{\mathrm{A}}(\alpha)$ entspringen. (Im transfiniten Term $\underset{+}{\mathrm{t}}$ dürfen dabei hochtypige Argumente vorkommen, nur der $\underset{+}{\mathrm{t}}$-Wert soll vom tiefsten nichtleeren Typ sein, also bei Spezialisierung der Argumente in $\underset{+}{\mathrm{t}}$ Ordinalzahlen liefern.)

**0.31** Angesichts einer dermaßen vielseitigen Abrundung nenne ich $\mathfrak{S}_1$ lieber normale „Sphäre“ über Grundzahlen (statt „Sprachschicht“).

**0.4** Im Unterschied zur Theorie der rekursiven Funktionen, wo das Phänomen der Rekursivität aus der Definition rekursiver Funktionen entfernt und auf die Anwendung der Definition verschoben wird ([11], p. 274), soll hier der Versuch gemacht werden, das konstruktive Prinzip selbst, nach dem man Definitionen und Beweise durch Induktion bildet, in einer möglichst ergiebigen Weise formal zu fixieren. Die kalkulatorisch vorgenommene Auswahl aus den formal durch keine (rekursive) Aufzählung auszuschöpfenden Möglichkeiten, Definitionen ohne unendlichen Regreß zu erfinden und Beweise durch vollständige Induktion zu führen, ist in ihrer Konzeption auf das Zustandebringen von Fortsetzungs- und Abschließungseigenschaften gerichtet.

Die Stärke des formalen Systems $\mathfrak{S}_1$ beruht grundsätzlich in seiner konsequenten Beschränkung auf Einsichten, die selbst erst aus dem Operieren mit Figuren nach Systemen von Regeln gewonnen worden sind, also in der Kodifizierung jener Hinwendung „zu den Sachen selber“, die den Mathematiker meist als zu schwach anmuten.

**0.5** Wer harte Formeldressur nicht scheut, kann $\mathfrak{S}_1$ im wesentlichen auf zweierlei Weise überspielen, ohne dabei den Bereich des einwandfrei operativ-dialogisch Deutbaren zu verlassen.

**0.51** Ausgehend von einem $\mathfrak{S}_1$ umgreifenden Figurbezirk konstruiere man zunächst $\mathfrak{S}_2$, und zwar derart, daß $\mathfrak{S}_1$ einerseits in $\mathfrak{S}_2$ mit bedingten Operatoren reproduzierbar ist und daß $\mathfrak{S}_1$ andererseits in $\mathfrak{S}_2$ aus einer (etwa primitiv rekursiven) Numerierung von $\mathfrak{S}_1$ auch f o r m a l völlig entschlüsselt werden kann. Iteriert man diese normale Sprachkonstruktion transfinit, solange Ordinalzahlen, die bereits deduktiv erreicht sind, als Iterationsindizes zur Verfügung stehen, so gelangt man (wie [21] S. 271) bis zu Regionen endlicher Höhe, welche die $\mathfrak{S}_\alpha$ mit $\alpha$ aus der vorangehenden Region bündeln. $\sigma_\alpha$ bezeichne wieder den Limes der in $\mathfrak{S}_\alpha$ erreichbaren Ordinalzahlen; die kritischen Stellen der Iteration, wo $\sigma_\alpha = \alpha$, lassen sich dann durch Einfädelung von Ordinalzahlmarken $\mu$ überwinden[9]. Und die Zufuhr von Ordinalzahlmarken kann ihrerseits noch gestaffelt werden.

Wer nur an der Ausdehnung des Ordinalzahlgegenstandes interessiert ist, mag Quantifikation, Induktion und Rekursion über die in $\mathfrak{S}_1 \rightleftharpoons {}^*\mathfrak{S}_1/\sigma_0^*$ erreichten (und darin bereits schwach wohlgeordneten) Ordinalzahlen $< \sigma_1$ zu $\mathfrak{S}_1$ hinzufügen;

---

[9] Ordinalzahlmarken $\mu$ werden, genau wie unbeschränkte Ordinalzahlvariablen $\alpha$, durch Markenpaare $\pi\rho$ dargestellt. Zur Unterscheidung hat man vorher in den so verwendeten $\pi\rho$ die Unterscheidungsindizes für $\mu$ gegenüber denen für $\alpha$ auszuzeichnen.

er erhält so $*\mathfrak{S}_1/\sigma_1 \rightleftharpoons *\mathfrak{S}_1/\sigma_1^*$ und daraufhin (wie [22]) die erste *expansible* Sprachschicht $*\mathfrak{S}_1/\sigma_\alpha^*$ über sich erweiternden Ordinalzahlabschnitten $< \sigma_\alpha^*$, welche jeweils einen transfinit fortsetzbaren Sektor aus $\mathfrak{S}_{\alpha+1}$ loslöst.

**0.52** Wer $\mathfrak{S}_1$ vorerst lieber nach „innen" engmaschiger auszuarbeiten wünscht, wird eine Verschlüsselung $\ulcorner . \urcorner \leftrightharpoons \mathrm{cf}_0$ von $\mathfrak{S}_1$ zur Definition eines Reflexionsprinzips benutzen, das $\mathfrak{S}_1 \rightleftharpoons \mathfrak{S}_1^0$ zu $\mathfrak{S}_1^1$ ergänzt; dieses Prinzip reflektiert auf chiffrierte Deduktionsformen, um dechiffrierte Aussageformen zu folgern:

$$\vdash_{\mathfrak{S}_1^1} \bigvee z\ \mathrm{dd}_{\mathfrak{S}_1^0}(z, \mathrm{cf}_0\, P0^{(x)}) \rightarrow Px.$$

Wer den Kode $\mathrm{cf}_0$, $\mathrm{dd}_{\mathfrak{S}_1^0}$ kennt, kann für jedes $m$ anstelle x aus dem vom Dialogpartner anzugebenden $n$ anstelle z eine Deduktion für $Pm$ in $\mathfrak{S}_1^0$ entnehmen; der Partner braucht die Entzifferung von $\mathrm{cf}_0$ nicht zu beherrschen.

$\mathfrak{S}_1^1$ leistet zwar keine Entschlüsselung der Chiffrierung $\mathrm{cf}_0$ (wie $\mathfrak{S}_2$ es tut), sorgt aber für eine bescheidene formale Ausschöpfung der unendlichen Induktion $\Rightarrow_n Pn \dot{\Rightarrow} Px$.

Durch transfinit iterierte Anwendung des Reflexionsprinzips auf schon durch Reflexion erzielte Ergänzungen von $\mathfrak{S}_1$ entwickelt sich, nach Maßgabe bereits erreichter Iterationsindizes, eine weitgehende Verdichtung von $\mathfrak{S}_1$:

$$\vdash_{\mathfrak{S}_1^{\cdot\backslash\cdot}} I(\pi, \rho) \dot{\rightarrow} \bigwedge x \bigvee z\ \mathrm{dd}_{\mathfrak{S}_1^{\pi\rho}}(z, \mathrm{cf}_{\pi\rho}\, P0^{(x)}) \rightarrow Px.$$

Ein derartiger Vervollkommnungsprozeß holt die chiffrierenden Prädikate $\mathrm{cf}_\alpha$ und $\mathrm{dd}_{\mathfrak{S}_1^\alpha}$, bis auf die von außen per definitionem zugeführte Erstverschlüsselung $\ulcorner . \urcorner \leftrightharpoons \mathrm{cf}_0$, aus $\mathfrak{S}_1$, wo sie einheitlich in $\pi$, $\rho$ unter $I(\pi, \rho)$ ausdrückbar sind.

$\mathfrak{S}_1^{\cdot\backslash\cdot}$ wird also (bis auf den Verschlüsselungsanfang und das Entschlüsselungsende) gewissermaßen noch von $\mathfrak{S}_1$ umspannt. $\mathfrak{S}_1^{\cdot\backslash\cdot}$ heiße darum auch die *koerzible Perfektion* von $\mathfrak{S}_1$.

**0.6** Um die Handhabung des als kanonisches System (Post) aufgestellten Regelwerks $\mathfrak{S}_1$ nicht vorzeitig mit der intendierten Deutung zu verquicken, kommen wir auf diese erst unter **1.5∗** zurück; und den Anschluß an die gebräuchliche Notationsweise finden wir bloß noch inoffiziell unter **1.6∗∗** wieder. In **1.632** lernen wir dabei verbal die Art der ausgewählten (relativen und bedingten) rekursiven Definitionen ihrer Syntax nach genauer kennen. Zum konstruktiven Inhalt der Deduktionstechnik wird unter **1.63∗** nachgetragen, was über **0.13∗** hinausgeht.

Das Hauptgewicht der Darlegung ruht angesichts der unter **3.∗∗** erörterten Spannweite von $\mathfrak{S}_1$ auf der minuziösen Detailarbeit in den Abschnitten **1.∗∗∗**. Unter **2.∗∗∗** wird in knapper, aber (bei einigen metamathematischen Vorkenntnissen) einwandfrei kontrollierbarer Form ein Kernstück der Deduktionsmöglichkeiten in diesem System der konstruktiven Arithmetik vorgeführt.

## 1. Konstitution des Kalküls

Zur Vorbereitung der Definition für die erste normale Sphäre über Grundzahlen stellen wir einige Verabredungen zusammen.

Ist ein endliches Alphabet $\boldsymbol{\alpha}$ aus $k_0(>0)$ Konstanten $\mathrm{c}_1, .., \mathrm{c}_{k_0}$ und aus $k_n(\geq 0)$ $n$-stelligen Funktoren ${}^n\mathrm{f}_1, .., {}^n\mathrm{f}_{k_n}$ $(n = 1, .., m; m > 0)$ gegeben, so seien Wörter w hergestellt nach den Regeln

$$\rightarrow {}_w\mathrm{c}_l \qquad (l = 1, .., k_0)$$
$${}_w\mathrm{w}_1, .., {}_w\mathrm{w}_n \rightarrow {}_w{}^n\mathrm{f}_l\, \mathrm{w}_1 .. \mathrm{w}_n \qquad (l = 1, .., k_n; n = 1, .., m).$$

Zu „verstehen" sind dabei $\rightarrow$ als Imperativ "aus .,..,. bilde .", die Kommata als Partition der Prämissen, die w als Mitteilung für schon hergestellte Wörter, ${}_w\mathrm{w}$ als "w ist Wort".

Ergänzen wir $\boldsymbol{\alpha}$ durch $k_0'(>0)$ Wortvariable $\mathrm{o}_1, .., \mathrm{o}_{k_0'}$, so seien Wortterme t gebildet gemäß

$$\rightarrow {}_{*w}\mathrm{o}_l \qquad (l = 1, .., k_0')$$
$$\rightarrow {}_{*w}\mathrm{c}_l \qquad (l = 1, .., k_0)$$
$${}_{*w}\mathrm{t}_1, .., {}_{*w}\mathrm{t}_n \rightarrow {}_{*w}{}^n\mathrm{f}_l\, \mathrm{t}_1 .. \mathrm{t}_n \qquad (l = 1, .., k_n; n = 1, .., m).$$

Die t sind als Mitteilung für schon gebildete Wortterme zu verstehen, ${}_{*w}\mathrm{t}$ als "t ist Wortterm". Aus t entstehen durch Substitution von $\mathrm{w}_l$ für die $\mathrm{o}_l$ in t immer wieder Wörter.

Ergänzen wir $\boldsymbol{\alpha}$ durch $k_0'$ Wortvariable und durch $k_n'(\geq 0)$ $n$-stellige Relatoren ${}^n\mathrm{r}_1, .., {}^n\mathrm{r}_{k_n'}$ $(n = 1, .., m'; m' > 0)$ zu $\boldsymbol{\alpha}'$, so seien Formeln W gebildet gemäß

$${}_{*w}\mathrm{t}_1, .., {}_{*w}\mathrm{t}_n \rightarrow {}_W{}^n\mathrm{r}_l\, \mathrm{t}_1 .. \mathrm{t}_n \qquad (l = 1, .., k_n'; n = 1, .., m').$$

${}_W\mathrm{W}$ lies "W ist Formel (Wort über Wörter)", wo W schon gebildete Formeln mitteilt. Aus jedem W ist eindeutig zu entnehmen, welche t dem Relator folgen; ebenso aus jedem t und aus jedem w, welche Wortterme beziehungsweise welche Wörter zu den Funktoren in t bzw. in w gehören — klammerfreie Łukasiewicz-Notation.

**1.1 Spezifikation.** Wir nehmen nun $k_0 = 7, k_1 = 1, k_2 = 7, k_0' = 19$, $k_1' = 4$, $k_2' = 9$, $k_3' = 1$, $k_4' = 1$ und für $\mathrm{c}_l$ speziell $\circ$, 0, $\xi$, $\Pi$, $\leq$, $\vee$, $\wedge$, für ${}^1\mathrm{f}_l$, ${}^2\mathrm{f}_l$ speziell $'$ bzw. ${}^{\cup}$, ${}^{\rightarrow}$, ${}^{\wedge}$, ${}^{\vee}$, ${}^{\bigwedge}$, ${}^{\bigwedge}$, ${}^{\rightarrow}$ für ${}^1\mathrm{r}_l$, ${}^2\mathrm{r}_l$, ${}^3\mathrm{r}_l$, ${}^4\mathrm{r}_l$ speziell ${}_k$, ${}_x$, ${}_{*k}$, ${}_A$ bzw. ${}_{\neq}$, ${}_2$, ${}_\pi$, ${}_{\cup *k}$, ${}_{\not k}$, ${}_{\cup x}$, ${}_P$, ${}_\angle$, ${}_{\leftrightarrows}$ bzw. ${}_+$ bzw. ${}_\|$. Die 7 Konstanten stehen auf der Zeile, die 8 Funktoren sind hoch gestellt und, vom einstelligen $'$ abgesehen, zweistellig, so daß die Wortbestandteile sich augenfällig wiederfinden lassen. Die 15 Relatoren sind tief gestellt und teils keine Primfiguren, sondern mnemotechnisch zusammengesetzt — im Hinblick auf die später intendierte Deutung, die aber ja nicht verstanden zu werden braucht. (Da jede Formel W mit einem einzigen Relator beginnt, ist beim Lesen unterhalb der Zeile klar, wo er endet. Wer die Stellenzahl der einzelnen Relatoren vergessen hat, kann sich an die Zahl der Wörter halten, die dem jeweiligen Relator folgen. Die Stellenzahl der Funktoren sollte man allerdings wissen.[10])

Zu verstehen sind in den nachfolgend postulierten Regeln von der Form $W_1, .., W_k \rightarrow W$, wobei $k = 0, 1, 2, 3, 4, 5, 6, 27$ sein wird, außer "$\rightarrow$" und "," noch die 19 Wortvariablen "$o_l$", nämlich als Mitteilung für Wörter, die den speziellen 7 Konstanten und 1 + 7 Funktoren entspringen: anstelle $o_1, .., o_{19}$ seien darum $w, w_1, .., w_5, v, v_1, .., v_5, u, u_1, .., u_6$ zur Mitteilung solcher Wörter in den Worttermen t (innerhalb der Formeln W) benutzt.

**1.2 Technische Vorbemerkungen.** Um Regeln mit denselben Prämissen zusammenzuziehen, wird statt '$W_1, .., W_k \rightarrow W^1; ..; W_1, .., W_k \rightarrow W^l$' kurz '$W_1, .., W_k \rightarrow W^1, .., W^l$' notiert; die rechts vom Imperativ auftretenden, als Partition der Konklusionen zu verstehenden Kommata stehen damit eine Reflexionsstufe höher als die (bisherigen) Kommata links vom Imperativ. Bei Regeln, die nicht in eine Zeile passen, seien aufeinanderfolgende Zeilen dadurch verbunden, daß das Zeichen am Zeilenende am nächsten Zeilenanfang wiederholt wird. Regeln (einschließlich Bezifferung) sind durch ';', Regelgruppen durch '⁏' getrennt.

Prämissen, die entbehrlich waren, obwohl sie der Eindämmung überflüssiger Produktionen dienlich gewesen wären, sind — namentlich bei den Relatoren $\not\downarrow$, $\|$ — weggelassen. Entbehrliche Prämissen sind — damit die Regelzahl, die jetzt = 98, nicht > 100 ansteige — auch dann nicht beibehalten worden, wenn dabei (wie bei den nachstehenden Regeln 8.2.1, 2, 12, 8.1.9, 10, 13, 11 und 10.1) in der Konklusion „lose" Wortmitteilungen eintreten, die nicht in der Weise gekoppelt sind, daß sie in einer der zugehörigen Prämissen vorkommen. Eine Neufassung des Kalküls $\mathfrak{S}_1$ ohne lose Wortmitteilungen würde gestatten, mit „Ur"mitteilungen zu arbeiten, die irgendein (ununterbrochenes) „Stück" aus einer schon produzierten Figur bezeichnen, statt, wie hier, die Mitteilungen u, v, w, .. auf einen vorher spezifizierten Wortbegriff zu beziehen. Mit genau 130 Regeln läßt sich diese Neufassung überdies so anlegen, daß im entscheidbaren Teil des Kalküls 1. jede Urmitteilung in den Prämissen einer Regel auch in der Konklusion vorkommt und 2. jede den Konklusionsrelator enthaltende Prämisse kürzer als die Konklusion ausfällt.

**1.3 Der pure Kalkül.** Doch will ich jetzt ohne Umschweife die Regeln des Sphärenkalküls $\mathfrak{S}_1$ in ihrer optisch knappsten Fassung präsentieren.

*Anmerkung.* Der Leser mache sich mit dem (später entstandenen) Zusatz S. 193 f. vertraut, wonach in 9.1.5 genauer $_{\bar{A}}u$ statt $_A u$ steht und für $_{\bar{A}}$ bloß 9.4.1 — 6, d. i. 9.2.1 — 6 mit $_{\bar{A}}$ anstelle $_A$, benutzt wird. Da diese Modifikation fortan unschwer zu verfolgen ist, kann sie im Text unberücksichtigt bleiben.

[10] Dies wäre nicht nötig bei der am Schluß von **1.2** erwähnten Version von $\mathfrak{S}_1$.

0.1 $\to {}_{k}0$ ;  0.2 ${}_{k}\mathrm{w} \to {}_{k}{'}\mathrm{w}$ ;·

1.1, 2 ${}_{k}\mathrm{w} \to {}_{\ddagger}0{'}\mathrm{w},\ {}_{\ddagger}{'}\mathrm{w}0$ ;  1.3 ${}_{\ddagger}\mathrm{vw} \to {}_{\ddagger}{'}\mathrm{v}{'}\mathrm{w}$ ;·

2.1 $\to {}_{2}00$ ;  2.2 ${}_{2}\mathrm{vw} \to {}_{2}{'}\mathrm{v}{''}\mathrm{w}$ ;·

3. ${}_{k}\mathrm{w} \to {}_{x}{}^{\cup}\xi\,\mathrm{w}$ ;·  4. ${}_{k}\mathrm{v},\ {}_{k}\mathrm{w} \to {}_{\pi}\mathrm{v}^{\cup\cup}\mathrm{v}\,\Pi\,\mathrm{w}$ ;·

5.1 $\to {}_{*k}0$ ;  5.2 ${}_{*k}\mathrm{w} \to {}_{*k}{'}\mathrm{w}$ ;  5.3 ${}_{x}\mathrm{w} \to {}_{*k}\mathrm{w}$ ;·

6.1 $\to {}_{\cup *k}0\circ$ ;  6.2 ${}_{*k}\mathrm{v},\ {}_{\cup *k}\mathrm{wu} \to {}_{\cup *k}{'}\mathrm{w}^{\cup}\mathrm{uv}$ ;·

7.1 ${}_{\cup *k}\mathrm{wu} \to {}_{+}\mathrm{u}\circ\mathrm{u}$ ;  7.2 ${}_{*k}\mathrm{v},\ {}_{+}\mathrm{u}_1\mathrm{u}_2\mathrm{u} \to {}_{+}\mathrm{u}_1{}^{\cup}\mathrm{u}_2\mathrm{v}^{\cup}\mathrm{uv}$ ;·

8.1.1 ${}_{\ddagger}\mathrm{vw} \to {}_{\dagger}{}^{\cup}\xi\,\mathrm{v}^{\cup}\xi\,\mathrm{w}$ ;

8.1.2 ${}_{x}\mathrm{w} \to {}_{\dagger}\mathrm{w}\circ$ ;  [8.1.0 ${}_{\dagger}\mathrm{wu},\ {}_{\dagger}\mathrm{wv} \to {}_{\dagger}\mathrm{w}^{\cup}\mathrm{uv}$];

8.3.1 $\to {}_{\cup x}0\circ$ ;  8.3.2 ${}_{x}\mathrm{v},\ {}_{\dagger}\mathrm{vu},\ {}_{\cup x}\mathrm{wu} \to {}_{\cup x}{'}\mathrm{w}^{\cup}\mathrm{uv}$ ;

8.1.3 ${}_{x}\mathrm{w} \to {}_{\dagger}\mathrm{w}0$ ;

8.2.1 ${}_{x}\mathrm{v} \to {}_{\|}\mathrm{vvv}_1\mathrm{v}_1$ ;  8.2.2 ${}_{\dagger}\mathrm{vu} \to {}_{\|}\mathrm{vuv}_1\mathrm{u}$ ;

8.2.3 ${}_{\|}\mathrm{vuv}_1\mathrm{u}_1 \to {}_{\|}\mathrm{v}{'}\mathrm{uv}_1{'}\mathrm{u}_1$ ;

8.1.8 ${}_{\dagger}\mathrm{vv}_1,\ {}_{\|}\mathrm{vuv}_1\mathrm{u} \to {}_{\dagger}\mathrm{vu}$ ;

8.2.4 ${}_{\|}\mathrm{wuw}_1\mathrm{u}_1,\ {}_{\|}\mathrm{wvw}_1\mathrm{v}_1 \to {}_{\|}\mathrm{w}^{\cup}\mathrm{uvw}_1{}^{\cup}\mathrm{u}_1\mathrm{v}_1$ ;

8.1.4, 5, 6, 7 ${}_{x}\mathrm{w} \to {}_{\dagger}\mathrm{w}\Pi,\ {}_{\dagger}\mathrm{w}\leq,\ {}_{\dagger}\mathrm{w}\vee,\ {}_{\dagger}\mathrm{w}\wedge$ ;

8.2.5, 6, 7 ${}_{\|}\mathrm{wuw}_1\mathrm{u}_1,\ {}_{\|}\mathrm{wvw}_1\mathrm{v}_1 \to {}_{\|}\mathrm{w}^{\to}\mathrm{uvw}_1{}^{\to}\mathrm{u}_1\mathrm{v}_1,\ {}_{\|}\mathrm{w}^{\wedge}\mathrm{uvw}_1{}^{\wedge}\mathrm{u}_1\mathrm{v}_1,$
$,\ {}_{\|}\mathrm{w}^{\vee}\mathrm{uvw}_1{}^{\vee}\mathrm{u}_1\mathrm{v}_1$ ;

8.1.9, 10 $\to {}_{\dagger}\mathrm{v}^{\vee}\mathrm{vu},\ {}_{\dagger}\mathrm{v}^{\wedge}\mathrm{vu}$ ;

8.2.9, 10 ${}_{\dagger}\mathrm{vw}_1,\ {}_{\dagger}\mathrm{vw},\ {}_{\|}\mathrm{wuw}_1\mathrm{u}_1 \to {}_{\|}\mathrm{w}^{\vee}\mathrm{vuw}_1{}^{\vee}\mathrm{vu}_1,\ {}_{\|}\mathrm{w}^{\wedge}\mathrm{vuw}_1{}^{\wedge}\mathrm{vu}_1$ ;

8.1.12 ${}_{\dagger}\mathrm{vu},\ {}_{\dagger}\mathrm{wu} \to {}_{\dagger}{}^{\cup}\mathrm{vwu}$ ;  8.1.13 $\to {}_{\dagger}\circ\mathrm{u}$ ;

8.1.16 ${}_{\ddagger}\mathrm{w}_1\mathrm{w}_2,\ {}_{\pi}\mathrm{w}_1\mathrm{u},\ {}_{\pi}\mathrm{w}_2\mathrm{v} \to {}_{\dagger}\mathrm{uv}$ ;

8.1.17 ${}_{k}\mathrm{v},\ {}_{\ddagger}\mathrm{w}_1\mathrm{w}_2 \to {}_{\dagger}{}^{\cup\cup}\mathrm{v}\Pi\mathrm{w}_1{}^{\cup\cup}\mathrm{v}\Pi\mathrm{w}_2$ ;

8.1.18, 19, 20, 21, 22, 23 ${}_{\pi}\mathrm{wv} \to {}_{\dagger}\mathrm{v}\circ,\ {}_{\dagger}\mathrm{v}0,\ {}_{\dagger}\mathrm{v}\xi,\ {}_{\dagger}\mathrm{v}\leq,\ {}_{\dagger}\mathrm{v}\vee,\ {}_{\dagger}\mathrm{v}\wedge$ ;

8.2.12 ${}_{\pi}\mathrm{wv} \to {}_{\|}\mathrm{vvv}_1\mathrm{v}_1$ ;

8.2.8 ${}_{\|}\mathrm{vuv}_1\mathrm{u}_1,\ {}_{\|}\mathrm{wu}_1\mathrm{w}_1\mathrm{u}_2 \to {}_{\|}{}^{\cup}\mathrm{vwu}^{\cup}\mathrm{v}_1\mathrm{w}_1\mathrm{u}_2$ ;

8.1.11 $\to {}_{\dagger}\mathrm{v}^{\to}\mathrm{vu}$ ;  8.2.11 ${}_{\dagger}\mathrm{vw}_1,\ {}_{\dagger}\mathrm{vw},\ {}_{\|}\mathrm{wuw}_1\mathrm{u}_1 \to {}_{\|}\mathrm{w}^{\to}\mathrm{vuw}_1{}^{\to}\mathrm{vu}_1$ ;

8.1.14, 15 ${}_{\dagger}{}^{\cup}\mathrm{vwu} \to {}_{\dagger}\mathrm{vu},\ {}_{\dagger}\mathrm{wu}$ ;·

9.1.1 $\to {}_{P}{''}0\leq$ ;  9.1.2, 3 ${}_{k}\mathrm{w} \to {}_{P}\mathrm{w}\vee,\ {}_{P}\mathrm{w}\wedge$ ;  9.1.4 ${}_{\pi}\mathrm{wu} \to {}_{P}\mathrm{wu}$ ;

9.2.1 ${}_{\cup *k}\mathrm{wv},\ {}_{P}\mathrm{wu} \to {}_{A}{}^{\cup}\mathrm{uv}$ ; 9.2.2, 3, 4 ${}_{A}\mathrm{u},\ {}_{A}\mathrm{v} \to {}_{A}{}^{\to}\mathrm{uv},\ {}_{A}{}^{\wedge}\mathrm{uv},\ {}_{A}{}^{\vee}\mathrm{uv}$ ;

9.3.1 $\quad {}_{\angle}uw,\ {}_{\angle}wv \rightarrow {}_{\angle}uv;$ 9.3.2, 11, 12 ${}_{A}u \rightarrow {}_{\angle}uu,\ {}_{\angle}{}^{\cup}\wedge\circ u,\ {}_{\angle}u{}^{\cup}\vee\circ;$

9.3.3, 4, 6, 7 $\quad {}_{A}u,\ {}_{A}v \rightarrow {}_{\angle}{}^{\wedge}uvu,\ {}_{\angle}{}^{\wedge}uvv,\ {}_{\angle}u{}^{\vee}uv,\ {}_{\angle}v{}^{\vee}uv;$

9.3.5 $\quad {}_{\angle}wu,\ {}_{\angle}wv \rightarrow {}_{\angle}w{}^{\wedge}uv;$ 9.3.8 $\quad {}_{\angle}uw,\ {}_{\angle}vw \rightarrow {}_{\angle}{}^{\vee}uvw;$

9.3.9 $\quad {}_{\angle}{}^{\wedge}wuv \rightarrow {}_{\angle}w{}^{\rightarrow}uv;$ 9.3.10 $\quad {}_{\angle}w{}^{\rightarrow}uv \rightarrow {}_{\angle}{}^{\wedge}wuv;$

9.3.17, 18 $\quad {}_{\cup *k}{}'wv \rightarrow {}_{\angle}{}^{\cup}\vee\circ{}^{\cup}\vee v,\ {}_{\angle}{}^{\cup}\wedge v{}^{\cup}\wedge\circ;$

9.3.19, 20, 21, 22 $\quad \rightarrow {}_{\angle}{}^{\cup}\vee\circ{}^{\cup}\leq{}^{\cup\cup}\circ 0{}^{\cup}\xi 0,$

$, {}_{\angle}{}^{\cup}\leq{}^{\cup\cup}\circ{}^{\cup}\xi 0{}^{\cup}\xi' 0{}^{\cup}\leq{}^{\cup\cup}\circ'{}^{\cup}\xi 0'{}^{\cup}\xi' 0,$

$, {}_{\angle}{}^{\cup}\leq{}^{\cup\cup}\circ'{}^{\cup}\xi 0'{}^{\cup}\xi' 0{}^{\cup}\leq{}^{\cup\cup}\circ{}^{\cup}\xi 0{}^{\cup}\xi' 0,$

$, {}_{\angle}{}^{\cup}\leq{}^{\cup\cup}\circ'{}^{\cup}\xi 00{}^{\cup}\wedge\circ;$

9.3.23 $\quad {}_{x}v,\ {}_{A}u,\ {}_{\|}vu'vu_1,\ {}_{\|}vu0u_2,\ {}_{\angle}uu_1 \rightarrow {}_{\angle}u_2u;$

9.2.5, 6 $\quad {}_{x}v,\ {}_{A}u \rightarrow {}_{A}{}^{\vee}vu,\ {}_{A}{}^{\wedge}vu;$

9.3.13, 15 $\quad {}_{x}v,\ {}_{*k}v_1,\ {}_{A}u,\ {}_{\|}vuv_1u_1 \rightarrow {}_{\angle}u_1{}^{\vee}vu,\ {}_{\angle}{}^{\wedge}vuu_1;$

9.3.14 $\quad {}_{x}v,\ {}_{\nmid}vw,\ {}_{\angle}uw \rightarrow {}_{\angle}{}^{\vee}vuw;$ 9.3.16 $\quad {}_{x}v,\ {}_{\nmid}vw, {}_{\angle}wu \rightarrow {}_{\angle}w{}^{\wedge}vu;$

10.1 $\quad \rightarrow {}_{\leftrightharpoons}{}^{\wedge}\circ uu;$ 10.2 $\quad {}_{x}v_1,\ {}_{\leftrightharpoons}{}^{\wedge}vuw \rightarrow {}_{\leftrightharpoons}{}^{\wedge\cup}vv_1u{}^{\wedge}v_1w;$

9.1.5 $\quad {}_{2}vw,\ {}_{\pi}vw_2,\ {}_{\cup x}vv_2,\ {}_{P}vv_1,\ {}_{P}ww_1,\ {}_{A}u,\ {}_{A}u_1,$

$, {}_{\pi}vw_3,\ {}_{\nmid}w_3{}^{\cup}u_1{}^{\cup}v_1w_1,\ {}_{\cup x}vv_3,\ {}_{\cup x}vv_4,$

$, {}_{\nmid}v_3v_4,\ {}_{\nmid}v_3u_1,\ {}_{\nmid}{}^{\cup}v_3v_4{}^{\cup}v_1w_1,\ {}_{+}v_4v_3v_5,$

$, {}_{\leftrightharpoons}{}^{\wedge}v_4{}^{\rightarrow\cup}v_1v_4{}^{\rightarrow\cup}w_1v_5{}^{\cup}w_3v_4u_2,\ {}_{\leftrightharpoons}{}^{\wedge}v_3{}^{\rightarrow\cup}v_1v_3{}^{\rightarrow}u_2{}^{\cup}w_3v_3u_3,$

$, {}_{\angle}u_1{}^{\rightarrow}u_3{}^{\rightarrow\cup}v_1v_3{}^{\cup}w_3v_3,$

$, {}_{\pi}vw_4,\ {}_{\pi}vw_5,\ {}_{\nmid}w_4w_5,\ {}_{\nmid}{}^{\cup}w_4w_5{}^{\cup}u{}^{\cup}u_1{}^{\cup}v_1w_1,\ {}_{\nmid}v_3u,$

$, {}_{\|}{}^{\cup}w_2v_2u{}^{\cup}w_4v_3u_4,\ {}_{\|}{}^{\cup}w_2v_2u{}^{\cup}w_5v_3u_5,$

$, {}_{\leftrightharpoons}{}^{\wedge}v_4{}^{\rightarrow\cup}v_1v_4{}^{\rightarrow\cup}w_1v_5{}^{\wedge\rightarrow\cup}w_4v_4{}^{\cup}w_5v_4{}^{\rightarrow\cup}w_5v_4{}^{\cup}w_4v_4u_6,$

$, {}_{\angle}u_1{}^{\rightarrow\cup}v_1v_3{}^{\rightarrow}u_6{}^{\wedge\rightarrow}u_4u_5{}^{\rightarrow}u_5u_4 \rightarrow$

$\rightarrow {}_{P}v{}^{\cup\cup}u_1{}^{\cup}v_1w_1{}^{\rightarrow)\cup}w_2v_2u;$

9.3.24, 25 $\quad {}_{P}v{}^{\cup\cup}u_1{}^{\cup}v_1w_1{}^{\rightarrow)\cup}w_2v_2u,\ {}_{\cup x}vv_3,\ {}_{\nmid}v_3{}^{\cup}u{}^{\cup}u_1{}^{\cup}v_1w_1,$

$, {}_{\|}{}^{\cup}v_2w_2u{}^{\cup}v_3{}^{\cup\cup}u_1{}^{\cup}v_1w_1{}^{\rightarrow)\cup}w_2v_2uu_2 \rightarrow$

$\rightarrow {}_{\angle}u_1{}^{\rightarrow\cup}v_1v_3{}^{\rightarrow}u_2{}^{\cup\cup\cup}u_1{}^{\cup}v_1w_1{}^{\rightarrow)\cup}w_2v_2uv_3,$

$, {}_{\angle}u_1{}^{\rightarrow\cup}v_1v_3{}^{\rightarrow\cup\cup\cup}u_1{}^{\cup}v_1w_1{}^{\rightarrow)\cup}w_2v_2uv_3u_2;$

9.3.26 $\quad {}_{\pi}{}'wu,\ {}_{\cup x}{}'wv,\ {}_{x}v_1,\ {}_{x}v_2,\ {}_{\nmid}v_2{}^{\cup}v_1v,$

$, {}_{\|}v_1{}^{\cup}uvv_2u_1 \rightarrow {}_{\angle}{}^{\wedge\cup}\leq{}^{\cup\cup}\circ v_1v_2{}^{\cup}\leq{}^{\cup\cup}\circ v_2v_1{}^{\rightarrow\cup}uvu_1;$

9.2.7, 8 $\quad {}_{\pi}wv,\ {}_{A}u \rightarrow {}_{A}{}^{\wedge}vu,\ {}_{A}{}^{\vee}vu;$

9.3.27, 29 $\quad {}_{\pi}wv,\ {}_{P}wv_1,\ {}_{A}u,\ {}_{\|}vuv_1u_1 \rightarrow {}_{\angle}{}^{\wedge}vuu_1,\ {}_{\angle}u_1{}^{\vee}vu;$

9.3.28 $\quad {}_{\pi}wv,\ {}_{\nmid}vu_1,\ {}_{\angle}u_1u \rightarrow {}_{\angle}u_1{}^{\wedge}vu;$

9.3.30 $\quad {}_{\pi}wv,\ {}_{\nmid}vu_1,\ {}_{\angle}uu_1 \rightarrow {}_{\angle}{}^{\vee}vuu_1.$

**1.4 Technischer Überblick.** Unter den 98 Regeln finden sich 10 Anfänge —Regeln mit Prämissenzahl $k = 0$, ohne lose Wortmitteilung — (: 0.1, 2.1, 5.1, 6.1, 8.3.1, 9.1.1, 9.3.19—22) und 5 Anfangsformen —Regeln mit $k = 0$, aber mit loser Wortmitteilung— (: 8.1.9, 10, 13, 11, 10.1). In der Regel 9.1.5 könnte $k$ um 2 auf 25 vermindert werden: man ziehe die 12., 13. und 23. Prämisse auf $_{\nmid}\mathrm{v}_3{}^{\cup\cup}\mathrm{v}_4\mathrm{u}_1\mathrm{u}$ zusammen. In 9.3.17, 18 genügt $_{\cup}x$ anstelle $_{\cup}*_k$.

Die Regel 8.1.0 macht zwar 8.3.1,2 schnell zugänglich, läßt sich aber streichen, weil sie eliminierbar ist, *falls* w noch $_x\mathrm{w}$ oder $_{\pi}.\mathrm{w}$ erfüllt oder $\circ$ mitteilt, — nämlich via 8.2.2, 8.2.4, 8.1.8 (mittels 8.1.1, 17, 13). Die Regelzahl 98 errechnet sich mithin, beziehentlich für 0. bis 7., 8.1.—3., 9.1.—3. und 10. gezählt, aus $2+3+2+1+1+ +3+2+2+(23+12+2)+(5+8+30)+2$.

Die Bezifferung der Regeln legt Wert auf zusammengehörige Gruppen, z.B. 8.1.2—7 analog 8.1.18—23, 8.1.9—11 im selben Takt wie 8.2.9—11. Vor der letzten Nummer einer Regelbezifferung steht die Bezifferung des Relators der Konklusionsformel; ist nur ein Punkt da, handelt es sich um unmittelbar definierte Relatoren (: 0., 2.) oder um sukzessive definierte Relatoren (: 1., 3., 4. stützen sich auf 0.; 5., 10. auf 3.; 6. auf 5.; 7. auf 5., 6. (wobei sich 6. mit zwei Regeln $\rightarrow_{+}\circ\circ\circ$ und $*_k\mathrm{v}$, $_{+}\mathrm{u}\circ\mathrm{u}_1 \rightarrow {}_{+}{}^{\cup}\mathrm{u}\,\mathrm{v} \circ {}^{\cup}\mathrm{u}_1\mathrm{v}$ anstelle 7.1 umgehen ließe)); sind zwei Punkte da, werden mehrere Relatoren simultan definiert oder im Anschluß an solche (: 8.1. stützt sich auf 1., 3., 4., 0. und via 8.1.8 auf 8.2.; 8.2. hinwieder stützt sich via 8.2.2, 9—11 auf 8.1. und sonst auf 3., 4.. 8.3. stützt sich auf 3. und 8.1.. 9.1. stützt sich auf 0., 4. und via 9.1.5 auf 9.2., 9.3. neben 2., 4., 7., 10., 8.1.—3.; 9.2. hinwieder stützt sich via 9.2.1 auf 9.1. und im übrigen auf 6., 3., 4.; 9.3. endlich stützt sich via 9.3.24, 25 auf 9.1. neben 8.1.—3., via 9.3.2, 11, 12, 4, 6, 7, 23, 13, 15 auf 9.2. neben 3., 8.2., 5., via 9.3.27, 29 auf 9.1. und 9.2. neben 4., 8.2., und im übrigen lediglich auf 6. (oder 8.3.), 3., 8.1., 4., 8.3., 8.2.; ganz 9. ist ohne lose Wortmitteilung).

Bis auf 3., 4. greifen alle definierten Relatoren auf sich selbst zurück (: 0.2, 1.3, 2.2, 5.2, 6.2, 7.2, 8.1.0, 8, 12, 14, 15, 8.2.3—11, 8.3.2, 9.1.5, 9.2.2—8, 9.3.1, 5, 8—10, 23, 14, 16, 28, 30, 10.2). Die produzierten Formeln nehmen im induktiven Prozeß allein via 8.1.8, 14, 15 und 9.3.1, 9, 10, 23 nicht an Länge zu.

Nichtsdestoweniger ist der Teilkalkül, der nach Streichung von 9.1.—3. übrigbleibt, entscheidbar. Dies übersieht man am leichtesten, wenn die Regeln an Hand der intendierten Deutung der bisher formal gehandhabten Konstanten, Funktoren und Relatoren rekapituliert werden. Dabei motiviert sich auch die gegenüber der Bezifferung durcheinandergewürfelte Reihenfolge beim Anschreiben der Regeln; dieses nimmt Bezug auf eine gestaffelte Ingangsetzung der Regeln — gelegentlich, bei 9.3.2, 11, 12, auf Kürzung durch Konklusionen für dieselben Prämissen.

**1.5 Intendierte Interpretation.** Der intuitive Sinn der zur Formulierung des Kalküls erforderlichen bedeutungsvollen Zeichen (: $\rightarrow$, u.. v.. w..), für die allein bisher in 1.* eine Erklärung gegeben wurde, gehört zum Bezirk der HILBERTschen Metamathematik, deren methodische Anforderungen unter dem Terminus "finit" — finitary, finitär — geläufig sind.[11] In finiter Sicht sind Extremalklauseln, daß *nur* die so und so zu konstruierenden Dinge unter einen genannten Begriff fallen, überflüssig; bei schrittweiser Erweiterung des Endlichen treten „umfassendere" Erfüllungen der Regeln gar nicht erst in den Kreis des Möglichen.

Die beabsichtigte Interpretation der mitgeteilten und, finit besehen, bedeutungsfreien Formeln bleibt im Bereich der konstruktiven Mathe-

[11] Vgl. beispielsweise [9], S. 171—173; dazu [7], S. 280—282.

matik; diese leugnet jedwede Vorstellung vom aktual Unendlichen, ohne deren primäre Wahrheit die klassische Mathematik (und die TARSKIsche Metamathematik) nicht voll arbeiten könnte, und faßt das Unendliche ausschließlich potentiell auf. Die operativ-dialogische Einstellung [13, 14] zur Konstruktivität wirkt einer neutralen Anerkennung oder Verwerfung formaler Postulate, bei der man die Inhalte dahinter mehr oder weniger gleichgültig taxiert, dadurch entgegen, daß das bloße „Inkonsistenz"risiko, das bei nicht-globaler Anwendungspraxis schwach bleibt, von einem starken „Verlust"risiko beim Verteidigen von Behauptungen abgelöst wird. Die hier vorgeschlagene konstruktive Arithmetik verzichtet gegenüber [15], wo der wesentliche Bestand der Analysis konstruktiv begründet wird, naturgemäß auf indefinite Variablen und Quantoren; sie grenzt ihre Aussagen vielmehr durch Aufzählung ab. Jedoch gestatten die benutzten Prädikatmarken auch andere Substitutionen als die systemgebundenen, und davon macht sogar die Deutung des Systems Gebrauch: beim Definieren eines Prädikats gewinnt sie die Einsicht in die (dialogische) Definitheit der Definition an Hand der formalen Deduktion gewisser Aussageschemata durch eine informale Anwendung. Entgegen [13—15] bleibt das tertium-non-datur hier ganz aus dem Spiel, weil es durch formale Übersetzungsverfahren lediglich partiell zu eliminieren ist und auch nach sonstigen konstruktiven Methoden nicht ohne weiteres formal gerechtfertigt werden kann.

Um gleichwohl die bezweckten Vollständigkeitseigenschaften zu ermöglichen, kommt es auf äußerste deduktive Ausgewogenheit der produzierbaren Beweis- und Definitionsmittel an. Die Quintessenz, die der rein äußerlich ins Auge springende technische Angelpunkt 9.1.5 des Kalküls $\mathfrak{S}_1$ im Verein mit 9.3.24, 25 als erste Antwort auf die formale Vollständigkeitsfrage bietet, läuft auf eine durchgreifende, konstruktiv noch akzeptable Verallgemeinerung des DEDEKINDschen Satzes über die Definition durch Induktion (1888) [3] hinaus: nämlich von absoluten unbedingten Definitionen durch gewöhnliche Induktion auf relative bedingte Definitionen durch entsprechende (schon konstruktiv deduzierte) Vorgängerinduktion; und von zu definierenden Funktionen, deren Definiensschema primitiv rekursiv ist, auf zu definierende Prädikate, deren Definiensschema hinsichtlich des aufgewiesenen Vorgängerprädikats konstruktiv deduzierbar (relativ bedingt) rekursiv „definierend", d.h. bis auf Bisubjunktion $\leftrightarrow$ bestimmend ist — ohne jedoch vom konstruktiven Sinn solcher Bestimmung erwarten zu dürfen, über jede Primaussage, die aus einem deduzierbar rekursiv „definierten" Prädikat (ohne freie Parameter) hervorgeht, sei irgendwo, -wann und -wie eine Entscheidung gefällt, auf die hin uns das formale tertium-non-datur orakelhaft erlaubt wäre.

Die intendierte Deutung der Konstanten, Funktoren und Relatoren sei jetzt kurz zusammengestellt.

Konstante: o (κενόν, das Leere), 0 (Null),
ξ (Variablenkern), Π (Prädikatmarkenkern),
$\leq$ (Ordnungsprädikat), ⋁ (verum), ⋀ (falsum).
Funktoren: ′ (Nachfolger); $^{\cup}$ (Juxtapositor, kurz Juxtor),
→ (Subjunktor), ∧ (Konjunktor), ∨ (Adjunktor),
⋁ (Partikularisator), ⋀ (Generalisator),
⇢ (Rekursifikator, kurz Rekursor).

Unter Zugrundelegung dieser Bedeutungen der 7 Konstanten und 1 + 7 Funktoren bezeichne *w* die Bedeutung des Wortes w, welche sich dem Aufbau von w parallel zusammensetzt.

Relatoren: $_k$w (*w* ist Grundzahl), $_x$w (*w* ist Variable für Grundzahlen),
$_{*k}$w (*w* ist Term für Grundzahlen),
$_A$w (*w* ist Aussageschema über Grundzahlen);
$_{\neq}w_1 w_2$ ($w_1$ ungleich $w_2$), $_2 w_1 w_2$ ($w_2$ verdoppelt $w_1$),
$_\pi w_1 w_2$ ($w_2$ ist Marke für $w_1$-stellige Prädikate),
$_{\cup *k} w_1 w_2$ ($w_2$ ist $w_1$-Tupel aus Termen),
$_{\not\downarrow} w_1 w_2$ ($w_1$ ist in $w_2$ unfrei, $w_1$ enthält keine in $w_2$ frei vorkommende 'Variable' oder 'Marke'),
$_{\cup x} w_1 w_2$ ($w_2$ ist $w_1$-Tupel aus verschiedenen Variablen),
$_P w_1 w_2$ ($w_2$ ist Prädikatschema über Grundzahlen mit $w_1$ Stellen),
$_\angle w_1 w_2$ ($w_1$ impliziert $w_2$),
$_{\leftrightarrows} w_1 w_2$ ($w_1$ ist eine Abkürzung für $w_2$, jedoch kann $w_1$ das längere oder das gegebene Wort sein);
$_+ w_1 w_2 w_3$ ($w_3$ verschmilzt $w_1$ mit $w_2$);
$_{\|} w_1 w_2 w_3 w_4$ ($w_1$ ist in $w_2$ frei zu konfusionsloser Substitution von $w_3$ (für $w_1$), *und* eine Substitution für $w_1$ in $w_2$ durch $w_3$ [von $w_3$ für $w_1$ in $w_2$] ergibt $w_4$;
das Ergebnis ist eindeutig, und zwar entsteht, falls $w_1$ und $w_3$ mit einem $^{\cup}$-Exemplar anfangen, das nicht zu einer 'Variable' oder 'Marke' gehört, dasjenige Ergebnis, welches — wenn $w_1$ für $^{\cup}w_{11} w_{12}$ und $w_3$ für $^{\cup}w_{31} w_{32}$ steht — die nacheinander ausgeführten Substitutionen von $w_{31}$ für $w_{11}$ und dann von $w_{32}$ für $w_{12}$ liefern).

Daß bei Deutung von $_{\not\downarrow}$, $_{\leftrightarrows}$, $_{\|}$ keine Kursivbuchstaben auftreten, weist auf den syntaktischen Gebrauch derartiger Begriffe hin, für den die Wörter einer Formel autonym zu verstehen sind: z.B. können weder Marken noch auch Variablen einfach als Namen für Substitute dienen, erst die Ergebnisse des formalen Substitutionsvorgangs werden manchmal gedeutet (: 9.3.23, 5. Prämisse, Konklusion; 9.3.13, 15, Konklusionen;

9.1.5, 27. Prämisse; 9.3.24, 25, Konklusionen; 9.3.26, Konklusion; 9.3.27, 29, Konklusionen).

**1.51** Daraufhin kann ein mit Formalismen vertrauter Mathematiker bis auf 9.1.5 und 9.3.24, 25 alles lesen, und für diesen Rest genügt es vorerst, einen näheren, wenn auch noch nicht den dialogischen Sinn mit dem dort mehrmals auftauchenden Wort $^{\cup\cup}\mathrm{u}_1{}^{\cup}\mathrm{v}_1\mathrm{w}_1{\rightarrow}{}^{\cup}\mathrm{w}_2\mathrm{v}_2\mathrm{u}$ zu verbinden: es bezeichnet das ($v$-stellige) Prädikatschema, das *rekursiv definiert* wird unter der Voraussetzung $u_1$ bei der Bedingung $v_1$ hinsichtlich der Fundierung $w_1$ *durch* das Aussageschema $u$ für die Prädikatmarke $w_2$ nach dem Variablen-$v$-tupel $v_2$, und zwar konstruktiv deduzierbar rekursiv definiert gemäß der 18. und der 27. Prämisse in 9.1.5. (Bei absoluten, unbedingten, expliziten Definitionen tritt beziehentlich $^{\cup}\vee\circ$ anstelle $\mathrm{u}_1$, $\vee$ anstelle $\mathrm{v}_1$, $\wedge$ anstelle $\mathrm{w}_1$.)

**1.52** Partikularisierte Marken gemäß 9.2.8, 9.3.29, 30 werden in **2.** entbehrlich sein; Existenzbehauptungen über Prädikate verlangen also dort einen direkten unnegierbaren Aufweis. Ohne die zitierten Regeln müßte man existentiale Formulierungen in **3.** gelegentlich unnötig abschwächen (durch negative Übersetzung).

Generalisierte Marken gemäß 9.2.7, 9.3.27, 28 wären in **2.** entbehrlich, solange es um eine deduktive Erreichung konstanter Ordinalzahlen sowie der Anordnungseigenschaften ihrer Repräsentanten geht.

Zur Substitution von passend stelligen Prädikatschemata für Prädikatmarken ist man ohne 9.3.27 auf das Kopieren von Deduktionen angewiesen; dabei mag die Konfusionslosigkeit der Substitute wieder zum Substitutionsvorgang geschlagen werden — deswegen ist die Originaldeduktion eventuell über kongruente, durch geeignete Umbenennung gebundener Variablen und Marken entstehende Aussageschemata zu leiten, nicht nur um die Substitution konfusionslos durchzuhalten, sondern auch um die Schlüssigkeit nicht durch Variablen und Marken zu stören, die im Substitut anonym frei vorkommen.

Zur Darstellung variabler Ordinalzahlen werden in **2.** generalisierte Marken innerhalb der hypothetischen Voraussetzung von Rekursoren genügen. Das Ziel, eine schwache Form transfiniter Induktion über alle (über Grundzahlen) darstellbaren Ordinalzahlen zu deduzieren, führt auch in Implikataussagen zu generalisierten Marken.

**1.53** Ohne quantifizierte Marken, ohne relative und ohne bedingte Definitionen, sogar allein mit Definitionen durch gewöhnliche (Vorgänger-)Induktion neben expliziten Definitionen lassen sich bereits hohe Ordinalzahlen erreichen: in diesem Teil von $\mathfrak{S}_1$ können zum Beispiel die beiden transfiniten Induktionen, die SCHÜTTE in bezug auf zwei simultan rekursiv definierte binäre Prädikate konstruktiv-inhaltlich nachweist [17] (vgl. S. 115—118, 108), formal nachvollzogen werden; dabei hat man anzuwenden, daß Prädikatmarken sich durch etliche Deduktionskopien variieren lassen (wie in **1.52** erwähnt), und daß Variablentupel variabler Gliederzahl mittels Gödels $\beta$-Funktion fortzuschaffen sind.

Bei solcher Beschränkung auf Rekursionen hinsichtlich der natürlichen Fundierung (Wohlordnung) der Grundzahlen muß man allerdings auf die Definition eines Niveauvergleichs (: [21], p. 268) zwischen den Repräsentanten deduktiv erreichter konstanter Ordinalzahlen verzichten.

**1.54** Bedingte Definitionen sind für die Ordinalzahltheorie in **2.** unvermeidlich, sobald Einengungen eines fundierenden Prädikats R als Ordinalzahlrepräsentanten PR (gemäß **0.21**) fungieren und untereinander sowie mit uneingeengten Repräsentanten $\vee$ R verglichen werden sollen. $\vee\wedge$ repräsentiert übrigens die Ordinalzahl 1, und erst $\wedge$R repräsentieren die Ordinalzahl 0.

Relative Definitionen stellen sich bei der Ordinalzahltheorie in **2.** ein, sobald unbeschränkte Ordinalzahltherme (gemäß **0.23**) untereinander verglichen werden sollen.

**1.6 Rekapitulation. Inoffizielle Notation.** Die Anwendung von 0. macht die Formeln $_k{}'\cdot\cdot{}'0$ herstellbar. Die formalen Grundzahlen $'\cdot\cdot{}'0$ seien inoffiziell durch k, l, .. mitgeteilt. Der Inhalt, der $k$ vor k auszeichnet, betrifft lediglich die Abstraktion von der Art der sukzessive aufeinanderfolgenden gleichförmigen Objekte (sowie von deren graphischer Ausführung), keinen ordinalen oder kardinalen Sinngehalt.

1. und 2. leiten $_{\neq}$kl bzw. $_{2}$kl ab, für die $k \neq l$ bzw. $k + k = l$. Die formalen Variablen $^{\cup}\xi$l und Marken $^{\cup\cup}$k$\Pi$l, die an Hand **3.** bzw. **4.** als (erste) Wörter hinter $_x$ bzw. als zweite Wörter hinter $_\pi$ entstehen, seien inoffiziell mit $\xi_l$ bzw. $^k\Pi_l$ notiert und durch x, y, .. bzw. $\pi$, $\rho$, .. mitgeteilt.

Die formalen Grundzahlterme k und $'\cdot\cdot{}'^{\cup}\xi$l (: konstante bzw. variable) gemäß 5. seien durch t, s, .. mitgeteilt. (Die Bildung der t entspricht der Herstellung von Wortermen, nur daß jetzt die Variablenliste nicht beschränkt ist.) Die formalen Term-$k$-tupel $^{\cup\cdot\cdot\cup}\circ\, \mathrm{t}_1 \,..\, \mathrm{t}_k$, die sich gemäß 6. als zweite Wörter hinter $_{\cup}{*}_k$ ergeben, mögen kurz durch $\overset{k}{\mathrm{t}}$ mitgeteilt werden. 7. liefert genau die Formeln

$$_+{}^{\cup\cdot\cdot\cup}\circ\, \mathrm{t}_1 \,..\, \mathrm{t}_k{}^{\cup\cdot\cdot\cup}\circ\, \mathrm{s}_1 \,..\, \mathrm{s}_l{}^{\cup\cdot\cdot\cup\cup\cdot\cdot\cup}\circ\, \mathrm{t}_1 \,..\, \mathrm{t}_k\, \mathrm{s}_1 \,..\, \mathrm{s}_l,$$

deren dritte Wörter sich inoffiziell durch $\overset{k}{\mathrm{t}}\,\mathrm{s}$ mitteilen lassen: fügt man $\overset{k}{\mathrm{t}}$ und $\overset{l}{\mathrm{s}}$ zu $^{\cup}\overset{k}{\mathrm{t}}\overset{l}{\mathrm{s}}$ nebeneinander, so entsteht weder ein gemäß 6. gebildetes Tupel aus $k + l$ Termen noch ein Wort, das in Formeln, die später gemäß 9.1.—3. herstellbar werden, als Teilwort auftreten könnte.

**1.61** Um das Geflecht der übrigen 82 Regeln ad 8.—10. zu entwirren, denke man sich vorerst die 30 Implikationsregeln 9.3. gestrichen und in 9.1.5 bloß die 7 ersten Prämissen, welche die erste Zeile anfüllen, beibehalten. Die so beschnittenen Regeln seien mit $\tilde{9}$.1.1—5, $\tilde{9}$.2.1—8 beziffert, die abgewandelten Relatoren ebenfalls mit einer Tilde versehen: $_{\tilde p}$w$_1$w$_2$ ($w_2$ ist Prädikatskelett über Grundzahlen mit $w_1$ Stellen), $_{\tilde A}$w ($w$ ist Aussageskelett über Grundzahlen).

$\tilde{9}$.1.1—4 liefern als zweite Wörter hinter $_{\tilde p}$ folgende Prädikatskelette: das 2-stellige $\leq$ und die $k$-stelligen $\vee$, $\wedge$, $^{\cup\cup}$k$\Pi$l. Teilen $\tilde{\mathrm P}$, $\tilde{\mathrm Q}$, $\tilde{\mathrm R}$, .. Prädikatskelette mit, deren figürliche Gestaltung sich später kraft $\tilde{9}$.1.5 wesentlich anreichern wird, so sind $^{\cup}\tilde{\mathrm{P}}\overset{k}{\mathrm{t}}$ die gemäß $\tilde{9}$.2.1 hergestellten Aussageskelette, falls $k$ die — oder (wie im Fall $\bigvee$, $\bigwedge$) eine — Stellenzahl von $\tilde{\mathrm{P}}$ ist. Teilen $\tilde{\mathrm{A}}$, $\tilde{\mathrm{B}}$, $\tilde{\mathrm{C}}$, .. Aussageskelette mit, so sind gemäß $\tilde{9}$.2.2—8 auch $^{\rightarrow}\tilde{\mathrm{A}}\tilde{\mathrm{B}}$, $^{\wedge}\tilde{\mathrm{A}}\tilde{\mathrm{B}}$, $^{\vee}\tilde{\mathrm{A}}\tilde{\mathrm{B}}$ und $^{\vee}\mathrm{x}\tilde{\mathrm{A}}$, $\bigwedge\mathrm{x}\tilde{\mathrm{A}}$, $\bigwedge\pi\tilde{\mathrm{A}}$, $\bigvee\pi\tilde{\mathrm{A}}$ solche. Inoffizielle Notation für die Konklusionen von $\tilde{9}$.2.1—8: $(\tilde{\mathrm{P}}\overset{k}{\mathrm{t}})$, $(\tilde{\mathrm{A}} \rightarrow \tilde{\mathrm{B}})$, $(\tilde{\mathrm{A}} \wedge \tilde{\mathrm{B}})$, $(\tilde{\mathrm{A}} \vee \tilde{\mathrm{B}})$, $\bigvee\mathrm{x}\tilde{\mathrm{A}}$, $\bigwedge\mathrm{x}\tilde{\mathrm{A}}$, $\bigwedge\pi\tilde{\mathrm{A}}$, $\bigvee\pi\tilde{\mathrm{A}}$. Oder, mit Punkten statt Klammern: $\tilde{\mathrm{P}}\overset{k}{\mathrm{t}}$, $\tilde{\mathrm{A}} \overset{(l)}{\rightarrow} \tilde{\mathrm{B}}$, $\tilde{\mathrm{A}} \overset{(l)}{\wedge} \tilde{\mathrm{B}}$, $\tilde{\mathrm{A}} \overset{(l)}{\vee} \tilde{\mathrm{B}}$, $\bigvee_{\mathrm{x}} \tilde{\mathrm{A}}.$, $\bigwedge_{\mathrm{x}} \tilde{\mathrm{A}}.$, $\bigwedge_{\pi} \tilde{\mathrm{A}}.$, $\bigvee_{\pi} \tilde{\mathrm{A}}.$; $l+1$-fach punktierte Junktoren ($\rightarrow$, $\wedge$, $\vee$) mögen schärfer trennen als $l$-fach punktierte und, zwecks Punkteinsparung, $l$-fach punktierte Subjunktoren schärfer als $l$-fach punktierte Konjunktoren und Adjunktoren; die Junktorenpunktation im Wirkungsstück eines Operators ($\bigvee_{*}$, $\bigwedge_{*}$, später $\rightarrowtail_{*}$) soll die Junktorenpunktation außerhalb solcher Formelteile nicht beeinflussen; am Formelende lassen sich die Endpunkte der Wirkungsstücke von Operatoren tilgen. $\tilde{\mathrm{P}}\overset{k}{\mathrm{t}}$, $\tilde{\mathrm{A}} \overset{(l)}{\rightarrow} \tilde{\mathrm{B}}$, $\tilde{\mathrm{A}} \overset{(l)}{\wedge} \tilde{\mathrm{B}}$, $\tilde{\mathrm{A}} \overset{(l)}{\vee} \tilde{\mathrm{B}}$ werden auch mit $\bigvee\mathrm{x}(\tilde{\mathrm{A}})$, $\bigwedge\mathrm{x}(\tilde{\mathrm{A}})$, $\bigwedge\pi(\tilde{\mathrm{A}})$, $\bigvee\pi(\tilde{\mathrm{A}})$ kombiniert — man unterscheide dazu $\bigvee_{\mathrm{x}}$ von $\bigvee\mathrm{x}$, usw. Von selbst versteht sich $\mathrm{t}_1 \leq \mathrm{t}_2$ anstelle $\leq \overset{2}{\mathrm{t}}$.

Inoffiziell tritt $\Rightarrow$ anstelle des imperativen $\rightarrow$, entsprechend Doppelkommata anstelle partitiver Kommata.

Zur Handhabung von $\tilde{9}$.1.5 bedarf es vorher 8.3., und für 8.3. genügt die Kenntnis von 8.1. in dem 8.3.1,2 vorangestellten bescheidenen Umfang 8.1.0—2. Bis auf weiteres darf $_{\not\downarrow}\mathrm{vu}$ sogar als "v enthält keine in u enthaltene 'Variable'" gedeutet werden. (Beachte, daß 8.1.0 nicht über das eliminierbare Ausmaß — vgl. **1.4** — hinaus Anwendung findet.) Unter den laut 8.1.0—2 herstellbaren $_{\not\downarrow}\mathrm{xw}$ befinden sich neben solchen mit w, die Tupel $\overset{k}{\mathrm{t}}$ aus (von x verschiedenen) $\mathrm{x}_1 \mathinner{..} \mathrm{x}_k$ bilden, auch zweckfreie Formeln wie $_{\not\downarrow}\ \mathrm{x}^{\cup}\mathrm{o}^{\cup\cup}\mathrm{o}^{\cup}\mathrm{ooo}$ oder $_{\not\downarrow}{}^{\cup}\xi 0\ {}^{\cup\cup}\xi''0\ \mathrm{o}$. 8.3. stellt sodann genau solche Variablen-$k$-tupel $^{\cup..\cup}\ \mathrm{o}\ \mathrm{x}_1 \mathinner{..} \mathrm{x}_k$ als zweite Wörter hinter $_{\cup}x$ her, in deren $k$ Variablen keine gleichen Unterscheidungsindizes auftreten. Die so ausgezeichneten $\overset{k}{\mathrm{t}}$ seien durch $\overset{k}{\mathrm{x}}$ mitgeteilt.

Nunmehr bildet $\tilde{9}$.1.5 aus $\overset{k}{\mathrm{x}}$, $k$-stelligen $\pi$, $\tilde{\mathrm{P}}$ und $k+k$-stelligem $\tilde{\mathrm{R}}$ sowie $\tilde{\mathrm{A}}$, $\tilde{\mathrm{C}}$ ein neues Prädikatskelett $\tilde{\mathrm{Q}}$ mit $k$ Stellen, nämlich $^{\cup\cup}\tilde{\mathrm{C}}^{\cup}\tilde{\mathrm{P}}\tilde{\mathrm{R}}^{\rightarrowtail}\ ^{\cup}\pi\overset{k}{\mathrm{x}}\tilde{\mathrm{A}}$. Inoffizielle Notation: mit Klammern $((\tilde{\mathrm{C}}(\tilde{\mathrm{P}}\tilde{\mathrm{R}})) \rightarrowtail (\pi\overset{k}{\mathrm{x}})\tilde{\mathrm{A}})$ bzw. $\tilde{\mathrm{C}}\tilde{\mathrm{P}}\tilde{\mathrm{R}} \rightarrowtail \pi\overset{k}{\mathrm{x}}\tilde{\mathrm{A}}$, welch letztere Notation dank der Klammern ad $\tilde{9}$.2. den Aufbau nach $\tilde{9}$.1.—2. noch eindeutig ermitteln läßt; oder mit Punkten $\tilde{\mathrm{C}} \underset{\tilde{\mathrm{P}}}{\overset{\tilde{\mathrm{R}}}{\rightarrowtail}}{}_{\pi\overset{k}{\mathrm{x}}} \tilde{\mathrm{A}}.$, auch Kombination mit $\tilde{\mathrm{C}} \underset{\tilde{\mathrm{P}}}{\overset{\tilde{\mathrm{R}}}{\rightarrowtail}} \pi\overset{k}{\mathrm{x}}\ (\tilde{\mathrm{A}})$, wobei der Anfang von $\tilde{\mathrm{C}}$ durch erhöhte Punktezahl über dem eventuell vorangehenden Junktor kenntlich gemacht wird, soweit der Anfang nicht schon aus der den Operatoren zugehörigen Punktverteilung ersichtlich ist. (Die den Operatoren zugehörige Punktanordnung bleibt übrigens bei Substitutionen erhalten, unter Einlagerung der Substitutpunktation.)

Aussageskelette, die sich gemäß $\tilde{9}$.2.1 ergeben, heißen *Prim*aussageskelette; $\tilde{\mathrm{A}}$, die nicht prim sind, bauen sich *logisch* an Hand $\tilde{9}$.2.2—8 aus primen auf. Prädikatskelette gemäß $\tilde{9}$.1.1—4 heißen *Prim*prädikatskelette; $\tilde{\mathrm{P}}$, die nicht prim sind, bauen sich *definitorisch* an Hand $\tilde{9}$.1.5 auf, unter Heranziehung zuvor erhaltener Aussage- und Prädikatskelette. $\tilde{\mathrm{A}}$, die prim sind, können hinwieder mit Hilfe von $\tilde{\mathrm{P}}$, die nicht prim sind, definitorisch aufgebaut sein.

**1.62** Um 8.1.0—2 dahin zu erweitern, daß ⅄xw außer ⅄x$\overset{k}{\mathrm{x}}$ auch ⅄x$\overset{k}{\mathrm{t}}$ erfaßt, genügt 8.1.3 und ⅄vu → ⅄v′u, was sich, analog 8.1.0, für x statt v via 8.2.2, 8.2.3, 8.1.8 (mittels 8.1.1) eliminieren läßt. Die 8.1. auf 8.2. stützende Rückverbindung 8.1.8 kann übrigens, unter Inkaufnahme einer Schar derartig eliminierbarer Regeln, ganz entfernt werden. $_{\|}$vu$v_1u_1$ darf bis auf weiteres sogar als "eine Auswechslung jedes in u enthaltenen Exemplars des Substituenden v gegen ein Exemplar des Substituts $v_1$ ergibt $u_1$" gedeutet werden. 8.2.1—4 allein umgreifen, neben allerlei überschüssigen Produktionsmöglichkeiten, die Auswechslung von Variablen gegen Terme in Variablen, Termen und Termtupeln.

Die Niederschrift der Regeln setzt hiernach die gestaffelte Ausweitung von ⅄ und $_{\|}$ fort, indem sie 8.1.2, 3 durch 8.1.4—7 auf alle von ξ verschiedenen 6 (= 7 — 1) Konstanten erweitert und 8.2.4 durch 8.2.5—7 von $^{\cup}$ auf die 3 Junktoren überträgt. Die Wörter w, für die ⅄xw (via 8.1.8) herleitbar wird, umfassen jetzt insbesondere die aus $\tilde{9}$.1.1—4 gemäß $\tilde{9}$.2.1—4 herzustellenden „kleinen" $\tilde{\mathrm{A}}$, in denen x nicht als Teilwort enthalten ist; in diesem Kreis ist die Tilde überhaupt gegenstandslos und jedes Skelett ein Schema laut 9.1.1—4, 9.2.1—4. Zu den inzwischen herleitbaren Formeln $_{\|}$vu$v_1u_1$ zählen speziell $_{\|}$xut$u_1$ für kleine Aussageschemata u, die zu x, u, t eindeutig ein $u_1$ ergeben, das wieder ein kleines Aussageschema bildet.

Mit 8.1.9, 10 und 8.2.9, 10 verlassen wir die bisherige engere Auslegung von ⅄ und $_{\|}$, obwohl sie sich anderwärts (bei 8.1.12—23, 8.2.8) noch eine Weile durchhalten ließe. In der vollen ⅄-Deutung könnte man allerdings den Passus ".. oder 'Marke'" vorläufig streichen. Nach Einführung von 8.1.9, 10 erfassen die w, für die ⅄xw (via 8.1.8 über 8.2.2, 3—7 *und* 9, 10 mittels 8.1.1) herleitbar wird, neben vielerlei Ausschuß jedes *primitive*, aus 9.1.1—4 gemäß 9.2.1—6 aufgebaute Aussageschema, in dem x nicht frei vorkommt, in dem x also höchstens gebunden vorkommt, d. h. als erstes Wort hinter einem Quantor $^{\vee}$, $^{\wedge}$ oder als im zweiten Wort dahinter — im Wirkungsstück — enthaltene Teilwörter; in diesem Wirkungsstück können solche x-Exemplare auch bereits gebunden sein, nämlich *durch* den jeweils innersten ein solches x bindenden Quantor. Die Hinzunahme von 8.2.9, 10 bewirkt, daß insbesondere die Formeln $_{\|}$xut$u_1$ für primitive Aussageschemata u erreicht werden, wobei sich zu x, u, t eindeutig ein $u_1$ ergibt, das wieder ein primitives Aussageschema bildet, nämlich dasjenige, welches durch Auswechslung jedes in u freien (d. h. in u nicht gebundenen) x-Exemplars gegen ein t-Exemplar entsteht, vorausgesetzt jedoch, daß diese Substitution konfusionslos ist, daß also kein in u freies x-Exemplar im Wirkungsstück eines Quantors liegt, der eine im Substitut t (frei) enthaltene Variable bindet.

Wer die beiden Konjuganden der $_{\|}$-Deutung getrennt definiert haben möchte, denke sich für den ersten Konjugand in 8.2.1—7, 9, 10 überall die vierten Wörter hinter $_{\|}$ fort und streiche die 2. Prämissen in 8.2.9, 10; für den zweiten Konjugand sind bisher bloß die 1. Prämissen in 8.2.9, 10 zu streichen.

Substitutionen für v in $^{\vee}$vu, $^{\wedge}$vu erledigt 8.2.2 via 8.1.9, 10; Konfusionslosigkeit in $^{\vee}$vu, $^{\wedge}$vu bei ⅄wu läuft über 8.2.2, sobald ⅄wv, ⅄wu → ⅄w$^{\vee}$vu, ⅄w$^{\wedge}$vu (analog 8.1.0) für x statt w via 8.2.2, 9 bzw. 10, 8.1.8 mittels 8.1.1 eliminiert worden ist. — Will man ohne die Rückverbindung 8.1.8 arbeiten, so sind die letztgenannten Regeln nicht zu entbehren; natürlich auch nicht die 8.1.0 analogen Regeln für die 3 Junktoren.

Die restlichen 16 Regeln ad 8.1.—2. greifen ziemlich ineinander, erneut verzahnt mit 8.2.1—7, 9, 10 und 8.1.8, wie auch 8.1.1—7, 9, 10 mit jenen.

8.1.12, 13 besorgen zunächst, daß ⅄xu für primitive Aussageschemata u auf ⅄$\overset{k}{\mathrm{x}}$u ausgedehnt wird — für u, in denen die in $\overset{k}{\mathrm{x}}$ enthaltenen Variablen nicht frei vorkommen. 8.1.13 kann ersetzt werden durch $\rightarrow {\nmid}\circ o$, ${\nmid}\circ 0$, ${\nmid}\circ\xi$, ${\nmid}\circ\Pi$, ${\nmid}\circ\leq$, ${\nmid}\circ\vee$, ${\nmid}\circ\wedge$ (: Elimination von 8.1.13 via 8.2.2, 3—7, 9—11, 8.1.8 mittels ${\nmid}\circ o$ gelingt zumindest, *falls* u ein $\overset{k}{t}$ oder $\tilde{P}$ oder $\tilde{A}$ ist, nämlich dank 8.1.2, 18, 12); 8.1.13 wäre auch mittels $\rightarrow {\nmid}\circ o$ eliminierbar, wenn man $\rightarrow {}_{\|}vuvu$ hinzunimmt. (8.1.12 ist nicht via 8.2.2, 8 und 8.1.8 eliminierbar, da man 8.1.8 erst mittels 8.1.12 anwenden kann.)

Beim Gebrauch von ${\nmid}$, ${}_{\|}$ in 9.1.—3. wird zwar für $\circ$ in u stets $\circ$ substituiert (: siehe 9.1.5, 24. und 25. Prämisse, sowie 9.3.24, 25, 4. Prämisse), aber 8.2.2 liefert mit 8.1.13 ohnehin ${}_{\|}\circ u v_1 u$, also u als Ergebnis.

8.1.16, 17 und 8.1.18—23 leisten für $\pi$, was 8.1.1 und 8.1.2—7 für x leisten. Bei ${\nmid}\pi\rho$ sind die beiden Stellen- oder die beiden Unterscheidungsindizes ungleich. Für die ${\nmid}$-Deutung würde zwar teils noch ".. oder in $w_2$ enthaltene 'Marke'" genügen — die volle ${\nmid}$-Deutung wird aber mit 8.1.10, 9 und 8.2.10, 9 für $\pi$ anstelle v akut. Das für x Erörterte überträgt sich auf $\pi$, nachdem auch 8.2.12 in Kraft ist, wodurch 8.2.1 von x auf $\pi$ übertragen wird. (Freilich verfügen wir hinsichtlich ${}_{\|}\pi u v_1 u_1$ bei primitiven Aussageschemata u, $u_1$ vorerst bloß über Primprädikatschemata aus 9.1.1—4 als Substitute $v_1$.)

8.2.8 führt simultane Substitutionen ein, die im allgemeinen von der Reihenfolge der Einzelsubstitutionen abhängen. Die Konfusionslosigkeit ließe sich daher nicht abtrennen, ohne auf die Zwischenergebnisse Bezug zu nehmen.

8.1.11, 8.2.11 entsprechen 8.1.9, 10, 8.2.9, 10. Das für die Quantoren $\vee$, $\wedge$ Gesagte überträgt sich auf den neuen Operator $\rightarrow$. $\rightarrow$ bindet in Skeletten $\tilde{P}$, $\tilde{A}$ Aussageschemata der Form $^{\cup}\pi\overset{k}{x}$ (gemäß 9.1.4, 9.2.1); darum besorgen 8.1.14, 15, die Inversionen zu 8.1.12, noch eine Verteilung der Bindung. 8.2.4 (und 8.1.0 im eliminierten Umfang) bewirken schließlich, daß ${\nmid}xu$, ${\nmid}\overset{k}{x}u$, ${\nmid}\pi u$, ${\nmid}^{\cup}\pi\overset{k}{x}u$ für jedes $\tilde{P}$, $\tilde{A}$ anstelle u herleitbar wird, in dem die im ersten Wort hinter ${\nmid}$ enthaltenen Variablen oder Marken nicht frei vorkommen. Man beachte, daß in $^{\cup\cup}\tilde{C}^{\cup}\tilde{P}\tilde{R}$ $\rightarrow^{\cup}\pi\overset{k}{x}\tilde{A}$ als Wirkungsstück des Rekursors allein $\tilde{A}$ fungiert, während $\tilde{C}$, $\tilde{P}$, $\tilde{R}$ von der $^{\cup}\pi\overset{k}{x}$-Bindung unberührt bleiben; gleichwohl formt $\rightarrow$ erst nach Anflicken von $\tilde{C}$, $\tilde{P}$, $\tilde{R}$ ein Prädikatskelett. Ebenso werden ${}_{\|}xutu_1$, ${}_{\|}\overset{k}{x}u\overset{k}{t}u_1$, ${}_{\|}\pi u\tilde{Q}u_1$, ${}_{\|}{}^{\cup}\pi\overset{k}{x}u^{\cup}\tilde{Q}\overset{k}{t}u_1$, ${}_{\|}{}^{\cup}\overset{k}{x}\pi u^{\cup}\overset{k}{t}\tilde{Q}u_1$ für jedes $\tilde{A}$ anstelle u herleitbar, wobei sich stets eindeutig ein $\tilde{A}_1$ als $u_1$ ergibt; z. B. im dritten Fall durch Auswechslung jedes in $\tilde{A}$ freien $\pi$-Exemplars gegen ein $\tilde{Q}$-Exemplar, vorausgesetzt, daß dabei kein in $\tilde{A}$ freies $\pi$-Exemplar im Wirkungsstück eines Operators $\vee$, $\wedge$, $\rightarrow$ liegt, der eine im Substitut $\tilde{Q}$ frei vorkommende Variable oder Marke bindet.

10.1, 2 liefern die Formeln ${}_{\leftrightarrows}\wedge^{\cup..\cup}\circ x_1 .. x_l u \wedge x_l .. \wedge x_1 u$. Die Abkürzung, das erste Wort hinter ${}_{\leftrightarrows}$, enthält 2 Primfiguren mehr als das abzukürzende zweite Wort; der Vorteil liegt in dem $\cdot\cdot$-freien Wort $\wedge\overset{k}{x}\tilde{A}$ anstelle $\wedge x_k .. \wedge x_1 \tilde{A}$.

Wer lose Wortvariablen störend findet, kann die Anfangsformen durch Anfänge für die Konstanten und Regeln zum Aufbau mit den Funktoren ersetzen. Man kann auch die Überproduktion eindämmen, z. B. ${}_{x}v$, ${}_{*k}v_1 \rightarrow {}_{\|}vvv_1v_1$ und ${}_{\pi}wv$, ${}_{P}wv_1 \rightarrow {}_{\|}vvv_1v_1$ statt 8.2.1, 12 fordern; dabei wird freilich 8. mit 9. verflochten, so daß die Bezifferung umzuändern wäre. Die vorliegende Variante sucht dem Verlangen nach „Rasanz" der Formulierung entgegenzukommen — die letzte Verbalfassung (1961) füllte 15 Seiten.

**1.63** Wir gehen jetzt von den Skeletten $\tilde{P}$, $\tilde{A}$ zu Prädikat- und Aussageschemata über und teilen Wörter u, für die ${}_{P}wu$ bzw. ${}_{A}u$, durch P, Q, R, .. bzw. A, B, C, .. mit. Fortan sind also alle 27 Prämissen von 9.1.5 in Kraft. Schemata ohne frei darin vorkommende Marken heißen Prädikat- bzw. Aussage*formen*. Formen ohne frei darin vorkommende Variablen heißen Prädikate bzw. Aussagen.

Die logischen Funktoren $^\rightarrow$, $^\wedge$, $^\vee$, $\bigvee$, $\bigwedge$ lese man nacheinander: dann (wenn .. so ..), und, oder (oder auch), es gibt (für mindestens ein, für einige), für alle.

9.3.1—10 bieten die Regeln der affirmativen Junktorenlogik, wobei es sich zunächst um kleine A handelt (ehe 9.2.5—8 und 9.1.5 mitspielen). 9.3.11, 12 fügen ex falso quodlibet und ex quodlibet verum hinzu. Für $\angle AB$ wird inoffiziell $A \angle B$ notiert.

Mit kleinen A lassen sich noch 9.3.17—23 vorwegnehmen, die eine Definition der Primprädikate $\vee$, $\wedge$, $\leq$ und eine Fassung der mathematischen (vollständigen) Induktion betreffen. $\vee$, $\wedge$, mit beliebiger, aber jeweils fester Stellenzahl, ermöglichen, P und R in $^{\cup\cup}C^{\cup}PR^{\rightarrow}{**}$ anfangs mit rekursorfreien Prädikaten zu besetzen. Die Aussagen $^{\cup}\vee\circ$, $^{\cup}\wedge\circ$ dürfen inoffiziell mit $\vee$ bzw. $\wedge$ notiert werden; wo es auf Unterscheidung von Prädikaten ankommt, sei $(\vee)$ bzw. $(\wedge)$ notiert.

Die Variablenspezialisierung in 9.3.19—22 (: $\vee \angle 0 \leq \xi_0$, $\xi_0 \leq \xi_1 \angle {}'\xi_0 \leq {}'\xi_1$, ${}'\xi_0 \leq {}'\xi_1 \angle \xi_0 \leq \xi_1$, ${}'\xi_0 \leq 0 \angle \wedge$) ist mit 9.3.16, 15 aufzuheben. Hierbei sind freilich schon nicht kleine, vorerst primitive, sogar ohne 9.2.7, 8 aufgebaute A im Spiel, und 9.3.13—16 ergänzen die effektive Junktorenlogik 9.3.1—12 für solche A zur effektiven Quantorenlogik (über Grundzahlen).

$\leq$ statt $=$ einzuführen, ist Geschmackssache: bei der unter **2.** innerhalb $\mathfrak{S}_1$ entwickelten Ordinalzahltheorie ist $\leq$ zwischen Ordinalzahlrepräsentanten definitorisch vorrangig.

Zur dialogischen Deutung von 9.3.1—23 (sowie nachher 9.3.26) sei auf [14] §§ 2, 4, 5 verwiesen, wegen 9.3.1 ist auch [14] S. 83/84 zu bedenken.

**1.631** 9.1.5 bewirkt den Übergang zu nicht primen P, wonach 9.3.1—23 auch für nicht primitive A zur Entfaltung kommen, ergänzt durch 9.3.24, 25. Auf diese zentralen Regeln gehen wir in **1.632** ausführlich ein.

9.3.26 drückt das Gleichheitsprinzip für Prädikatmarken aus. Für Prädikatformen anstelle der Marken ist das Gleichheitsprinzip eliminierbar. 9.3.26 erzwingt, daß bei informalen Deduktionskopien als Substitute für Marken nur arithmetische Prädikate in Frage kommen, die mit der Grundzahlgleichheit verträglich sind.

9.2.7, 8 und 9.3.27, 28, 29, 30 (analog 9.2.6, 5 bzw. 9.3.15, 16, 13, 14) sind auf die Prädikatschemata von $\mathfrak{S}_1$ selbst zugeschnitten: 2. Prämisse in 9.3.27, 29. Wieso man dabei formalistisch mit gebundenen Prädikatmarken statt Prädikatvariablen hantieren kann, wird in **1.64** erörtert.

Schon jetzt sei darauf hingewiesen, daß 9.3.26—30 ebenso wie 9.3.1—25 auf dem Wege über die Rückkopplung 9.1.5 die Spirale Prädikat-Aussage-Implikation-Prädikat immer weitertreiben.

**1.632** Die 20 Prämissen in Zeile 2 bis 9 von 9.1.5 werden von $\angle$ verursacht, und zwar gehören die 8.—15. Prämisse in den Zeilen 2 u. 3 zu der (mit Abkürzungen laut 16. und 17. Prämisse ausgedrückten) 18. Prämisse in Zeile 5; die 19.—25. Prämisse in den Zeilen 6 u. 7 gehören, zusammen mit der 10.—15. Prämisse in den Zeilen 2 u. 3, zu der (mit Abkürzung laut 26. Prämisse ausgedrückten) 27. Prämisse in Zeile 9.

Die 18. Prämisse, inoffiziell notiert

$$(1) \qquad C \angle \bigwedge \overset{k}{y}(P\overset{k}{y} \dot{\rightarrow} \bigwedge \overset{k}{z}(P\overset{k}{z} \dot{\rightarrow} R\overset{k}{z}\overset{k}{y} \rightarrow \rho\overset{k}{z}) \rightarrow \rho\overset{k}{y}) \dot{\rightarrow} P\overset{k}{y} \rightarrow \rho\overset{k}{y},$$

ist an Hand der 13. bzw. 9. Prämisse, wonach $\overset{k}{y}$ und $\rho$ in C nicht frei vorkommen, für $\overset{k}{y}$ wie für $\rho$ generalisierbar; die 9. und 14. Prämisse sorgen noch dafür, daß

$\rho, \overset{k}{y}, \overset{k}{z}$ nicht mit anonymen Parametermarken bzw. -variablen in P, R kollidieren können.

(1) bedeutet, daß die Definitionsvoraussetzung $C$ die $P$-bedingte (Vorgänger-) Induktion bezüglich $R$ impliziert, für $\varrho$ nach $\overset{k}{y}$.

Die 27. Prämisse, inoffiziell notiert

$$\text{(2)}\qquad C \angle P\overset{k}{y} \overset{\cdot\cdot}{\rightarrow} \wedge \overset{k}{z}(P\overset{k}{z} \overset{\cdot\cdot}{\rightarrow} R\overset{k}{z}\overset{k}{y} \overset{\cdot}{\rightarrow} \pi_1\overset{k}{z} \leftrightarrow \pi_2\overset{k}{z}) \overset{\cdot}{\rightarrow} A(\overset{k}{y}, \pi_1) \leftrightarrow A(\overset{k}{y}, \pi_2),$$

ist an Hand der 13. bzw. 22. Prämisse, wonach $\overset{k}{y}$ und $\pi_1, \pi_2$ in C nicht frei vorkommen, für $\overset{k}{y}$ wie für $\pi_1, \pi_2$ generalisierbar; die 22. und 23. Prämisse sorgen dafür, daß $\pi_1, \pi_2$ auch nicht mit anonymen Parametern in P, R und $\pi_1, \pi_2, \overset{k}{y}$ nicht mit solchen in $A(\overset{k}{x}, \pi)$ kollidieren können. (Der 13. und 14. Prämisse zufolge kommen $\overset{k}{y}$ in C und $\overset{k}{y}, \overset{k}{z}$ in P, R bereits nicht frei vor.)

(2) bedeutet: die Definitionshypothese $C$ impliziert die $P$-bedingte „Unabhängigkeit" des Definiensschemas $A(\overset{k}{x}, \pi)$ hinsichtlich des Vorgängerprädikats $R$ (für $\pi$ nach $\overset{k}{x}$), d.h. daß $A$ im Argument $\overset{k}{y}$ nur in solchen Argumenten $\overset{k}{z}$ von $\pi$ bis auf Bisubjunktion abhängt, die $\overset{k}{y}$ bezüglich $R$ bei der Bedingung $P$ vorangehen.

Die $\tilde{A}_1$, die sich aus ${}_{\|}xutu_1$, ${}_{\|}\overset{k}{x}u\overset{k}{t}u_1$, ${}_{\|}\pi uQu_1$, ${}_{\|}{}^{\cup}\pi\overset{k}{x}u{}^{\cup}Q\overset{k}{t}u_1$, ${}_{\|}{}^{\cup}\overset{k}{x}\pi u{}^{\cup}\overset{k}{t}Qu_1$ für jedes A anstelle u eindeutig als $u_1$ ergeben (weil jedes A ein $\tilde{A}$ und jedes Q ein $\tilde{Q}$ ist), sind jetzt Aussageschemata $A_1$, da gegebenenfalls immer 9.1.5 statt $\tilde{9}$.1.5 bemüht werden muß.

Bei der 24. und 25. Prämisse in 9.1.5 ist die Substitutionsreihenfolge irrelevant, da Substituend $\overset{k}{x}$ und Substitut $\overset{k}{y}$ je aus verschiedenen Variablen bestehen — ${}_{\nmid}v_2v_3$ ist aber wohlgemerkt nicht verlangt — und da in den Substituten $\pi_1, \pi_2$ für $\pi$ keine Variablen anonym frei vorkommen. Bei der 4. Prämisse in 9.3.24, 25 hingegen kommt es darauf an, zuerst $v_3$ für $v_2$ in u zu substituieren und hernach ${}^{\cup\cup}u_1{}^{\cup}v_1w_1{}^{\rightarrow\cup}w_2v_2u$ für $w_2$, während, andersherum, Variablen aus $v_2$, die eventuell in $u_1, v_1, w_1$ frei vorkommen, durch solche aus $v_3$ ersetzt würden.

9.3.24, 25 bewirken Umformungen, die, inoffiziell notiert,

$$\text{(3)}\qquad C \angle P\overset{k}{y} \overset{\cdot}{\rightarrow} A(\overset{k}{y}, CPR \rightarrow \pi\overset{k}{x}A) \leftrightarrow (CPR \rightarrow \pi\overset{k}{x}A\overset{k}{y})$$

postulieren; dabei sorgt die 3. Prämisse dafür, daß die Variablen aus $\overset{k}{y}$ in A, C, P, R nicht frei vorkommen, also insbesondere (hinter C) generalisierbar sind und auch nicht mit anonymen Parametern in P, R, A kollidieren — unbeschadet nachträglicher substitutioneller (oder definitorischer) Gleichsetzungen. Eine Umbenennung von $\overset{k}{y}$ in $\overset{k}{x}$ via 9.3.15 hängt davon ab, ob die Variablen aus $\overset{k}{x}$ nicht in C, P, R frei vorkommen, wo sie der Rekursor-Bindung von $\pi\overset{k}{x}$ entgehen. Sollte $\pi$ in C, P, R frei vorkommen, so kann die Substitution von $CPR \rightarrow \pi\overset{k}{x}A$ für $\pi$ in $A(\overset{k}{y}, \pi)$ zu keinerlei Kollisionen führen, weil jedes in $A(\overset{k}{y}, \pi)$ freie $\pi$-Exemplar durch $CPR \rightarrow \pi\overset{k}{x}A$ ersetzt wird.

(3) bedeutet: die Voraussetzung $C$ impliziert, daß in einem Argument $\overset{k}{y}$, das der Bedingung $P$ unterliegt, der Definiend $CPR \rightarrow \pi\overset{k}{x}A$ durch $A(\overset{k}{y}, CPR \rightarrow \pi\overset{k}{x}A)$

definiert ist. Nach der Bedeutung von (2) wäre dieses Definiens $A(\overset{k}{y}, CPR \rightarrow \pi \overset{k}{x} A)$ selbst unter $C$ definiert (bis auf Bisubjunktion), *falls* unter $C$ der Definiend schon für alle $\overset{k}{z}$, die $P\overset{k}{z}$ und $R\overset{kk}{zy}$ unterliegen, mit Hilfe des Definiensschemas $A(\overset{k}{x}, \pi)$ definiert ist. Daraus folgt nach der Bedeutung von (1), daß unter $C$ der Definiend $CPR \rightarrow \pi\overset{k}{x}A$ im Argument $\overset{k}{y}$ tatsächlich definiert ist.

Hinter der (relativen bedingten) Vorgängerinduktion bezüglich $R$ für "$CPR \rightarrow \pi\overset{k}{x}A$ ist in $\overset{k}{y}$ definiert" nach $\overset{k}{y}$ verbirgt sich kein imprädikatives Vorgehen. Zunächst hat man sich für eine voraussichtlich passende Definitionsfundierung $R$ zu entscheiden, um mit R die Ableitung von (1) und (2) zu versuchen. Ist dies geglückt, dann antizipiert die „unbestimmte" Marke ρ in der Ableitung von (1) nicht allein (in $\mathfrak{S}_1$ später als A, C, P, R zugängliche) „bestimmte" Prädikate oder Prädikatformen, die in Ableitungskopien zu (1) als Substitute für ρ fungieren, sondern — auf der Ebene konkreter, (konstruktiv-)inhaltlicher Beweiskopien aus der Ableitung von (1) — auch Induktionsaussageformen $\mathfrak{A}(\overset{k}{y})$ anstelle des abstrakt-formalen Induktionsaussageschemas $\rho\overset{k}{y}$, die den jeweils formal aufgezählten Aussagebezirk übersteigen. Für die Deutung von (3) innerhalb $\mathfrak{S}_1$ wird jedoch kein uferloser Bereich solcher $\mathfrak{A}$ beansprucht, sondern ein wohlabgestreckter Bezirk, der sich wiederum aufzählen läßt.

In (3) gehen einem konstanten Argument transfiniter „Höhe" — z. B. 1 bezüglich $0R2R4 \ldots R1R3 \ldots$ — nicht etwa Ausdrücke variabler und im allgemeinen transfiniter Länge voran, jedenfalls keine solchen, die nicht auf Aussageformen fester endlicher Länge hinauslaufen und somit der effektiven Quantorenlogik über Grundzahlen anheimfallen. Schlußformen wie die unendliche Induktion oder Definitionen wie die für $\in O$ und $<_O$, welche operativ betrachtet uneigentlich sind (: [13] S. 172), da in deren Prämisse eine generelle Implikation eingeht, führen zu Erzeugnissen, über die zwar im besonderen Einzelfall Einigkeit herrschen mag, über die aber hier und jetzt keine wohlbegrenzte allgemeine Reflexion einwandfrei denkbar ist. Weder intuitionistisch noch dialogisch-konstruktiv kann man heute allen künftig gültig werdenden Prämissen, die etwas "für jedes $n$" ansetzen, zuschauen; jene Ordinalzahlen, die zur schrittweisen Exhaustion des Definiens von $O$ nötig wären, um eine inklusions-isotone transfinite Folge von Grundzahlmengen $O_\alpha$ zu indizieren — vereinigt $O_\alpha$, was der Definiens-Term für $O$ in Anwendung auf vorgängige $O_\xi$, $<_{O_\xi}$ erzeugt, so bricht die Folge $O_\alpha$ nach der naiven oder axiomatischen Mengenlehre für ein $\lambda$ aus der zweiten Zahlklasse ab, indem außer $O_\lambda \subseteq O_{\lambda+1}$ auch $O_{\lambda+1} \subseteq O_\lambda$ zutrifft —, diese Ordinalzahlen wollen vielmehr unabhängig konstruiert sein, also mit einer eigentlichen Definition. (Die erste nicht-konstruktive Ordinalzahl ist folglich höchstens dann wohlbestimmt, wenn sie relativ zu einem formalen System verstanden wird.)

**1.633** Ein $P$-bedingt definiertes Prädikat hat im Fall $P\overset{k}{y} \angle \wedge$ überhaupt keine bestimmte Definition und fällt für solche $\overset{k}{y}$ nicht, wie eingeengte Prädikate, durchweg falsch aus.

Prädikatschemata Q, in denen Marken frei vorkommen, sind dank (1), (2) mit Gewißheit konstruktiv sinnvoll; denn die Substitution von Prädikatformen für alle in Q freien Marken liefert wieder Prädikatformen $Q_1$, und eine Substitution konstruktiv-inhaltlich sinnvoller Prädikatformen in $Q$ hat wieder ebenso sinnvolle Bedeutung. Prädikatskelette $\tilde{Q}$ hingegen sind meist ohne Sinn, zumindest konstruktiv beurteilt.

**1.634** Die dialogische Deutung von (3) überträgt sich aus der in **0.133** mit der einfachen Klausel, daß der Angreifer zuvor C und dann auch noch $\mathrm{P}\overset{k}{\mathrm{y}}$ übernommen hat. Es sei hier darauf verzichtet, diese Gewinnstrategien selber von ihrem Endstück aus wieder mit Hilfe eines Kalküls herzustellen; mit einer solchen „adäquaten" Umformulierung von $\mathfrak{S}_1$ gedenke ich mich anderwärts zu befassen.

**1.64** Bei engem Kontakt mit der Deutung würden anstelle $\wedge$-, $\vee$- bzw. $\rightarrow$-gebundener Marken ${}^k\Pi_l$ zwei Sorten von Prädikatvariablen angebracht sein — sagen wir „logische" ${}^k\Xi_l$ und „definitorische" ${}^k\mathrm{H}_l$, mitgeteilt durch X bzw. Y. Formalistisch dürfen wir aus folgenden Gründen die drei (unbeschränkten) Listen für $\pi$, X, Y zusammenschieben: aus dem Aufbau eines P oder A geht eindeutig hervor, welche $\pi$-Exemplare in P [in A] frei und damit „echte" Marken sind und welche $\pi$-Exemplare in P [in A] *durch* welchen Generalisator, welchen Partikularisator bzw. welchen Rekursor gebunden sind; überdies zeigen 9.1.5 und 9.3.24, 25, daß die Einführung von $*{*}{*}\rightarrow \mathrm{Y}{*}{*}$ und der Umgang damit ohne Verknüpfung mit $\wedge \mathrm{X}*$, $\vee \mathrm{X}*$ formulierbar ist.

**1.65** Dieser gesamten Deutung von $\mathfrak{S}_1$ in einer konstruktiven Theorie der Grundzahlen ist zu entnehmen, daß $\angle^{\cup}\vee\circ^{\cup}\wedge\circ$ unableitbar ist. $\mathfrak{S}_1$ ist ja so eingerichtet, daß jede abgeleitete Implikation — nach Substitution von Grundzahlen für freie Variablen und von systeminternen oder externen Prädikaten für freie Marken — dialogisch zu gewinnen ist, wenn die Partner metadialogische Überlegungen an Hand $\mathfrak{S}_1$ akzeptieren, seien letztere nun trivial (wie bei 9.3.1) oder nicht (wie bei 9.3.24, 25 im Verein mit 9.1.5).

Vom operativen Standpunkt aus ist $\mathfrak{S}_1$ in dem Sinne widerspruchsfrei, daß jede mittels $\mathfrak{S}_1$ abgeleitete Gleichheit zwischen konstanten Grundzahlen bereits direkt mit $\Rightarrow 0 = 0$ und $k = l \Rightarrow {}'k = {}'l$ herzustellen ist; sämtliche Definitions- und Beweismanipulationen von $\mathfrak{S}_1$ sind relativ eliminierbar — diese Einsicht ließe sich gegenüber [13] durch formalistische Fixierung jedes benutzten Metakalküls präzisieren.

**1.7 Rückblick.** Den Ansatz (1) „ (2) $\Rightarrow$ (3) meiner Version von $\mathfrak{S}_1$ gab ich, für den absoluten unbedingten Fall, 1960 in Stanford bekannt [22, 23], vorgeformt in [21], Etappe I und II; die erhaltenen Definiensschemata nannte ich 1960 „logisch" rekursiv. LORENZEN [13] will eine Fundierung aus einem (in endlich viele Bisubjunktionen separierten) Definitionsschema „ablesen" und spezialisiert die Fundierung $R$ infolgedessen auf Prädikate, die explizit durch ein (partikularisiertes) Adjugat von Gleichungskonjugaten definierbar sind; daß für abgelesene $R$, falls die diesbezügliche Vorgängerinduktion logisch ableitbar ist, (2) gilt, wird dort weder formuliert noch bewiesen.

Um jede Unsicherheit, die in teils verbalen und teils formalen Äußerungen verborgen sein könnte (: s. Fußnote 5 zu **0.131**), bei meinem Ansatz für $\mathfrak{S}_1$ auszuschließen, erschien es mir ratsam, den Einbau von (1) „ (2) $\Rightarrow$ (3) in ein formales System nicht der Vorstellung des Lesers zu überlassen; ich war vielmehr bemüht, ein (höchst wahrscheinlich) lückenloses System kalkulatorischer Vorschriften aus einem Guß an die Spitze zu stellen.

**1.8 Vorschau.** Ich glaube, daß die formale Widerspruchsfreiheit von $\mathfrak{S}_1$ sich noch auf einen (relativ vollständigen) Kalkül finit-direkter Kalkulation $\mathfrak{K}_1$ umrechnen läßt, dessen Formeln allein bei Anwendung einer Regel der „direkten" Induktion (mit logikfreiem Induktionsschritt) formal abgebaut werden. Um selbst ohne Gleichheit mit ihren Rückschlüssen wie ${}'0 = {}'0 \angle 0 = 0$ zu arbeiten, studiere man an (dyadischen) Gödelnummern für geregelte Ableitungen mit laufender Gebrauchsanweisung die Kalkulationskerne von Verfahren zur Elimination von Zusatzregeln, deren Reflexionstyp im allgemeinen dadurch kenntlich gemacht sei, daß man eine Nummer für eine Ableitung nach festgesetzten Regeln mit einer Zusatzchiffrierung versiegelt.

**1.9 Schlußbemerkung zur Definition.** Einem Gespräch mit A. HEYTING (1957) entnehme ich, daß intuitionistisch gewisse Bedenken gegen „transfinit“ viele Quantoren bestehen, wie sie durch gewisse rekursiv definierte Prädikate — etwa durch den Niveauvergleich zwischen Fundierungen (: [21], Etappe I; siehe hier **2.3**) — eingeschmuggelt zu werden scheinen. Ich hoffe allerdings, daß, nachdem **2.** ab **2.3** klarstellt, welcher in **1.3** finit „legalisierte“ Umgang mit transfiniter Rekursion mir damals als Werkzeug diente, der Intuitionismus weitherzig genug ist, solch eine sinngerechte Übertragung vom Zeichen $=$ bei Funktionen auf $\leftrightarrow$ bei Prädikaten zu gestatten, und bereit, die darin schlummernden Möglichkeiten auszubeuten — zumal die unterschiedlich weite Kluft zwischen "es gibt" und "für alle" im definiten Bezirk der Grundzahlen einerseits und im indefiniten Bereich der Prädikate über Grundzahlen andererseits hier nicht mit ein und derselben formalen Asymmetrie behandelt wird, sondern bei Prädikaten durch Rekursifikation bzw. durch freie Marken möglichst subtil in (Wörter-)Rechnung gestellt ist.

## 2. Partielle Stabilisierung. Konstruktive Ordinalzahltheorie

Dem Titel der Arbeit folgend wenden wir uns jetzt parenthetisch den Produktionen des definierten Regelsystems $\mathfrak{S}_1$ zu. Der Kürze halber wird vorausgesetzt, die Deduktionstechnik in $\mathfrak{S}_1$ sei durch metamathematische Theoreme, die einer Rekapitulation von Verfahren und Methoden aus HILBERT-BERNAYS (1934/39) [10] sowie KLEENE (1952) [11] entstammen, so weit gediehen, daß die Geschmeidigkeit verbaler Beweise durch Verweis auf die Mitteilung vorfabrizierter Deduktionsbausteine und auf eliminierbare Regeln auch formal erreichbar ist, jedoch mit den Vorzügen, die die Prägnanz einer formalen Ausdrucksweise vor der sprachlichen, nur teilweise formalen Darbietung logischer und definitorischer Gedankenabläufe auszeichnet — namentlich wenn diese beanspruchen, konstruktiv zu sein.

**2.1** Zur Übertragung von $\vdash$ in $\mathfrak{S}_1$ ist $\vdash \mathrm{A}$ als $\mathfrak{S}_1$-Ableitbarkeit von $\vee \angle \mathrm{A}$ zu verstehen.[12] Bei relativer Ableitbarkeit $\mathrm{A}_1, .. \vdash \mathrm{A}_0$ sind $\Rightarrow \vee \angle \mathrm{A}_1, ..$ zu $\mathfrak{S}_1$ hinzuzufügen. $\mathrm{A} \angle \mathrm{B}$ ist ableitbar, genau wenn $\vee \angle \mathrm{A} \rightarrow \mathrm{B}$ ableitbar ist; daher könnte man statt mit $_\angle$ auch mit einem einstelligen Relator $_S\mathrm{w}$ ($w$ ist Satzschema) arbeiten.

**2.11 Theorem 1.** *Deduzierbar rekursiv definierte Prädikatschemata mit demselben Definiens stimmen, bei derselben Bedingung und unter derselben Hypothese, (relativ bedingt) bisubjunktiv überein:*

$$\vdash \mathrm{C} \ddot{\rightarrow} \mathrm{P}\overset{k}{\mathrm{y}} \dot{\rightarrow} (\mathrm{C\,P\,R} \rightarrow \pi \overset{k}{\mathrm{x}} \mathrm{A} \overset{k}{\mathrm{y}}) \leftrightarrow (\mathrm{C\,P\,R}_1 \rightarrow \pi \overset{k}{\mathrm{x}} \mathrm{A} \overset{k}{\mathrm{y}});$$

die Variablengebote seien aus 9.3.24, 25 übernommen (hinsichtlich beider Fundierungen $\mathrm{R}, \mathrm{R}_1$).

---

[12] Man beachte den Wechsel im Gebrauch des Symbols $\vdash$ : das offizielle $\vdash_{\mathfrak{S}_1} \angle {}^{\cup}\vee \circ \mathrm{u}$ (aus **0.12**) geht über in das inoffizielle $\vdash_{\mathfrak{S}_1} \mathrm{u}$ (aus **0.21**).

**Beweis.** Man greife hinsichtlich R und C, P, A auf 9.1.5 zurück, um das $\dot{\rightarrow}$-Sukzedens in Theorem 1 aufgrund (1) in **1.632** formal-induktiv abzuleiten. Dazu kopiere man die Deduktion von (1) für das $\dot{\rightarrow}$-Sukzedens als Induktionsaussageform anstelle $\rho\overset{k}{y}$, „unterwegs" eventuell auf kongruenten Umleitungen — am Ende erscheint aber die Induktionsaussageform selbst, da sie in (1) konfusionslos substituierbar ist. Dann führe man den Induktions*schritt* $C \angle P\overset{k}{y} \dot{\rightarrow} \bigwedge_z^k * \cdot \rightarrow *$; nämlich indem sich aus der Induktions*annahme* $P\overset{k}{y} \wedge \bigwedge_z^k *$ dank (2) in **1.632**, wieder notfalls auf kongruenten Umleitungen, $A(\overset{k}{y}, CPR \rightharpoonup \pi\overset{k}{x}A) \leftrightarrow A(\overset{k}{y}, CPR_1 \rightharpoonup \pi\overset{k}{x}A)$ ergibt, also nach (3) in **1.632**, da $P\overset{k}{y}$, auch die Induktions*behauptung* $(CPR \rightharpoonup \pi\overset{k}{x}A\overset{k}{y}) \leftrightarrow (CPR_1 \rightharpoonup \pi\overset{k}{x}A\overset{k}{y})$. Der Induktionsschritt beweist nun die Induktions*voraussetzung* $C \angle \bigwedge_y^k *$, diese mit (1) die Induktions*folgerung* $C \angle P\overset{k}{y} \rightarrow *$, w. z. z. w.

**2.12** Derlei einfache Induktionen durch Induktionsschritt unterdrücken wir fortan als trivial. Sie tauchen naturgemäß bereits auf, wenn z.B. die Umbenennung für gebundene Marken rekapituliert wird.

Wegen (2) kann in (3) für $CPR \rightharpoonup \pi\overset{k}{x}A$ in $A(\overset{k}{y}, CPR \rightharpoonup \pi\overset{k}{x}A)$ auch der „Abschnitt" $CP^{R_y^k}R \rightharpoonup \pi\overset{k}{x}A$ mit $P^{R_y^k} \leftrightharpoons (\bigvee) \bigvee \bigwedge \rightharpoonup \pi\overset{k'}{z}(P\overset{k}{z} \wedge R\overset{k}{z}\overset{k}{y})$ eingesetzt werden. ($\leftrightharpoons$ wird über 10.1, 2 hinaus als Abbreviationssymbol benutzt.) Solche Abschnitte sind mit ebensolchen Abschnitten definierbar; aber Abschnitte von Abschnitten sind nicht ohne weiteres wieder Abschnitte, denn R braucht ja nicht transitiv zu sein. Der Ansatz (3) richtet sich im Interesse formaler Einfachheit sofort auf $CPR \rightharpoonup \pi\overset{k}{x}A$; ein auf die zugehörige Abschnittsform ausgehender Ansatz würde auf Implikationsformen (1), (2) fußen, die zwar auch beweisbar sind — aber (relative bedingte) explizite Definitionen, ohne die $CPR \rightharpoonup \pi\overset{k}{x}A$ selbst nicht aus der zugehörigen Abschnittsform hervorgeht, blieben dann gesondert anzusetzen.

**2.2 Definition 1.** $\neg A$ (: lies 'nicht A') stehe für $(A \rightarrow (\bigwedge))$. Die bekannte stabilisierende Übersetzung, die A auf °A abbildet, indem $°(P\overset{k}{t})$, $°(A \rightarrow B)$, $°(A \wedge B)$, $°(A \vee B)$, $°\bigvee x\, A$, $°\bigwedge x\, A$ nacheinander für $(P\overset{k}{t})$, $(°A \rightarrow °B)$, $(°A \wedge °B)$, $\neg(\neg °A \wedge \neg °B)$ $[\leftrightarrow \neg\neg(°A \vee °B)]$, $\neg\bigwedge x \neg °A$ $[\leftrightarrow \neg\neg\bigvee x\, °A]$, $\bigwedge x\, °A$ stehen, sei abgeändert und ergänzt durch die Vorschrift, daß $°(P\overset{k}{t})$ für $(°P\overset{k}{t})$ steht und $°\leq$, $°\bigvee$, $°\bigwedge$, $(°\pi\overset{k}{t})$, $°(CPR \rightharpoonup \pi\overset{k}{x}A)$[13], $°\bigwedge\pi A$, $°\bigvee\pi A$ nacheinander für $\leq$, $\bigvee$, $\bigwedge$, $\neg\neg(\pi\overset{k}{t})$, $(CPR \rightharpoonup \pi\overset{k}{x}\, °A)$, $\bigwedge\pi\, °A$, $\neg\bigwedge\pi\neg °A$ $[\leftrightarrow \neg\neg\bigvee\pi\, °A]$.

[13] Ohne Klammer säße ° auf C.

**Definition 2.** $^{\bullet\circ}A$ entstehe aus $^{\circ}A$, indem jedes $(^{\circ}\pi\overset{k}{t})$ nur dann mit $\neg\neg(\pi\overset{k}{t})$ übersetzt wird, wenn 1. A für ein $(\pi\overset{k}{t})$ steht, oder wenn 2. das betreffende $\pi$-Exemplar in A nicht oder „mittelbar" $\rightarrowtail$-gebunden ist *und* in A unmittelbar mit $\wedge$ oder $\bigwedge x$ oder $\bigwedge\pi$ zusammengesetzt oder unmittelbar $\rightarrow$-Sukzedens ist oder sich (mit seinem $\overset{k}{t}$) allein im Wirkungsstück eines (nicht $\pi$ bindenden) Rekursors befindet — sonst jedoch mit $(\pi\overset{k}{t})$ übersetzt wird. Ein $\pi$-Exemplar heiße dabei in A mittelbar $\rightarrowtail$-gebunden, wenn es im Definiens desjenigen Rekursors, durch welchen es gebunden wird, noch im Wirkungsstück eines anderen Rekursors oder mehrerer solcher liegt.

**Definition 3.** $^{\rho\bullet\circ}A$ entstehe aus $^{\bullet\circ}A$, indem $\neg\neg$ unmittelbar vor jedem nicht gebundenen Exemplar der speziellen Marke $\rho$ getilgt wird — soweit diese $\neg\neg$ von $^{\bullet\circ}$ verursacht sind —, *wenn* nicht das betreffende $\rho$-Exemplar im Wirkungsstück (Definiens) wenigstens eines Rekursors liegt.

In Definition 1 kann '$^{\circ}$' durch '$^{\bullet\circ}$' ersetzt werden, *außer* daß $^{\bullet\circ}(CPR\rightarrowtail\pi\overset{k}{x}A)$ für $(CPR\rightarrowtail\pi\overset{k}{x}\,^{\pi\bullet\circ}A)$ steht; daß $^{\bullet\circ}(A\rightarrow B)$ für $(A\rightarrow{}^{\bullet\circ}B)$ steht, falls A für ein $(\pi\overset{k}{t})$ steht; daß $^{\bullet\circ}(A\vee B)$ für $\neg\neg(A\vee{}^{\bullet\circ}B)$, $\neg\neg(^{\bullet\circ}A\vee B)$, $\neg\neg(A\vee B)$ steht, falls beziehentlich A für $(\pi\overset{k}{t})$, B für $(\rho\overset{l}{s})$, A für $(\pi\overset{k}{t})$ und B für $(\rho\overset{l}{s})$ steht; und daß $^{\bullet\circ}\bigvee xA$ für $\neg\neg\bigvee xA$ sowie $^{\bullet\circ}\bigvee\pi A$ für $\neg\neg\bigvee\pi A$ steht, falls A für $(\rho\overset{k}{t})$ steht.

**Definition 4.** A heiße hinsichtlich C, P, R *selbständig* deduzierbar rekursiv definierend, wenn (1) in **1.632** und

$$(^{\bullet\circ}2)\qquad C\angle P\overset{k}{y}\overset{\cdot\cdot}{\rightarrow}\bigwedge\overset{k}{z}(P\overset{k}{z}\overset{\cdot\cdot}{\rightarrow}R\overset{k}{z}\overset{k}{y}\overset{\cdot}{\rightarrow}\pi_1\overset{k}{z}\leftrightarrow\pi_2\overset{k}{z})\overset{\cdot}{\rightarrow}$$
$$\overset{\cdot}{\rightarrow}{}^{\pi_1\bullet\circ}A(\overset{k}{y},\pi_1)\leftrightarrow{}^{\pi_2\bullet\circ}A(\overset{k}{y},\pi_2)$$

in $\mathfrak{S}_1$ ableitbar sind; die Variablen- und Markengebote seien aus 9.1.5 übernommen.

Indem C, P, R in $((CPR\rightarrowtail\pi\overset{k}{x}A)\overset{k}{t})$ von $^{\circ}$ wie von $^{\bullet\circ}$ nicht affiziert werden, kann (1) für beide Übersetzungen von $C\overset{R}{\underset{P}{\rightarrowtail}}{}_{\pi\overset{k}{x}}A$ unverändert übernommen werden. Man darf schon im absoluten unbedingten Fall nicht erwarten, etwa von $I(\bigvee, R)$ [14] auf $I(\bigvee, {}^{\circ}R)$ schließen zu können, sofern nicht R in $\mathfrak{S}_1$ entscheidbar ist; und $^{\bullet\circ}I(\bigvee, R)$ schwächt im allgemeinen nicht nur R sondern auch die Induktionsfolgerung so ab $(:\neg\neg\rho\overset{k}{y})$, daß sie im konstruktiven Sinn von 9.1.5 als Definitionsprämisse unbrauchbar ist.

[14] Zu I vgl. **0.21** und **0.24**.

Daß ein durch Übersetzung geschwächtes Induktionsschema als Konklusion noch zur konstruktiven Auslegung einer Induktionsaxiomform (mit aufgezählten Induktionsaussageformen) aus einem Formalismus klassischer transfiniter Arithmetik verwendbar sein mag, steht auf einem anderen Blatt.

Die Definitionsprämisse (2) wird für die ${}^{\bullet\circ}$-Übersetzung von $\mathrm{C} \overset{\mathrm{R}}{\underset{\mathrm{P}}{\rangle\!\!\rangle}}{}^{k}_{\pi\mathrm{x}}\mathrm{A}$, nämlich $\mathrm{C} \overset{\mathrm{R}}{\underset{\mathrm{P}}{\rangle\!\!\rangle}}{}^{k}_{\pi\mathrm{x}}{}^{\pi\bullet\circ}\mathrm{A}$, in Gestalt der „Selbständigkeits"eigenschaft (${}^{\bullet\circ}$2) schlechthin postuliert. Eine Deduktion von (${}^{\bullet\circ}$2) nach Vorlage einer solchen von (2) mißlingt nur, wenn die Rekursion „unselbständig" fundiert ist, indem R oder P, wofern unentscheidbar in $\mathfrak{S}_1$, oder von P, R abhängige Prädikate in das Definiens A eingehen.

Kopiert man eine Deduktion von (${}^{\bullet\circ}$2) unter Substitution von $\neg\neg(\pi_1 *)$, $\neg\neg(\pi_2 *)$ für $(\pi_1 *)$, $(\pi_2 *)$, so entsteht eine Deduktion von

$$\mathrm{C} \angle \mathrm{P}\overset{k}{\mathrm{y}} \overset{..}{\rightarrow} \wedge\overset{k}{\mathrm{z}}(\mathrm{P}\overset{k}{\mathrm{z}} \overset{..}{\rightarrow} \mathrm{R}\overset{k}{\mathrm{z}}\overset{k}{\mathrm{y}} \dot{\rightarrow} \neg\neg\pi_1\overset{k}{\mathrm{z}} \leftrightarrow \neg\neg\pi_2\overset{k}{\mathrm{z}}) \dot{\rightarrow}$$
$$\dot{\rightarrow} {}^{\bullet\circ}\mathrm{A}(\overset{k}{\mathrm{y}}, \pi_1) \leftrightarrow {}^{\bullet\circ}\mathrm{A}(\overset{k}{\mathrm{y}}, \pi_2),$$

nachdem etwa entstandene $\neg\neg\neg\neg(\pi_{1,2} *)$ im Wirkungsstück von Rekursoren wieder durch $\neg\neg(\pi_{1,2} *)$ ersetzt worden sind — weil ja $\vdash \neg\neg\neg\mathrm{D} \leftrightarrow \neg\mathrm{D}$ infolge $\vdash \mathrm{D} \rightarrow \neg\neg\mathrm{D}$ und 'Kontraposition' $\vdash (\mathrm{D}_1 \rightarrow \mathrm{D}_2) \rightarrow (\neg\mathrm{D}_2 \rightarrow \neg\mathrm{D}_1)$. Das jetzige $\dot{\rightarrow}$-Antezedens vor ${}^{\bullet\circ}$A folgt noch aus dem homologen $\dot{\rightarrow}$-Antezedens in (${}^{\bullet\circ}$2), und daher gilt, *wenn* (${}^{\bullet\circ}$2) abgeleitet ist, auch die Variante

(${}^{\bullet\circ}$2′)
$$\mathrm{C} \angle \mathrm{P}\overset{k}{\mathrm{y}} \overset{..}{\rightarrow} \wedge\overset{k}{\mathrm{z}}(\mathrm{P}\overset{k}{\mathrm{z}} \overset{..}{\rightarrow} \mathrm{R}\overset{k}{\mathrm{z}}\overset{k}{\mathrm{y}} \dot{\rightarrow} \pi_1\overset{k}{\mathrm{z}} \leftrightarrow \pi_2\overset{k}{\mathrm{z}}) \dot{\rightarrow}$$
$$\dot{\rightarrow} {}^{\bullet\circ}\mathrm{A}(\overset{k}{\mathrm{y}}, \pi_1) \leftrightarrow {}^{\bullet\circ}\mathrm{A}(\overset{k}{\mathrm{y}}, \pi_2).$$

**2.21 Theorem 2.** *(Partielle Stabilisierung.)* 1. $\vdash {}^{\bullet\circ}\mathrm{A} \leftrightarrow \mathrm{A}$ *gilt für Aussageformen* A *jedenfalls dann, wenn* A *selbst und die Definientes der am definitorischen Aufbau von* A *beteiligten deduzierbar rekursiv definierten Prädikatschemata, die überdies sämtlich als -formen vorausgesetzt seien, einen logischen Aufbau ohne* $\vee$, $\vee\mathrm{x}$, $\wedge\pi$, $\vee\pi$ *zeigen.* 2. *Für jedes Aussageschema* A, *das sich definitorisch durchgängig — im ganzen Stammbaum — aus* $\leq$, $\vee$, $\wedge$, $\pi$ *und absolut unbedingt selbständig deduzierbar rekursiv definierten Prädikatschemata aufbaut, ist* °A *stabil, d.h.* $\vdash \neg\neg{}^{\circ}\mathrm{A} \rightarrow {}^{\circ}\mathrm{A}$. 3. *Für die* ad 2. *genannten* A *gilt* $\vdash {}^{\bullet\circ}\mathrm{A} \leftrightarrow {}^{\circ}\mathrm{A}$. 4. *Schränkt man den Deduktionsprozeß* 9.3.1—30 *auf die* ad 2. *genannten* A *ein, fügt* $\vee \angle \mathrm{A} \vee \neg\mathrm{A}$ *hinzu und bezeichnet den so modifizierten, sich bei ausdrücklich gesperrter Rückkopplung via* 9.1.5 *entfaltenden Ableitbarkeitsbegriff mit* $\vdash'$, *so gilt* ${}^{\circ}\mathrm{A}_1, .., {}^{\circ}\mathrm{A}_n \vdash {}^{\circ}\mathrm{A}$, *wenn* $\mathrm{A}_1, .., \mathrm{A}_n \vdash' \mathrm{A}$ *gilt; insbesondere für* $n = 0$. (2.—4. relativ bedingt zu formulieren, würde weitere vorbereitende Definitionen erheischen.)

**Beweise.** Ad 1. ${}^{\bullet\circ}$A steht in den genannten Fällen für A: zunächst bei Primaussageformen aus den Primprädikaten $\leq$, $\vee$, $\wedge$; dann beim

Aufbau mit $\rightarrow$, $\wedge$, $\bigwedge x$, falls schon bei den Teilaussageformen; endlich bei nicht primitiven Primaussageformen $((\mathrm{CPR} \rightharpoonup \pi \overset{k}{x} \mathrm{B}) \overset{k}{t})$, falls schon bei den am logischen Aufbau von B eventuell beteiligten nicht primitiven Primaussageformen, denn ad 1. steht ${}^{\pi\bullet\circ}\mathrm{B}$ für B. (Die ad 1. am definitorischen Aufbau von A beteiligten Prädikatformen sind automatisch selbständig deduzierbar rekursiv definiert.)

Ad 2., 3. simultan. Jede Primaussageform aus $\leq$, $\bigvee$, $\bigwedge$ ist stabil; bei solchen A steht °A für A. $\neg\neg(\pi \overset{k}{t})$ ist stabil wegen $\vdash \neg\neg\neg \mathrm{D} \rightarrow \neg \mathrm{D}$; °- wie •°-Übersetzung von $(\pi \overset{k}{t})$ heißen $\neg\neg(\pi \overset{k}{t})$. $^\circ(\mathrm{A} \rightarrow \mathrm{B})$ ist stabil, falls schon °B stabil ist, denn $\vdash \neg\neg(\mathrm{D}_1 \rightarrow \mathrm{D}_2) \leftrightarrow (\mathrm{D}_1 \rightarrow \neg\neg \mathrm{D}_2)$; wenn 3. schon für A, B gilt, dann auch für (A $\rightarrow$ B), wobei, falls A für ein $(\pi \overset{k}{t})$ steht, zwar 3. für B genügt, aber die Stabilität von °B sowie $\vdash (\neg\neg \mathrm{D}_1 \rightarrow \neg\neg \mathrm{D}_2) \leftrightarrow (\mathrm{D}_1 \rightarrow \neg\neg \mathrm{D}_2)$ heranzuziehen ist. Die Stabilität vererbt sich von °A, °B auf $^\circ(\mathrm{A} \wedge \mathrm{B})$, denn $\vdash \neg\neg (\mathrm{D}_1 \wedge \mathrm{D}_2) \leftrightarrow$ $\leftrightarrow \neg\neg \mathrm{D}_1 \wedge \neg\neg \mathrm{D}_2$; 3. vererbt sich von A, B auf (A $\wedge$ B). Analog bei $\bigwedge x$, $\bigwedge \pi$ anstelle $\wedge$, obwohl lediglich $\vdash \neg\neg \bigwedge x \, \mathrm{D} \rightarrow \bigwedge x \neg\neg \mathrm{D}$ und entsprechend für $\pi$ statt x. $^\circ(\mathrm{A} \vee \mathrm{B})$ ist wieder stabil, weil $\vdash \neg\neg\neg \mathrm{D} \rightarrow \neg \mathrm{D}$; 3. vererbt sich von A, B auf (A $\vee$ B), wobei, falls A für $(\pi \overset{k}{t})$ oder B für $(\rho \overset{l}{s})$ steht, oder beides, $\vdash \neg(\mathrm{D}_1 \vee \mathrm{D}_2) \leftrightarrow$ $\leftrightarrow \neg \mathrm{D}_1 \wedge \neg \mathrm{D}_2$ und die Stabilität von $\neg(\pi \overset{k}{t}), \neg(\rho \overset{l}{s})$ zu beachten ist. Ähnlich bei $\bigvee x$, unter Beachtung von $\vdash \neg \bigvee x \, \mathrm{D} \leftrightarrow \bigwedge x \neg \mathrm{D}$; desgleichen für $\pi$ statt x. (° und •° beeinträchtigen eine deduktive Variation von Variablen und Marken in 2., 3. zu Ersetzungszwecken nicht.)

Bei einem nicht primitiven Primaussageschema $(((\bigvee) \bigvee \mathrm{R} \rightharpoonup \pi \overset{k}{x} \mathrm{A}) \overset{k}{t})$, das einem selbständig deduzierbar rekursiv definierten Prädikatschema entspringt, sei 2., 3. schon bewiesen für die am logischen Aufbau von A eventuell beteiligten nicht primitiven (ad 2., 3. eo ipso absolut unbedingt selbständig deduzierbar rekursiv definierten) Primaussageschemata, ebenso für A selbst. Aus (•°2) für $(\bigvee)$, $\bigvee$ anstelle C, P folgt, dank 3. für A, über (•°2′) hinaus

$$(^\circ 2) \qquad \vdash \bigwedge \overset{k}{z} (\mathrm{R} \overset{k}{z} \overset{k}{y} \dot{\rightarrow} \pi_1 \overset{k}{z} \leftrightarrow \pi_2 \overset{k}{z}) \dot{\rightarrow} {}^\circ\mathrm{A}(\overset{k}{y}, \pi_1) \leftrightarrow {}^\circ\mathrm{A}(\overset{k}{y}, \pi_2),$$

also die Definitionsprämisse (2) für $(\bigvee) \bigvee \mathrm{R} \rightharpoonup \pi \overset{k}{x}\, {}^\circ\mathrm{A}$. Nunmehr liefert (1) für $(\bigvee)$, $\bigvee$ anstelle C, P, zusammen mit (°2), an Hand 9.3.24, 25 und Definition 1:

$$(^\circ 3) \qquad \vdash {}^\circ\mathrm{A}(\overset{k}{y}, {}^\circ((\bigvee) \bigvee \mathrm{R} \rightharpoonup \pi \overset{k}{x} \mathrm{A})) \leftrightarrow {}^\circ(((\bigvee) \bigvee \mathrm{R} \rightharpoonup \pi \overset{k}{x} \mathrm{A}) \overset{k}{y}),$$

wonach sich 2. für $(((\bigvee) \bigvee \mathrm{R} \rightharpoonup \pi \overset{k}{x} \mathrm{A}) \overset{k}{t})$ aus 2. für $\mathrm{A}(\overset{k}{x}, \pi)$ — so ursprünglich, vor Substitutionen — ergibt, indem man die Ableitung der °A-Stabilität durch passende Kopie, notfalls erst kongruent umleitend, derart umformt, daß $\overset{k}{t}$, $^{\circ}((\vee)\vee R \rightarrow \pi \overset{k}{x} A)$ für $\overset{k}{x}$, $\pi$ in $^{\circ}A(\overset{k}{x}, \pi)$ substituiert ist.

Bleibt 3. für $(((\vee)\vee R \rightarrow \pi \overset{k}{x} A)\,\overset{k}{t})$ zu zeigen, und zwar durch absolute unbedingte Induktion bezüglich R für die behauptete Bisubjunktion zwischen $^{\bullet\circ}$- und $^{\circ}$-Übersetzung. (1), ($^{\bullet\circ}$2) mit $(\vee)$, $\vee$ anstelle C, P liefern gemäß 9.3.24, 25 und Definitionen 2, 3:

$$(^{\bullet\circ}3) \qquad \vdash {}^{\pi\bullet\circ}A(\overset{k}{y}, {}^{\bullet\circ}((\vee)\vee R \rightarrow \pi \overset{k}{x} A)) \leftrightarrow {}^{\bullet\circ}(((\vee)\vee R \rightarrow \pi \overset{k}{x} A)\,\overset{k}{y}).$$

Die Induktionsannahme

$$\bigwedge\nolimits_z^k R\overset{k}{z}\overset{k}{y} \dot{\rightarrow} {}^{\bullet\circ}(((\vee)\vee R \rightarrow \pi \overset{k}{x} A)\,\overset{k}{z}) \leftrightarrow {}^{\circ}(((\vee)\vee R \rightarrow \pi \overset{k}{x} A)\,\overset{k}{z})$$

führt wegen des letzten Beweisschrittes ad 2. zunächst zu

$$\bigwedge\nolimits_z^k R\overset{k}{z}\overset{k}{y} \dot{\rightarrow} \neg\neg\, {}^{\bullet\circ}(((\vee)\vee R \rightarrow \pi \overset{k}{x} A)\,\overset{k}{z}) \leftrightarrow {}^{\bullet\circ}(((\vee)\vee R \rightarrow \pi \overset{k}{x} A)\,\overset{k}{z}),$$

dies mittels ($^{\bullet\circ}$2) durch passende Kopie (und $\neg\neg\neg\neg$-Reduktion, wo nötig) zu

$$^{\bullet\circ}A(\overset{k}{y}, {}^{\bullet\circ}((\vee)\vee R \rightarrow \pi \overset{k}{x} A)) \leftrightarrow {}^{\pi\bullet\circ}A(\overset{k}{y}, {}^{\bullet\circ}((\vee)\vee R \rightarrow \pi \overset{k}{x} A));$$

durch passende Kopie aus 3. für A folgt

$$^{\bullet\circ}A(\overset{k}{y}, {}^{\bullet\circ}((\vee)\vee R \rightarrow \pi \overset{k}{x} A)) \leftrightarrow {}^{\circ}A(\overset{k}{y}, {}^{\bullet\circ}((\vee)\vee R \rightarrow \pi \overset{k}{x} A)),$$

und schließlich führt die Induktionsannahme mittels ($^{\circ}$2) noch, passend kopiert, zu

$$^{\circ}A(\overset{k}{y}, {}^{\bullet\circ}((\vee)\vee R \rightarrow \pi \overset{k}{x} A)) \leftrightarrow {}^{\circ}A(\overset{k}{y}, {}^{\circ}((\vee)\vee R \rightarrow \pi \overset{k}{x} A)).$$

An Hand der 3 letzten Bisubjunktionen und laut ($^{\bullet\circ}$3), ($^{\circ}$3) ist damit der Induktionsschritt geleistet, wonach auch

$$\vdash {}^{\bullet\circ}(((\vee)\vee R \rightarrow \pi \overset{k}{x} A)\,\overset{k}{t}) \leftrightarrow {}^{\circ}(((\vee)\vee R \rightarrow \pi \overset{k}{x} A)\,\overset{k}{t})$$

leicht erzielt wird.

Ad 4. Der Deduktionszusammenhang wird bei $^{\circ}$-Abbildung einer $\vdash'$-Ableitung nirgends zerrissen, da $\vdash \neg\neg(D \vee \neg D)$. Sonst handelt es sich um Anwendung derselben Regeln 9.3.1—30 auf $^{\circ}$-geschwächte Aussageschemata, bei 9.3.6, 7, 13, 29 ergänzt durch $\vdash D \rightarrow \neg\neg D$ für das Konklusionsimplikat, bei 9.3.8, **14**, 30 durch doppelte Kontraposition der Konklusion mit nachträglicher Anwendung von 2. für das Implikat, bei 9.3.24, 25 durch Rückgriff auf ($^{\circ}$3) ad 2., 3.; bei 9.3.26, das keine $^{\circ}$-geschwächte Anwendung vorsieht, folgt das $^{\circ}$-Bild durch doppelte Kontraposition des Konklusionsimplikats. Bei 9.3.27, 29 ist noch zu berücksichtigen, daß für die ad 2. genannten A und Q das $^{\circ}$-Bild des Ergebnisses der Substitution von Q für $\pi$ in A, $^{\circ}(A(Q))$, mit dem Ergebnis der Substitution von °Q für $\pi$ in °A, °A(°Q)), bisubjunktiv übereinstimmt: $°Q\overset{k}{t}$ ist nach 2. stabil, so daß die $\neg\neg$ vor $\pi$ in °A und damit vor °Q in °A(°Q) wegfallen dürfen — genau wie in °(A(Q)). Dies beendet die Beweise zu Theorem 2.[15]

2.22 Wie weit eine derartige formale Rechtfertigung des tertium-non-datur auch trägt, für den Intuitionismus bleibt dessen Wahrheit auf je Endliches begrenzt und da, wo es darüber hinaus formal ohne Widerspruch anwendbar ist, inhaltlich eine sekundäre Scheinwahrheit. Im Unterschied zu einfacheren arithmetischen Formalismen, wo das tertium-non-datur total eliminierbar ist, macht das differierende Wahrheitsniveau sich hier auch formal bemerkbar — die Rekursorbestandteile C, P, R entziehen sich der Stabilisierung und 9.1.5 muß für $\vdash'$ unterbunden werden.

2.3 Auf dem Weg zu einem (finit kontrollierbaren) konstruktiven Widerspruchsfreiheitsbeweis für die (formalisierte) axiomatische Mengenlehre [1, 2, 6] kommt es vor allem darauf an, in beträchtlichem Ausmaß mit Ordinalzahlen zu operieren, ohne dabei den Rahmen dessen zu überschreiten, was in konstruktive Wahrheit übersetzt oder umgedeutet werden kann. Nachdem Ordinalzahlen einmal zur Numerierung von (konstruktiblen) Mengen dienen [6], kann man $\in$ auch rein ordinalzahltheoretisch definieren [19, 20][16]; der Vorteil besteht, von $\mathfrak{S}_1$ aus beurteilt, darin, daß sich die Eigenschaften der Ordinalzahlwohlordnung $\leq$ noch für variable transfinite Terme konstruktiv weitgehend nachvollziehen lassen, während eine Darstellung von Mengen mit $\in_x A(x)$ dafür rasch zu höheren Konstruktionen $\mathfrak{S}_\alpha$ veranlaßt.

Wir prüfen nun, wie die in [21] (Etappe I) vorgeschlagene Darstellung der Ordinalzahlwohlordnung $\leq$, die mit Hilfe eines Niveauvergleichs zweier Ordinalzahlrepräsentanten $\vdash I(P, R)$, $\vdash I(Q, S)$ arbeitet, zur konstruktiven Deduktion von Sätzen hinführt, in denen eine übersetzte Wiedergabe einiger zentraler Axiome der Ordinalzahltheorie steckt.

2.31 Der Zusammendrängung halber sei $\bigwedge x_P A$ bzw. $\bigwedge_{\underset{P}{x}} A$ anstelle $\bigwedge x(Px \to A)$ bzw. $\bigwedge_x Px \to A$ notiert, ebenso $\bigvee x_P A \leftrightarrows \bigvee x(Px \wedge A)$ und $\bigvee_{\underset{P}{x}} A \leftrightarrows \bigvee_x Px \wedge A$. Der Übersicht wegen schreiben wir gelegentlich sogar p statt $x_P$, q statt $y_P$, ..; wo solche bedingten Variablen frei

---

[15] Anläßlich des 4. Kolloquiums für mathematische Logik und Grundlagenforschung am 18. bis 20. November 1954 in Marburg/Lahn habe ich einen Vortrag „Über die Eliminierbarkeit des tertium non datur bei induktiven Definitionen" gehalten, der sich auf den Inhalt dieses Theorems richtete.

[16] Als ich [22, 23] schrieb, hatte ich keine Kenntnis von [19, 20]. Ich danke G. TAKEUTI für zahlreiche Informationen seit der Begegnung in Stanford.

vorkommen, ist $A(p) \leftrightharpoons Px \to A(x), \ldots$ zu lesen (genauer [11] p. 420, Zeile 13—20). Da eventuell $P \leftrightharpoons \wedge$, wollen wir hier jedoch nicht mit $p, q, \ldots$ selbst deduktiv umgehen, sondern halten uns an die ausführliche Lesart der Abkürzungen; dies unterbindet Schlüsse, die, wie $\wedge p\, D(p) \to \to \vee p\, D(p)$, eventuell falsch werden. Indizes seien verschieden bedingten Variablen vorbehalten, also $p_1$ statt $x_{P_1}, \ldots$

Um den Niveauvergleich zweier Ordinalzahlrepräsentanten $\vdash I(P, R)$ und $\vdash I(Q, S)$ ganz allgemein, auch wenn diese Ableitungen hypothetisch sind, zu formulieren, definiere ich ${}^{R}_{P}{\leq}^{S}_{Q}$, [21] (Etappe I, 2. Absatz; Etappe II, 2. Absatz) entsprechend, so, daß (3) in **1.632** für ${}^{R}_{P}{\leq}^{S}_{Q}$ als ein 2-stelliges Prädikat(schema) folgendermaßen aussieht:

(D1)
$$I(P, R) \wedge I(Q, S) \angle P \times Q x_0 y_0 \dot{\to} x_0\, {}^{R}_{P}{\leq}^{S}_{Q}\, y_0 \leftrightarrow \leftrightarrow \bigwedge_{\substack{x\\P}} Rxx_0 \to \bigvee_{\substack{y\\Q}} Syy_0 \wedge x\, {}^{R}_{P}{\leq}^{S}_{Q}\, y,$$

wobei $P \times Q x_0 y_0 \leftrightarrow Px_0 \wedge Qy_0$; eine Fundierung von (D1) geschieht (laut [21], p. 269, Zeile 21—22) hinsichtlich $R \times S xyx_0y_0 \leftrightarrow Rxx_0 \wedge Syy_0$. Wie man die $\times$ via $(\vee)\, \frac{\wedge}{\vee}\rangle_{*}{*}$ und ${}^{R}_{P}{\leq}^{S}_{Q}$ via $C\frac{R \times S}{P \times Q}\rangle_{*}{*}$ offiziell ausdrückt, ist klar. Auf die zu ${}^{R}_{P}{\leq}^{S}_{Q}$ gehörigen Definitionsprämissen (1), (2) kommen wir in **2.32** zurück.

Der Umstand, daß (D1) im allgemeinen unselbständig fundiert ist (wofern nicht gerade P, R, Q, S deduzierbar stabil sind), und der Wunsch, möglichst viel an klassischer Ordinalzahltheorie im Übersetzungsweg zu „retten", geben Veranlassung, die Definientes für Prädikate, Prädikatformen oder -schemata, die eine Ordinalzahl, einen beschränkten oder einen unbeschränkten Ordinalzahlterm repräsentieren, von der ${}^{\bullet\circ}$- und ${}^{\circ}$-Stabilisierung in Definitionen wie Deduktionen auszunehmen. Dementsprechend stelle ich (D1) die folgende (ebenfalls bei $P \times Q$ hinsichtlich $R \times S$ fundierte) relative bedingte, gemäß **2.32** gleichfalls deduzierbar rekursive Definition für einen „halb"stabilisierten Niveauvergleich ${}^{R}_{P}{\overset{\cap}{\leq}}^{S}_{Q}$ zur Seite:

(${}^{\cap}$D1)
$$I(P, R) \wedge I(Q, S) \angle P \times Q x_0 y_0 \dot{\to} x_0\, {}^{R}_{P}{\overset{\cap}{\leq}}^{S}_{Q}\, y_0 \leftrightarrow \leftrightarrow \bigwedge_{\substack{x\\P}} Rxx_0 \to \neg\neg \bigvee_{\substack{y\\Q}} Syy_0 \wedge x\, {}^{R}_{P}{\overset{\cap}{\leq}}^{S}_{Q}\, y.$$

**2.311** Nachdem der Niveauvergleich (D1) erfolgt ist, wird laut [21] (p. 268) ein Ordinalzahlvergleich $\leq$ als 0-stelliges Prädikat(schema) explizit —hier relativ explizit— definiert, dessen Definiens dem in (D1) ähnelt:

(D2)
$$I(P, R) \wedge I(Q, S) \angle PR \leq QS \leftrightarrow \bigwedge_{\substack{x\\P}} \bigvee_{\substack{y\\Q}} x\, {}^{R}_{P}{\leq}^{S}_{Q}\, y.$$

Daneben tritt, gleichlautend von (${}^{\cap}$D1) her übertragen, ein halbstabilisierter Ordinalzahlvergleich $\overset{\frown}{\leq}$:

$$(^{\frown}\mathrm{D2}) \qquad \mathrm{I}(\mathrm{P},\mathrm{R}) \wedge \mathrm{I}(\mathrm{Q},\mathrm{S}) \angle \mathrm{PR} \overset{\frown}{\leq} \mathrm{QS} \leftrightarrow \underset{\mathrm{P}}{\bigwedge}{}_{\mathrm{x}} \neg\neg \underset{\mathrm{Q}}{\bigvee}{}_{\mathrm{y}}\, \mathrm{x}^{\mathrm{R}}_{\mathrm{P}} \overset{\frown}{\leq} {}^{\mathrm{S}}_{\mathrm{Q}}\mathrm{y}.$$

In (D2) wird (D1) und in ($^{\frown}$D2) wird ($^{\frown}$D1) nicht über die Bedingung $\mathrm{P}\times\mathrm{Q}$ hinaus beansprucht, da in (D2) ein weiteres Px hinter $\bigvee \mathrm{y}$ geschafft und von dort wieder beseitigt werden kann, in ($^{\frown}$D2) ebenso dank $\vdash \mathrm{D} \rightarrow \neg\neg \mathrm{D}$, $\vdash \neg\neg \mathrm{D}_1 \wedge \neg\neg \mathrm{D}_2 \leftrightarrow \neg\neg (\mathrm{D}_1 \wedge \mathrm{D}_2)$.

Die Frage, ob das im allgemeinen instabile (und nicht stabilisierbare) Fundierungsmaterial, das $\mathfrak{S}_1$ zur Ordinalzahldarstellung produziert, von $\overset{\frown}{\leq}$ ausgehend mit einer konstruktiv gestützten halbstabilisierten „Fassade" so verkleidet werden kann, daß sich formal auf ihr transfinite klassische Arithmetik treiben läßt, findet bald eine bejahende Antwort. $\overset{\frown}{\leq}$ liefert insbesondere die natürliche (schwache) Halbwohlordnung zwischen Ordinalzahlrepräsentanten, die der natürlichen Wohlordnung der von ihnen (klassisch) dargestellten Ordinalzahlen entspricht — „derselben" Ordinalzahl entsprechen viele Repräsentanten.

2.32 Die Unabhängigkeitsprämisse (2) besagt im Fall (D1)

$$\mathrm{P}\times\mathrm{Q}\,\mathrm{x}_0\mathrm{y}_0 \overset{..}{\rightarrow} \underset{\mathrm{P}\times\mathrm{Q}}{\bigwedge}{}_{\mathrm{xy}} \mathrm{R}\times\mathrm{S}\,\mathrm{xyx}_0\mathrm{y}_0 \overset{.}{\rightarrow} \pi_1\mathrm{xy} \leftrightarrow \pi_2\mathrm{xy} \,.\, \overset{.}{\rightarrow}$$
$$\overset{.}{\rightarrow} \underset{\mathrm{P}}{\bigwedge}{}_{\mathrm{x}} \mathrm{Rxx}_0 \rightarrow \underset{\mathrm{Q}}{\bigvee}{}_{\mathrm{y}} \mathrm{Syy}_0 \wedge \pi_1\mathrm{xy} \,..\!\leftrightarrow \underset{\mathrm{P}}{\bigwedge}{}_{\mathrm{x}} \mathrm{Rxx}_0 \rightarrow \underset{\mathrm{Q}}{\bigvee}{}_{\mathrm{y}} \mathrm{Syy}_0 \wedge \pi_2\mathrm{xy}.$$

Die letzte Bisubjunktion läßt sich, dem logischen Aufbau des Definiens $\mathrm{A}(\mathrm{x}_0, \mathrm{y}_0, \pi)$ von $\pi$ aus parallel, durch Ersetzungen folgern, die dem $\overset{.}{\rightarrow}$-Antezedens entspringen. (Variablen- und Markengebote bei Repräsentanten beschränkter oder unbeschränkter Ordinalzahlterme, die mit anonymen Variablen bzw. Marken in P, R, Q, S behaftet sind, trage man an Hand 9.1.5 und 9.3.24, 25 nach.) Im Fall ($^{\frown}$D1) ergibt sich (2) analog.

Die Definitionsprämisse (1) für (D1) und ($^{\frown}$D1) folgt — vgl. [21] (p. 269, Zeile 22—25) — aus

$$(\mathrm{o}\,1) \qquad \vdash \mathrm{I}(\mathrm{P},\mathrm{R}) \vee \mathrm{I}(\mathrm{Q},\mathrm{S}) \rightarrow \mathrm{I}(\mathrm{P}\times\mathrm{Q}, \mathrm{R}\times\mathrm{S}).$$

*Beweis.* Es genügt, $\vdash \mathrm{I}(\mathrm{P},\mathrm{R}; \hat{\pi}, \mathrm{x}) \rightarrow \mathrm{I}(\mathrm{P}\times\mathrm{Q}, \mathrm{R}\times\mathrm{S}; \pi, \mathrm{xy})$ zu zeigen,[14] wozu $\hat{\pi}\mathrm{t} \leftrightharpoons \underset{\mathrm{Q}}{\bigwedge}{}_{\mathrm{y}}\, \pi\mathrm{ty}$ gesetzt sei. (Symmetrisch dazu kann man auch mit $\mathrm{I}(\mathrm{Q},\mathrm{S}; \underset{\mathrm{P}}{\bigwedge}{}_{\mathrm{x}}\, \pi\mathrm{x}{*}, \mathrm{y})$ anfangen.) Die Induktionsannahme bezüglich R „auf" P für $\hat{\pi}$ nach x impliziert insbesondere $\mathrm{P}\times\mathrm{Q}\,\mathrm{x}_0\mathrm{y}_0 \overset{.}{\rightarrow}$ $\overset{.}{\rightarrow} \underset{\mathrm{P}\times\mathrm{Q}}{\bigwedge}{}_{\mathrm{xy}} \mathrm{R}\times\mathrm{S}\,\mathrm{xyx}_0\mathrm{y}_0 \rightarrow \pi\mathrm{xy}$, also die Annahme bezüglich $\mathrm{R}\times\mathrm{S}$ auf $\mathrm{P}\times\mathrm{Q}$ für $\pi$ nach xy. Die diesbezügliche Induktionsvoraussetzung liefert $\pi\mathrm{x}_0\mathrm{y}_0$ für jedes $\mathrm{y}_0$, das $\mathrm{Qy}_0$, mithin die Induktionsbehauptung

$\hat{\pi} x_0$. Dieser Induktionsschritt von $I(P, R; \hat{\pi}, x)$ führt zur Induktionsfolgerung $Px \rightarrow \hat{\pi} x$, also zur Folgerung von $I(P \times Q, R \times S; \pi, xy)$.

**2.321** *Bemerkung.* Der indirekte klassische Schluß vom Gegenbeispiel auf einen unendlichen Regreß ebensolcher (FERMATS descente infinie) oder die heuristisch noch wirksamere Widerlegung der Existenz eines minimalen Gegenbeispiels wird konstruktiv vom Induktionsschritt vertreten, um allgemein möglichst starke Induktionsfolgerungen zu erhalten. Demgegenüber wird der Schluß von einem unendlichen Regreß auf einen anderen, bereits widerlegten konstruktiv durch Transformation der Induktionsaussageform und Ausnutzung der Induktionsvoraussetzungen — statt bloß -annahmen — zu bewältigen gesucht, ohne dabei $\pi *$ durch $\neg\neg$ abzuschwächen.

**Satz 1.** *Es sei* $P \subseteq Q \leftrightharpoons \bigwedge_x Px \rightarrow Qx$, $R \subseteq S \leftrightharpoons \bigwedge_{xy} Rxy \rightarrow Sxy$, $R^P xy \leftrightharpoons Rxy \wedge Px \wedge Py$, *ferner* $\overset{*}{R}_P xy \leftrightharpoons \bigvee_z \overset{'z}{R}_P xy$, *wo* $\overset{z}{R}_P xy$ *nach* z *durch*

$$\overset{z}{R}_P xy \dot{\leftrightarrow} z = 0 \wedge x = y \dot{\vee} \bigvee_{z_0} z = {}'z_0 \wedge \underset{P}{\bigvee}_z \overset{z_0}{R}_P xz \wedge Rzy$$

*rekursiv definiert sei; dann gilt*

(o 2) $\vdash P \subseteq Q \wedge I(Q, R) \rightarrow I(P, R)$, (o 4) $\vdash I(P, R^P) \leftrightarrow I(P, R)$,

(o 3) $\vdash R \subseteq S \wedge I(P, S) \rightarrow I(P, R)$, (o 5) $\vdash I(P, R) \leftrightarrow I(P, \overset{*}{R}_P)$.

($\subseteq$ ist, falls P, Q, R, S bei in $\mathfrak{S}_1$ unentscheidbaren Bedingungen (: $\nvdash P_0 x \vee \neg P_0 x$) definiert sind, im allgemeinen erst mit bedingten Quantoren ausdrückbar, worauf hier nicht näher eingegangen wird.)

**Beweise.** (o 2): substituiere $Pt \rightarrow \pi t$ für $\rho t$ in $I(Q, R; \rho, x)$ durch Kopie und wende, um auf $I(P, R; \pi, x)$ zu kommen, mehrmals Antezedenstausch sowie Absorption von Q durch $P \subseteq Q$ an. (o 3): die Annahme in $I(P, S; \pi, x)$ impliziert die Annahme in $I(P, R; \pi, x)$ wegen $R \subseteq S$; daraufhin folgt aus der Induktionsvoraussetzung in $I(P, R; \pi, x)$ die in $I(P, S; \pi, x)$; nunmehr ergibt sich $I(P, S; \pi, x) \rightarrow I(P, R; \pi, x)$. (o 4): $\leftarrow$ ergibt sich aus (o 3); für $\rightarrow$ beweise man $I(P, R^P; \pi, x) \rightarrow$ $\rightarrow I(P, R; \pi, x)$ durch Importation von $Px$ in der Annahme von $I(P, R^P; \pi, x)$ sowie durch Versetzung eines (wiederholten) $Px_0$ hinter $\bigwedge_x$ in der Voraussetzung von $I(P, R; \pi, x)$, mit anschließender Importation von $Px$ und Exportation nach Anwendung von $\vdash D_1 \wedge (D_2 \rightarrow D_3) \rightarrow$ $\rightarrow (D_1 \wedge D_2 \rightarrow D_3)$ innerhalb der dortigen Annahme.

(o 5): $\leftarrow$ ergibt sich aus (o 3) und (o 4), da $R^P xy \rightarrow \overset{*}{R}_P xy$; für $\rightarrow$ muß $\pi$ in $I(P, R; \pi, x)$ variiert werden, nämlich zu $\hat{\pi} t \leftrightharpoons \bigwedge_z z = t \vee$ $\vee \overset{*}{R}_P zt \rightarrow \pi z$. Die Annahme $Px_0 \dot{\rightarrow} \underset{P}{\bigwedge}_x Rxx_0 \rightarrow \hat{\pi} x$ in $I(P, R; \hat{\pi}, x)$ impliziert $Px_0 \dot{\rightarrow} \bigwedge_z \overset{*}{R}_P z x_0 \rightarrow \pi z$, denn $\overset{*}{R}_P z x_0 \rightarrow \underset{P}{\bigvee}{}_x R x x_0 \dot{\wedge} z = x \vee \overset{*}{R}_P z x$ und für solche x gilt annahmegemäß $\hat{\pi} x$, mithin für z, für die $z = x \vee \overset{*}{R}_P z x$ (wegen $\overset{*}{R}_P z x_0$), $\pi z$. Da $\overset{*}{R}_P z x_0$ auch $Pz$ nach sich zieht, impliziert die Annahme in $I(P, R; \hat{\pi}, x)$ die in $I(P, \overset{*}{R}_P; \pi, x)$. Nunmehr leistet die Induktionsvoraussetzung in $I(P, \overset{*}{R}_P; \pi, x)$ den Induktionsschritt von $I(P, R; \hat{\pi}, x)$, zu dem ja bloß noch $\pi x_0$ (zwecks $z = x_0 \rightarrow \pi z$) fehlt, und die Induktionsfolgerung $Px \rightarrow \hat{\pi} x$ aus $I(P, R; \hat{\pi}, x)$ enthält speziell, für $z = x$, die Induktionsfolgerung $Px \rightarrow \pi x$ von $I(P, \overset{*}{R}_P; \pi, x)$. (Beachte, daß bei $\overset{*}{R}_P$, $\overset{z}{R}_P$ ungeschwächte $\vee$, $\bigvee_*$ wesentlich sind.)

**2.33 Satz 2.**

(o 6) $\vdash I(P, R) \wedge I(Q, S) \ddot{\rightarrow} P \times Q x y \dot{\rightarrow} \neg\neg x\, {}^{R}_{P} \overset{\cap}{\leq} {}^{S}_{Q}\, y \leftrightarrow x\, {}^{R}_{P} \overset{\cap}{\leq} {}^{S}_{Q}\, y$,

(o 7) $\vdash I(P, R) \wedge I(Q, S) \dot{\rightarrow} \neg\neg PR \overset{\cap}{\leq} QS \leftrightarrow PR \overset{\cap}{\leq} QS$,

(o 8) $\vdash I(P, R) \wedge I(Q, S) \ddot{\rightarrow} P \times Q x y \dot{\rightarrow} x\, {}^{R}_{P} \leq {}^{S}_{Q}\, y \rightarrow x\, {}^{R}_{P} \overset{\cap}{\leq} {}^{S}_{Q}\, y$,

(o 9) $\vdash I(P, R) \wedge I(Q, S) \dot{\rightarrow} PR \leq QS \rightarrow PR \overset{\cap}{\leq} QS$.

**Beweise.** Die $P \times Q$-bedingte Stabilität von ${}^{R}_{P} \overset{\cap}{\leq} {}^{S}_{Q}$ und die Stabilität von $\overset{\cap}{\leq}$ resultieren aus $\vdash \neg\neg \bigwedge x_P D_0 \rightarrow \bigwedge x_P \neg\neg D_0$, indem man $\vdash \neg\neg \bigwedge x D \rightarrow \bigwedge x \neg\neg D$ kraft $\vdash \neg\neg(Px \rightarrow D_0) \leftrightarrow (Px \rightarrow \neg\neg D_0)$ auf bedingte Variablen überträgt. (o 8), (o 9) findet man leicht, aus (D 1), ($^{\cap}$D 1) per Induktion und (D 2), ($^{\cap}$D 2).

Übrigens gilt auch $\vdash \neg \bigvee y_Q D_0 \leftrightarrow \bigwedge y_Q \neg D_0$, weil $\vdash \neg \bigvee y D \leftrightarrow \leftrightarrow \bigwedge y \neg D$, $\vdash \neg (Qy \wedge D_0) \leftrightarrow (Qy \rightarrow \neg D_0)$.

**2.34** $\leq$, $\overset{\cap}{\leq}$ sind Halbordnungen, d.h. reflexiv und transitiv. Zum Beweis genügt nach (o 8), (o 9) der für

**Satz 3.** (o 10) $\vdash I(P, R) \dot{\rightarrow} Px \rightarrow x\, {}^{R}_{P} \leq {}^{R}_{P}\, x$,

(o 11) $\vdash I(P, R) \rightarrow PR \leq PR$,

(o 12) $\vdash I(P_1, R_1) \wedge I(P_2, R_2) \wedge I(P_3, R_3) \ddot{\rightarrow}$
$\ddot{\rightarrow} P_1 \times P_2 \times P_3 x_1 x_2 x_3 \dot{\rightarrow} x_1\, {}^{R_1}_{P_1} \leq {}^{R_2}_{P_2}\, x_2 \wedge x_2\, {}^{R_2}_{P_2} \leq {}^{R_3}_{P_3}\, x_3 \rightarrow x_1\, {}^{R_1}_{P_1} \leq {}^{R_3}_{P_3}\, x_3$,

(o 13) $\vdash I(P_1, R_1) \wedge I(P_2, R_2) \wedge I(P_3, R_3) \dot{\rightarrow}$
$\dot{\rightarrow} P_1 R_1 \leq P_2 R_2 \wedge P_2 R_2 \leq P_3 R_3 \rightarrow P_1 R_1 \leq P_3 R_3$;

(o 14), (o 15) *wie* (o 12), (o 13), *bloß mit* $\overset{\cap}{\leq}$ *anstelle* $\leq$, *auch bei* $\underset{*}{\overset{*}{}} \leq \underset{*}{\overset{*}{}}$.

**Beweise.** (o 10) resultiert durch P-bedingten Induktionsschritt bezüglich R für $x\, {}^{R}_{P} \leq {}^{R}_{P}\, x$, da ja bei *gleich* (und zwar mittels $Px \wedge Rxx_0$) bedingten Variablen p, q gilt $\vdash \bigwedge p D(p, p) \rightarrow \bigwedge p \bigvee q D(p, q)$; dies, mittels Bedingung P, auf (o 10) angewendet, führt zu (o 11).

(o 12) resultiert durch $P_1 \times P_2 \times P_3$-bedingten Induktionsschritt bezüglich $R_1 \times R_2 \times R_3$ für das $\dot{\rightarrow}$-Sukzedens, indem bei *verschieden* bedingten Variablen $p_1$, $p_2$, $p_3$ gilt

$$\vdash \bigwedge p_1 \bigvee p_2 D_{12}(p_1, p_2) \wedge \bigwedge p_2 \bigvee p_3 D_{23}(p_2, p_3) \rightarrow$$
$$\rightarrow \bigwedge p_1 \bigvee p_2 \bigvee p_3 (D_{12}(p_1, p_2) \wedge D_{23}(p_2, p_3)), \quad \text{falls } {\not\,}p_3 D_{12},\ {\not\,}p_1 D_{23},$$

wonach die Induktionsannahme auf eine Aussage $\bigwedge p_1 \bigvee p_3 D_{13}(p_1, p_3)$ führt. Analoge Schlüsse wie bei diesem Induktionsschritt liefern (o 13) aus (o 12).

Für (o 14), (o 15) haben wir uns, ganz entsprechend, lediglich noch von

$$\vdash \bigwedge p_1 \neg\neg \bigvee p_2 D_{12}(p_1, p_2) \wedge \bigwedge p_2 \neg\neg \bigvee p_3 D_{23}(p_2, p_3) \rightarrow$$
$$\rightarrow \bigwedge p_1 \neg\neg \bigvee p_2 \bigvee p_3 (D_{12}(p_1, p_2) \wedge D_{23}(p_2, p_3)), \text{ falls } {\not\,}p_3 D_{12}, {\not\,}p_1 D_{23},$$

zu überzeugen. Nun gilt $\vdash D_1(p_2) \dot{\rightarrow} \bigwedge p_2 D_2(p_2) \rightarrow \bigvee p_2(D_1(p_2) \wedge D_2(p_2))$. also $\vdash \neg\neg \bigvee p_2 D_1(p_2) \dot{\rightarrow} \bigwedge p_2 D_2(p_2) \rightarrow \neg\neg \bigvee p_2 (D_1(p_2) \wedge D_2(p_2))$,

mithin insbesondere

$$\vdash \neg\neg \bigvee p_2 D_{12}(p_1, p_2) \wedge \bigwedge p_2 \neg\neg \bigvee p_3 D_{23}(p_2, p_3) \rightarrow$$
$$\rightarrow \neg\neg \bigvee p_2 (D_{12}(p_1, p_2) \wedge \neg\neg \bigvee p_3 D_{23}(p_2, p_3)),$$
$$\rightarrow \neg\neg \bigvee p_2 \neg\neg (D_{12}(p_1, p_2) \wedge \bigvee p_3 D_{23}(p_2, p_3)),$$
$$\rightarrow \neg\neg \bigvee p_2 \neg\neg \bigvee p_3 (D_{12}(p_1, p_2) \wedge D_{23}(p_2, p_3)), \text{ da } {\not\,}p_3 D_{12},$$
$$\rightarrow \neg\neg \bigvee p_2 \bigvee p_3 (D_{12}(p_1, p_2) \wedge D_{23}(p_2, p_3))$$

und schließlich die anfängliche Behauptung, da ${\not\,}p_1 D_{23}$.

**2.35** Um die schwache Konnektivität $\vdash I(P, R) \wedge I(Q, S) \rightarrow \rightarrow \neg\neg(PR \overset{\cap}{\lesssim} QS \vee QS \overset{\cap}{\lesssim} PR)$ über schärfere Zwischenergebnisse zu erhalten, sei definiert

$$({}^{\supset}D3) \quad I(P, R) \wedge I(Q, S) \angle P \times Q\, x_0 y_0 \dot{\rightarrow} x_0 {}^{R}_{P} \overset{\supset}{>} {}^{S}_{Q} y_0 \leftrightarrow$$
$$\leftrightarrow \bigvee_{\substack{x\\P}} R\,x\,x_0 \wedge \bigwedge_{\substack{y\\Q}} S\,y\,y_0 \rightarrow \neg\neg x {}^{R}_{P} \overset{\supset}{>} {}^{S}_{Q} y,$$

$$({}^{\supset}D4) \quad I(P, R) \wedge I(Q, S) \angle PR \overset{\supset}{>} QS \leftrightarrow \bigvee_{\substack{x\\P}} \bigwedge_{\substack{y\\Q}} \neg\neg x {}^{R}_{P} \overset{\supset}{>} {}^{S}_{Q} y.$$

(D3), (D4) sollen aus $({}^{\supset}D3)$, $({}^{\supset}D4)$ durch Tilgung der ${}^{\supset}$ und von $\neg\neg$ entstehen. Für $({}^{\supset}D3)$, (D3) kann man sich die Definitionsprämissen (1), (2) geradeso überlegen wie für $({}^{\cap}D1)$, (D1) in **2.32**; auch die Anmerkung zu (D2), $({}^{\cap}D2)$ in **2.311** läßt sich auf (D4), $({}^{\supset}D4)$ übertragen, nur kommt $Px$ „unter" $\bigwedge_y$ und, bei ${}^{\supset}$, (mit $Qy$) hinter $\neg\neg$.

Im allgemeinen ist keines der für Ordinalzahlrepräsentanten neu definierten Prädikate (schemata) ${}^{R}_{P} \overset{\supset}{>} {}^{S}_{Q}$, $\overset{\supset}{>}$, ${}^{R}_{P} > {}^{S}_{Q}$, $>$ stabil. $\overset{*}{\underset{*}{}} > \overset{*}{\underset{*}{}}$ und $>$ haben auch nichts mit einer Übertragung des konversen asymmetrischen

Prädikatteils (: $R^{<-}$ gemäß $R^{<}xy \leftrightharpoons Rxy \wedge \neg Ryx$, $S^{-}xy \leftrightharpoons Syx$) auf die Prädikatenprädikate $\underset{*}{\overset{*}{}}\leq\underset{*}{\overset{*}{}}$, $\leq$ zu tun. Effektive junktorenlogische Konsequenzen aus Satz 2 und 3 für Abkürzungen wie $PR < QS \leftrightharpoons \leftrightharpoons PR \leq QS \wedge \neg QS \leq PR$ (und ebenso mit $^{\cap}$) werden hier als selbstverständlich behandelt; erst recht derartige Konsequenzen für Abkürzungen wie $PR \overset{\cap}{\equiv} QS \leftrightharpoons PR \overset{\cap}{\leq} QS \wedge QS \overset{\cap}{\leq} PR$ (und ebenso ohne $^{\cap}$), die dem symmetrischen Prädikatteil $R^{=}xy \leftrightharpoons Rxy \wedge Ryx$ nachgebildet sind.

Nicht nur, damit die erstgenannte Abkürzung nicht (D4) widerstreite, sondern auch mit Rücksicht auf die etwas verwickelten und bei Umstellung ungleichförmig erscheinenden Definientes in den übrigen Definitionen ($^{[\cap]}$D1), ($^{[\cap]}$D2), ($^{[\supset]}$D3), ($^{\supset}$D4) wollen wir jedwedes „Umklappen“ von $\downarrow\ddagger \overset{\leftarrow}{\leq} \ddagger\uparrow$, $\downarrow \overset{\leftarrow}{\leq} \uparrow$, $\uparrow\ddagger \overset{\rightarrow}{>} \ddagger\downarrow$, $\uparrow \overset{\rightarrow}{>} \downarrow$ in $\uparrow\ddagger \overset{\rightarrow}{\geq} \ddagger\downarrow$, $\uparrow \overset{\rightarrow}{\geq} \downarrow$, $\downarrow\ddagger \overset{\leftarrow}{<} \ddagger\uparrow$, $\downarrow \overset{\leftarrow}{<} \uparrow$ unterlassen.

**Satz 4.**

(o 16) $\vdash I(P,R) \wedge I(Q,S) \ddot{\rightarrow} P\times Q\,xy \dot{\rightarrow} \neg x\,{}^{R}_{P} \overset{\supset}{>} {}^{S}_{Q}\,y \leftrightarrow x\,{}^{R}_{P} \overset{\cap}{\leq} {}^{S}_{Q}\,y$,

(o 17) $\vdash I(P,R) \wedge I(Q,S) \dot{\rightarrow} \neg PR \overset{\supset}{>} QS \leftrightarrow PR \overset{\cap}{\leq} QS$,

(o 18) $\vdash I(P,R) \wedge I(Q,S) \ddot{\rightarrow} P\times Q\,xy \dot{\rightarrow} x\,{}^{R}_{P} > {}^{S}_{Q}\,y \rightarrow x\,{}^{R}_{P} \overset{\supset}{>} {}^{S}_{Q}\,y$,

(o 19) $\vdash I(P,R) \wedge I(Q,S) \dot{\rightarrow} PR > QS \rightarrow PR \overset{\supset}{>} QS$.

**Beweise.** (o 16) resultiert durch $P\times Q$-bedingten Induktionsschritt bezüglich $R\times S$ für das $\dot{\rightarrow}$-Sukzedens, indem man zweimal $\neg \bigvee p D \leftrightarrow \leftrightarrow \bigwedge p \neg D$ (für verschieden bedingte p) anwendet, bevor die Induktionsannahme benutzt wird. (o 17) mit entsprechenden Schlüssen aus (o 16). (o 18), (o 19) folgt aus (D3), ($^{\supset}$D3) per Induktion und (D4), ($^{\supset}$D4).

**Satz 5.**

(o 20) $\vdash I(P_1, R_1) \wedge I(P_2, R_2) \wedge I(P_3, R_3) \ddot{\rightarrow}$

$$\ddot{\rightarrow} P_1\times P_2\times P_3\,x_1 x_2 x_3 \dot{\rightarrow} x_1\,{}^{R_1}_{P_1} \overset{\supset}{>} {}^{R_2}_{P_2}\,x_2 \wedge x_2\,{}^{R_2}_{P_2} \overset{\supset}{>} {}^{R_3}_{P_3}\,x_3 \rightarrow x_1\,{}^{R_1}_{P_1} \overset{\supset}{>} {}^{R_3}_{P_3}\,x_3,$$

(o 21) $\vdash I(P_1, R_1) \wedge I(P_2, R_2) \wedge I(P_3, R_3) \dot{\rightarrow}$

$$\dot{\rightarrow} P_1 R_1 \overset{\supset}{>} P_2 R_2 \wedge P_2 R_2 \overset{\supset}{>} P_3 R_3 \rightarrow P_1 R_1 \overset{\supset}{>} P_3 R_3;$$

(o 24) $\vdash I(P_1, R_1) \wedge I(P_2, R_2) \wedge I(P_3, R_3) \ddot{\rightarrow}$

$$\ddot{\rightarrow} P_1\times P_2\times P_3\,x_1 x_2 x_3 \dot{\rightarrow} x_1\,{}^{R_1}_{P_1} \overset{\cap}{\leq} {}^{R_2}_{P_2}\,x_2 \wedge x_3\,{}^{R_3}_{P_3} \overset{\supset}{>} {}^{R_2}_{P_2}\,x_2 \rightarrow x_3\,{}^{R_3}_{P_3} \overset{\supset}{>} {}^{R_1}_{P_1}\,x_1,$$

(o 25) $\vdash I(P_1, R_1) \wedge I(P_2, R_2) \wedge I(P_3, R_3) \dot{\rightarrow}$

$$\dot{\rightarrow} P_1 R_1 \overset{\cap}{\leq} P_2 R_2 \wedge P_3 R_3 \overset{\supset}{>} P_2 R_2 \rightarrow P_3 R_3 \overset{\supset}{>} P_1 R_1;$$

(o 26), (o 27), (o 28), (o 29) *wie* (o 20), (o 21), (o 24), (o 25), *bloß ohne* $^{\supset}$ *und* $^{\cap}$, (o 30), (o 31) *wie* (o 24), (o 25) *ohne* $^{\supset}$ *und* $^{\cap}$, *aber mit Indizes* $_{23,\ 21,\ (31)}$ *anstelle* $_{12,\ 32,\ (31)}$;

(o 32) $\vdash \mathrm{I}(\mathrm{P},\mathrm{R}) \wedge \mathrm{I}(\mathrm{Q},\mathrm{S}) \ddot{\rightarrow} \mathrm{P}\times\mathrm{Q}\,\mathrm{x}\,\mathrm{y} \dot{\rightarrow} \mathrm{x}{}^{\mathrm{R}}_{\mathrm{P}} > {}^{\mathrm{S}}_{\mathrm{Q}}\mathrm{y} \rightarrow \mathrm{y}{}^{\mathrm{S}}_{\mathrm{Q}} \leq {}^{\mathrm{R}}_{\mathrm{P}}\mathrm{x}$,

(o 33) $\vdash \mathrm{I}(\mathrm{P},\mathrm{R}) \wedge \mathrm{I}(\mathrm{Q},\mathrm{S}) \dot{\rightarrow} \mathrm{PR} > \mathrm{QS} \rightarrow \mathrm{QS} \leq \mathrm{PR}$.

**Korollar.**

(o 22) $\vdash \mathrm{I}(\mathrm{P},\mathrm{R}) \wedge \mathrm{I}(\mathrm{Q},\mathrm{S}) \rightarrow \neg\neg(\mathrm{PR} \overset{\cap}{\leq} \mathrm{QS} \vee \mathrm{QS} \overset{\cap}{\leq} \mathrm{PR})$,

(o 23) $\vdash \mathrm{I}(\mathrm{P},\mathrm{R}) \wedge \mathrm{I}(\mathrm{Q},\mathrm{S}) \dot{\rightarrow} \neg \mathrm{PR} \overset{\cap}{\leq} \mathrm{QS} \leftrightarrow \mathrm{QS} \overset{\cap}{<} \mathrm{PR}$;

(o 34) $\vdash \mathrm{I}(\mathrm{P},\mathrm{R}) \wedge \mathrm{I}(\mathrm{Q},\mathrm{S}) \dot{\rightarrow} \mathrm{PR} > \mathrm{QS} \rightarrow \mathrm{QS} < \mathrm{PR}$.

**Beweise.** (o 20), (o 21) und (o 24), (o 25) gehen, ähnlich wie beim Beweis zu (o 14), (o 15), auf Anwendungen von

$$\vdash \bigvee p_1 \bigwedge p_2 \neg\neg D_{12}(p_1, p_2) \wedge \bigvee p_2 \bigwedge p_3 \neg\neg D_{23}(p_2, p_3) \rightarrow$$
$$\rightarrow \bigvee p_1 \bigvee p_2 \bigwedge p_3 \neg\neg(D_{12}(p_1, p_2) \wedge D_{23}(p_2, p_3)), \text{ falls } {}_{\nmid}p_3 D_{12},\ {}_{\nmid}p_1 D_{23},$$

bzw.

$$\vdash \bigwedge p_1 \neg\neg \bigvee p_2 D_{12}(p_1, p_2) \wedge \bigvee p_3 \bigwedge p_2 \neg\neg D'_{32}(p_3, p_2) \rightarrow$$
$$\rightarrow \bigvee p_3 \bigwedge p_1 \neg\neg \bigvee p_2(D_{12}(p_1, p_2) \wedge D'_{32}(p_3, p_2)), \text{ falls } {}_{\nmid}p_3 D_{12},\ {}_{\nmid}p_1 D'_{32},$$

zurück. (o 26) bis (o 31) ergeben sich nach demselben Muster: die $\neg\neg$ darf man tilgen (ohne $\vdash$ zu verletzen), und danach stört beim zweiten $\vdash$ auch die Indexumordnung nicht mehr. (o 32): Induktionsschritt durch Alteration bedingter Quantoren (: $\vdash \bigvee p_1 \bigwedge p_2 D \rightarrow \bigwedge p_2 \bigvee p_1 D$) vor Rückgriff auf die Induktionsannahme. (o 33) analog, mit (o 32).

(o 22): $\neg\neg(\mathrm{PR} \overset{\supset}{>} \mathrm{QS} \vee \mathrm{PR} \overset{\cap}{\leq} \mathrm{QS})$ nach (o 17), $\vdash \neg\neg(D \vee \neg D)$; $\overset{\supset}{>}$ irreflexiv nach (o 17) und (o 11) (mit (o 9)), also asymmetrisch wegen (o 21); daher $\mathrm{PR} \overset{\supset}{>} \mathrm{QS} \rightarrow \mathrm{QS} \overset{\cap}{\leq} \mathrm{PR}$ nach (o 17). (o 23): $\leftarrow$ per definitionem; für $\rightarrow$ fehlt $\neg \mathrm{PR} \overset{\cap}{\leq} \mathrm{QS} \rightarrow \mathrm{QS} \overset{\cap}{\leq} \mathrm{PR}$, was aus $\vdash \neg\neg(D_1 \vee D_2) \leftrightarrow \leftrightarrow (\neg D_1 \rightarrow \neg\neg D_2)$ mit (o 22), (o 7) folgt. (o 34): (o 33) liefert $\mathrm{QS} \leq \mathrm{PR}$; mit $\overset{\supset}{>}$ ist auch $>$ irreflexiv infolge (o 19), so daß $\neg \mathrm{PR} \leq \mathrm{QS}$ über (o 29) oder (o 31).

(Während $\overset{\cap}{\leq}$ und $\overset{\supset}{>}$ antitransitiv sind, sind allgemein weder $\leq$ noch $>$ antitransitiv — womit die Eigenschaft 'non-$*$ ist transitiv' gemeint sei.)

**2.36** Zur Vorbereitung der schwachen Induktion bezüglich $\overset{\cap}{<}$ (unter **2.37**) bauen wir in gewisser Hinsicht (o 32), (o 33) nach verschiedenen Richtungen aus. Interessant ist, daß die (unter **2.38** angehängte) „Abschnitts"darstellung von QS, die $\overset{\cap}{<}$ PR, auf der in der naiven Mengenlehre die Normaldarstellung jeder Ordinalzahl $\alpha$ durch $W_\alpha = \in_\beta \beta < \alpha$ fußt, welche ihrerseits ja die Trichotomie $<$ - $=$ - $>$ trägt, hier erst dann benötigt wird, wenn auch Rekursionen hinsichtlich $\overset{\cap}{<}$, etwa [20] 1.3.13, in $\mathfrak{S}_1$ übersetzt werden sollen.

**Satz 6.**

(o 35) $\vdash I(P,R) \wedge I(Q,S) \overset{..}{\to} P\times Q\,xy \overset{.}{\to} x\,{}^{R}_{P} \overset{\supset}{>} {}^{S}_{Q}\,y \leftrightarrow \bigvee_{\substack{z\\P}} Rzx \wedge y\,{}^{S}_{Q} \overset{\cap}{\leq} {}^{R}_{P}\,z$,

(o 37) $\vdash I(P,R) \overset{..}{\to} P\times P\,xy \overset{.}{\to} Rxy \to y\,{}^{R}_{P} > {}^{R}_{P}\,x$;

(o 38) $\vdash I(P,R) \overset{.}{\to} {}_{R}P_{z}x \to x\,{}^{R}_{P} \leq {}^{R}_{{}_{R}P_{z}}\,x \wedge x\,{}^{R}_{{}_{R}P_{z}} \leq {}^{R}_{P}\,x$,

(o 39) $\vdash I(P,R) \overset{.}{\to} {}_{R}P_{z}x \to z\,{}^{R}_{P} > {}^{R}_{P}\,x$,

(o 40) $\vdash I(P,R) \overset{.}{\to} Pz \to PR > {}_{R}P_{z}R$,

(o 41) $\vdash I(P,R) \wedge I(Q,S) \overset{.}{\to} PR \overset{\supset}{>} QS \leftrightarrow \bigvee_{\substack{z\\P}} QS \overset{\cap}{\leq} {}_{R}P_{z}R$,

(o 43) $\vdash I(P,R) \overset{..}{\to} {}_{R}P_{z}y \overset{.}{\to} {}_{RR}P_{zy}x \leftrightarrow {}_{R}P_{y}x$,

*wo* ${}_{R}P_{z}x \leftrightharpoons Pz \wedge \overset{*}{R}_{P}xz$ *sei (vgl. Satz 1)*;

(o 36), (o 42) *wie* (o 35), (o 41), *bloß ohne* $^{\supset}$ *und* $^{\cap}$.

**Beweise.** (o 35) ergibt sich durch $P\times Q$-bedingten Induktionsschritt bezüglich $R\times S$ für das $\overset{.}{\to}$-Sukzedens:

$$\begin{aligned} x_0\,{}^{R}_{P} \overset{\supset}{>} {}^{S}_{Q}\,y_0 &\leftrightarrow \bigvee_{Rxx_0} x_P \bigwedge_{Syy_0} y_Q \neg\neg\, x\,{}^{R}_{P} \overset{\supset}{>} {}^{S}_{Q}\,y \quad (:\,({}^{\supset}D3))\\ &\leftrightarrow \bigvee_{Rxx_0} x_P \bigwedge_{Syy_0} y_Q \;\neg\neg \bigvee_{Rzx} z_P\;\; y\,{}^{S}_{Q} \overset{\cap}{\leq} {}^{R}_{P}\,z \quad (:\text{ per Annahme})\\ &\leftrightarrow \bigvee_{Rxx_0} x_P\; y_0\,{}^{S}_{Q} \overset{\cap}{\leq} {}^{R}_{P}\,x \quad (:\,({}^{\cap}D1)); \end{aligned}$$

x ist ein $z_0$, wie (o 35) es behauptet. (o 36) entsprechend, ohne $^{\supset}$, $^{\cap}$ und $\neg\neg$. (o 37) ergibt sich durch $P\times P$-bedingten Induktionsschritt bezüglich $R\times R$ für das $\overset{.}{\to}$-Sukzedens:

$$\begin{aligned} Rx_0y_0 &\to \bigwedge_{Rxx_0} x_P\;\, x_0\,{}^{R}_{P} > {}^{R}_{P}\,x \quad (:\text{ per Annahme}),\\ &\to \bigvee_{Ryy_0} y_P \bigwedge_{Rxx_0} x_P\;\, y\,{}^{R}_{P} > {}^{R}_{P}\,x \quad (:\,x_0 \text{ ist solch ein } y)\\ &\leftrightarrow y_0\,{}^{R}_{P} > {}^{R}_{P}\,x_0 \quad (:\,(D3)). \end{aligned}$$

(o 38) ergibt sich durch $P\times {}_{R}P_{z}$-bedingten Induktionsschritt bezüglich $R\times R$ für $x\,{}^{R}_{P} \leq {}^{R}_{{}_{R}P_{z}}\,x$ und ${}_{R}P_{z}\times P$-bedingten Induktionsschritt bezüglich $R\times R$ für $x\,{}^{R}_{{}_{R}P_{z}} \leq {}^{R}_{P}\,x$, indem $Px_0 \wedge {}_{R}P_{z}x_0 \leftrightarrow {}_{R}P_{z}x_0 \overset{.}{\to}$ $\overset{.}{\to} Px \wedge Rxx_0 \leftrightarrow {}_{R}P_{z}x \wedge Rxx_0$, da ${}_{R}P_{z}x_0 \leftrightarrow Pz \wedge \overset{*}{R}_{P}x_0z\,(\to Px_0)$ und $\overset{'z_0}{R_{P}xz} \leftrightarrow Px \wedge \bigvee_{\substack{x_0\\P}} Rxx_0 \wedge \overset{z_0}{R_{P}x_0z}$ — auch, umarrangiert, als Definiensadjugand (in Satz 1) verwendbar —, und da die Umkehrung hinter dem letzten $\overset{.}{\to}$ für sich trivial.

(o 39) ergibt sich durch (unbedingte) gewöhnliche Induktion für $\bigwedge_{xz} Pz \wedge \overset{'y}{R_{P}xz} \to z\,{}^{R}_{P} > {}^{R}_{P}\,x$ nach y: im Fall $y = 0$ gilt $Pz \wedge \overset{'0}{R_{P}xz} \to$

$\to Pz \wedge Px \wedge Rxz$, also $z\,{}^{R}_{P}\!>{}^{R}_{P}\,x$ laut (o 37); im Fall $y = y_0$ gilt $Pz \wedge \overset{'y_0}{R}_P xz \to Pz \wedge \underset{P}{\bigvee}_{z_0} \overset{y_0}{R}_P x z_0 \wedge R z_0 z\ (\to Px)$, woraus annahmegemäß und laut (o 37) folgt $z_0\,{}^{R}_{P}\!>{}^{R}_{P}\,x \wedge z\,{}^{R}_{P}\!>{}^{R}_{P}\,z_0$, also, laut (o 26), $z\,{}^{R}_{P}\!>{}^{R}_{P}\,x$, womit der gewöhnliche Induktionsschritt geleistet ist. Dann aber gilt sofort $Pz \wedge \bigvee y_{>0} \overset{y}{R}_P xz \to z\,{}^{R}_{P}\!>{}^{R}_{P}\,x$.

Ohne Induktion folgt jetzt (o 40), (o 41):

$$Pz \to \underset{{}_RP_z}{\bigwedge}_x z\,{}^{R}_{P}\!>{}^{R}_{P}\,x \quad (:(\text{o }39)) \leftrightarrow \underset{{}_RP_z}{\bigwedge}_x z\,{}^{R}_{P}\!>{}^{R}_{{}_RP_z}\,x \quad (:(\text{o }38), (\text{o }28)),$$

$$\to \underset{P}{\bigvee}_x \underset{{}_RP_z}{\bigwedge}_y x\,{}^{R}_{P}\!>{}^{R}_{{}_RP_z}\,y \quad (:\text{nimm } z \text{ als } x) \leftrightarrow PR > {}_RP_z R \quad (:(D\,4));$$

$$PR \overset{\supset}{\gtrsim} QS \leftrightarrow \bigvee x_P \bigwedge y_Q \neg\neg x\,{}^{R}_{P} \overset{\supset}{\gtrsim} {}^{S}_{Q}\,y \quad (:({}^{\supset}D\,4))$$

$$\leftrightarrow \bigvee x_P \bigwedge y_Q \neg\neg \bigvee z_P (Rzx \wedge y\,{}^{S}_{Q} \overset{\cap}{\lesssim} {}^{R}_{P}\,z) \quad (:(\text{o }35))$$

$$\leftrightarrow \bigvee z_P \bigwedge y_Q \neg\neg \bigvee x_P (Rxz \wedge y\,{}^{S}_{Q} \overset{\cap}{\lesssim} {}^{R}_{{}_RP_z}\,x) \quad (:\text{Umbenennung},$$

$$Pz \wedge Px \wedge Rxz \to {}_RP_z\,x, (\text{o }38), (\text{o }8), (\text{o }14))$$

$$\leftrightarrow \bigvee z_P QS \overset{\cap}{\lesssim} {}_RP_z R \quad (:\to ({}^{\cap}D\,2), \leftarrow (\text{o }40), (\text{o }19), (\text{o }25)).$$

(o 42) entsprechend, ohne $^{\supset}$, $^{\cap}$ und $\neg\neg$, mit (o 36) statt (o 35), ohne (o 8) und (o 19), mit (o 12), (o 29) statt (o 14), (o 25).

Nun noch (o 43): in der Tat, ${}_{RR}P_{zy}\,x \leftrightharpoons {}_RP_z\,y \wedge \overset{*}{R}_{{}_RP_z} xy$; ferner ${}_RP_z\,y \dot{\to} \overset{*}{R}_{{}_RP_z} xy \leftrightarrow \overset{*}{R}_P xy$, indem ${}_RP_z \subseteq P$ und $\overset{z_0}{R}_{{}_RP_z} xy \to \overset{z_0}{R}_P xy$ ad $\to$ (ohne $\dot{\to}$-Antezedens), und $Pz \wedge \overset{*}{R}_P yz \dot{\to} \bigwedge_x \overset{*}{R}_P xy \to Pz \wedge \overset{*}{R}_P xz$ sowie ${}_RP_z\,y \dot{\to} \overset{z_0}{R}_P xy \to \overset{z_0}{R}_{{}_RP_z} xy$ ad $\leftarrow$. Also gilt ${}_RP_z y \dot{\to} {}_{RR}P_{zy}\,x \leftrightarrow \overset{*}{R}_{{}_RP_z} xy \leftrightarrow \overset{*}{R}_P xy \leftrightarrow \overset{*}{R}_P xy \wedge Py$, da $\overset{*}{R}_P yz \to Py$.

**2.37** Gestützt auf die Vorarbeit in den Sätzen 2—6 erreicht man den folgenden, diese konstruktiven Schlußketten vorläufig abschließenden

**Satz 7.** *(Unbeschränkte schwache transfinite Induktion.) Es gilt*

$$\vdash \bigwedge_{\alpha_0} \bigwedge_{\alpha_1} \alpha_1 < \alpha_0 \to \neg\neg \underset{+}{\pi}\,\alpha_1 \,.\!\to \neg\neg \underset{+}{\pi}\,\alpha_0 \,.\!\to \neg\neg \underset{+}{\pi}\,\alpha,$$

*repräsentiert in der Form*

$$(\text{o }44)\ \vdash \underset{I(\pi_0,\rho_0)}{\bigwedge_{\pi_0\rho_0}} \underset{I(\pi_1,\rho_1)}{\bigwedge_{\pi_1\rho_1}} \pi_1\rho_1 \overset{\cap}{<} \pi_0\rho_0 \to \neg\neg \underset{+}{\pi}(\pi_1\rho_1) \,.\!\to \neg\neg \underset{+}{\pi}(\pi_0\rho_0) \,.\dot{\to}$$

$$\dot{\to} I(\pi, \rho) \to \neg\neg \underset{+}{\pi}(\pi\rho),$$

*wo* ${}^1\underset{+}{\Pi}_l(\pi\rho)$ *für* $I(\pi, \rho) \vee \wedge \to {}^0\Pi_m {}^0\Pi_l$ *(mit* $l \neq m$*) steht. Erst recht gilt* — (o 3) *mittels* (o 17), (o 23) *und mittels* (o 19) *auf das Aussageschema* $\underset{+}{I}(I, \overset{\cap}{<})$ *von* (o 44) *in übertragener, aber in* $\mathfrak{S}_1$ *durchaus ableitbarer Form angewendet* — (o 44) *bezüglich* $\overset{\supset}{>}{}^{-}$ *und bezüglich* $>^{-}$ *anstelle* $\overset{\cap}{<}$.

**Beweis.** Das $\dot{\rightarrow}$-Sukzedens in (o 44) wird durch Vermittlung von $I(\pi, \rho; \hat{\pi}, x)$, wo $\hat{\pi} t \leftrightharpoons \underset{I(\pi_1, \rho_1)}{\bigwedge_{\pi_1 \rho_1}} \pi_1 \rho_1 \overset{\cap}{\leq} {}_\rho\pi_t \rho \rightarrow \neg\neg \underset{+}{\pi}(\pi_1 \rho_1)$, dargetan.

Dazu genügt es, aus der Induktionsvoraussetzung in (o 44), dem dortigen $\dot{\rightarrow}$-Antezedens, und aus $I(\pi, \rho)$ die Induktionsvoraussetzung in $I(\pi, \rho; \hat{\pi}, x)$ zu folgern: danach wird das $\dot{\rightarrow}$-Antezedens in (o 44) nämlich $I(\pi, \rho) \dot{\rightarrow} \pi x \rightarrow \hat{\pi} x$ implizieren, also auch die Induktionsannahme $I(\pi, \rho) \rightarrow \underset{I(\pi_1, \rho_1)}{\bigwedge_{\pi_1 \rho_1}} \pi_1 \rho_1 \overset{\cap}{<} \pi \rho \rightarrow \neg\neg \underset{+}{\pi}(\pi_1 \rho_1)$ in (o 44); denn aus $I(\pi, \rho) \wedge I(\pi_1, \rho_1) \wedge \neg \pi \rho \overset{\cap}{\leq} \pi_1 \rho_1$ folgt nach (o 17), (o 41) $\neg\neg \underset{\pi}{\bigvee}_x \pi_1 \rho_1 \overset{\cap}{\leq} {}_\rho\pi_x \rho$, aus

$$I(\pi, \rho) \wedge \pi x \wedge I(\pi_1, \rho_1) \wedge \pi_1 \rho_1 \overset{\cap}{\leq} {}_\rho\pi_x \rho \rightarrow \neg\neg \underset{+}{\pi}(\pi_1 \rho_1)$$

(: infolge $I(\pi, \rho) \dot{\rightarrow} \pi x \rightarrow \hat{\pi} x$) aber

$$I(\pi, \rho) \wedge I(\pi_1, \rho_1) \dot{\rightarrow} [\neg\neg] \underset{\pi}{\bigvee}_x \pi_1 \rho_1 \overset{\cap}{\leq} {}_\rho\pi_x \rho \,.\rightarrow \neg\neg \underset{+}{\pi}(\pi_1 \rho_1).$$ [17]

Geht man noch mit jener implizierten Annahme in das $\dot{\rightarrow}$-Antezedens von (o 44), so ergibt sich das $\dot{\rightarrow}$-Sukzedens $I(\pi, \rho) \dot{\rightarrow} \neg\neg \underset{+}{\pi}(\pi \rho)$, und (o 44) wäre bewiesen.

Nun folgt aus $I(\pi, \rho)$ und per Annahme in $I(\pi, \rho; \hat{\pi}, x)$ allein schon $\underset{I(\pi_1, \rho_1)}{\bigwedge_{\pi_1 \rho_1}} \pi_1 \rho_1 \overset{\cap}{<} {}_\rho\pi_{x_0} \rho \rightarrow \neg\neg \underset{+}{\pi}(\pi_1 \rho_1)$. Denn aus $\neg {}_\rho\pi_{x_0} \rho \overset{\cap}{\leq} \pi_1 \rho_1$ folgt dabei nach (o 17), (o 41) zunächst $\neg\neg \underset{\rho\pi_{x_0}}{\bigvee_z} \pi_1 \rho_1 \overset{\cap}{\leq} {}_{\rho\rho}\pi_{x_0 z} \rho$, also $\neg\neg \underset{\rho\pi_{x_0}}{\bigvee_z} \pi_1 \rho_1 \overset{\cap}{\leq} {}_\rho\pi_z \rho$ laut (o 43). Sodann ergibt sich

$$I(\pi, \rho) \wedge I(\pi_1, \rho_1) \dot{\rightarrow} [\neg\neg] \bigvee_z {}_\rho\pi_{x_0} z \wedge \pi_1 \rho_1 \overset{\cap}{\leq} {}_\rho\pi_z \rho \,.\rightarrow \neg\neg \underset{+}{\pi}(\pi_1 \rho_1),$$

weil ${}_\rho\pi_{x_0} z$ einerseits $\pi x_0 \wedge \overset{*}{\rho}_\pi z x_0$, mithin $\pi x_0 \wedge \underset{\pi}{\bigvee}_x \rho x x_0 \mathbin{\dot{\wedge}} \overset{*}{\rho}_\pi z x \vee z = x$, also, annahmegemäß, $\hat{\pi} x$ für solch ein $x$ nach sich zieht, andererseits ${}_\rho\pi_z \rho \leq {}_\rho\pi_x \rho$ (im Fall $z = x$ aus (o 10), im Fall $\overset{*}{\rho}_\pi z x$ aus ${}_\rho\pi_x z$, nämlich über ${}_\rho\pi_z \rho < {}_\rho\pi_x \rho$ (dank (o 40), (o 43)) mit (o 33)) — wonach, wegen $\pi_1 \rho_1 \overset{\cap}{\leq} {}_\rho\pi_z \rho$, mit (o 9), (o 15) das $\rightarrow$-Antezedens in $\hat{\pi} x$ und somit $\neg\neg \underset{+}{\pi}(\pi_1 \rho_1)$ deduziert ist.

Daraufhin fehlt am Induktionsschritt zu $\hat{\pi} x_0$ in $I(\pi, \rho; \hat{\pi}, x)$ bloß noch $\underset{I(\pi_1, \rho_1)}{\bigwedge_{\pi_1 \rho_1}} \pi_1 \rho_1 \overset{\cap}{\doteqdot} {}_\rho\pi_{x_0} \rho \rightarrow \neg\neg \underset{+}{\pi}(\pi_1 \rho_1)$, wenn man berücksichtigt, daß mittels $\vdash \neg\neg D_{12} \leftrightarrow \neg\neg ((D_{12} \wedge D_{21}) \vee (D_{12} \wedge \neg D_{21}))$ und (o 7) folgt $\pi_1 \rho_1 \overset{\cap}{\leq} {}_\rho\pi_{x_0} \rho \leftrightarrow \neg\neg (\pi_1 \rho_1 \overset{\cap}{\doteqdot} {}_\rho\pi_{x_0} \rho \vee \pi_1 \rho_1 \overset{\cap}{<} {}_\rho\pi_{x_0} \rho)$, und daß doppelte Negation vor dem $\rightarrow$-Antezedens von $\hat{\pi} x_0$ tilgbar ist. Die fehlende Behauptung ergibt sich nunmehr aus der Induktionsvoraussetzung in

[17] Formelteile in eckigen Klammern dürfen nachträglich getilgt werden.

(o 44): für jedes $\pi_0 \rho_0$ mit $I(\pi_0, \rho_0)$ und $\pi_0 \rho_0 \stackrel{\cap}{\equiv} {}_{\rho}\pi_{x_0}\rho$ gilt aufgrund des vorigen Absatzes $\bigwedge_{\substack{\pi_1 \rho_1 \\ I(\pi_1, \rho_1)}} \pi_1 \rho_1 \stackrel{\cap}{<} \pi_0 \rho_0 \to \neg\neg \underset{+}{\pi}(\pi_1 \rho_1)$, denn der transitive Schluß von $\stackrel{\cap}{<}$ und $\stackrel{\cap}{\leq}$ auf $\stackrel{\cap}{<}$ ist ja (via (o 15)) selbstverständlich gültig; die so bewiesene Induktionsannahme in (o 44) liefert $\neg\neg \underset{+}{\pi}(\pi_0 \rho_0)$, folglich sind, diesen Schluß über $\pi_0 \rho_0$ generalisierend, Induktionsschritt und -voraussetzung in $I(\pi, \rho; \hat{\pi}, x)$ erzielt — und laut vorvorigem Absatz auch (o 44) selbst.

**2.38** Als Anfang einer weiteren (hier nicht mehr abgehandelten) konstruktiven Schlußkette, die im Satz über unbeschränkte transfinite Rekursion bei Definiensaufbau mit beschränkten Operatoren über Ordinalzahlen gipfelt, bringe ich noch den folgenden, eine schwache Verschärfung von (o 41) bietenden

**Satz 8.**

(o 45) $\quad \vdash I(P, R) \dot{\to} Px \to x \overset{R}{\underset{P}{}} \leq \overset{\overset{*}{R}_P}{\underset{P}{}} x \wedge x \overset{\overset{*}{R}_P}{\underset{P}{}} \leq \overset{R}{\underset{P}{}} x,$

(o 46) $\quad \vdash I(P, R) \to PR \equiv P\overset{*}{R}_P,$

(o 47) $\quad \vdash I(P, R) \wedge I(Q, S) \dot{\to} \neg PR \stackrel{\cap}{\leq} QS \leftrightarrow \neg\neg \bigvee_{\substack{z \\ P}} QS \stackrel{\cap}{\equiv} {}_R P_z R.$

**Beweise.** (o 45) durch $P \times P$-bedingte Induktionsschritte bezüglich $R \times \overset{*}{R}_P$ bzw. $\overset{*}{R}_P \times R$, denn einerseits $Px \dot{\to} Rxx_0 \to \overset{*}{R}_P xx_0$ und andererseits $\overset{*}{R}_P xx_0 \to \bigvee_{\substack{y \\ P}} Ryx_0 \dot{\wedge} \overset{*}{R}_P xy \vee x = y$, worauf $Py \wedge \overset{*}{R}_P xy \to$ $\to y \overset{R}{\underset{P}{}} > \overset{R}{\underset{P}{}} x$ (: (o 39)) mit (o 32), (o 26) die zweite Induktionsbehauptung liefert. (o 46) aus (o 45).

(o 47) durch P-bedingten Induktionsschritt bezüglich $\overset{*}{R}_P$ für $\bigwedge z_P \neg QS \stackrel{\cap}{\equiv} {}_R P_z R \dot{\to} Px \to \neg\neg \bigvee y_Q \, x \overset{R}{\underset{P}{}} \stackrel{\cap}{\leq} \overset{S}{\underset{Q}{}} y$ nach x; dieser wird dann sofort zu $PR \stackrel{\cap}{\leq} QS$ führen, und mit Kontraposition zu (o 47)-$\to$, während (o 47)-$\leftarrow$ aus (o 41), (o 17) folgt. Annahmegemäß gelte das $\dot{\to}$-Sukzedens für x mit $Px \wedge \overset{*}{R}_P xx_0 \wedge Px_0 \leftrightarrow {}_R P_{x_0} x$; also, falls $\dot{\to}$-Antezedens, $\bigwedge_{\substack{x \\ {}_RP_{x_0}}} \neg\neg \bigvee_{\substack{y \\ Q}} x \overset{R}{\underset{{}_RP_{x_0}}{}} \stackrel{\cap}{\leq} \overset{S}{\underset{Q}{}} y$ (: (o 38), (o 8), (o 14)) $\leftrightarrow {}_R P_{x_0} R \stackrel{\cap}{\leq} QS$ (: ($^{\cap}$D 2)), und daher $\neg QS \stackrel{\cap}{\leq} {}_R P_{x_0} R$ (: $\neg QS \stackrel{\cap}{\equiv} {}_R P_z R \leftarrow Pz$). Dies besagt

$$\neg \bigwedge_{\substack{y \\ Q}} \neg\neg \bigvee_{\substack{x \\ {}_RP_{x_0}}} y \overset{S}{\underset{Q}{}} \stackrel{\cap}{\leq} \overset{R}{\underset{{}_RP_{x_0}}{}} x \,..\leftrightarrow$$

$$\leftrightarrow \neg\neg \bigvee_{\substack{y_0 \\ Q}} \bigwedge_{\substack{x \\ {}_RP_{x_0}}} \neg\neg y_0 \overset{S}{\underset{Q}{}} \stackrel{\supset}{>} \overset{R}{\underset{{}_RP_{x_0}}{}} x \quad (: (o\ 16))$$

$$\leftrightarrow \neg\neg \bigvee_{\substack{y_0 \\ Q}} \bigwedge_{\substack{x \\ {}_RP_{x_0}}} \neg\neg \bigvee_{\substack{y \\ Q}} Syy_0 \wedge x \overset{R}{\underset{{}_RP_{x_0}}{}} \stackrel{\cap}{\leq} \overset{S}{\underset{Q}{}} y \quad (: (o\ 35))$$

$$\leftrightarrow \neg\neg \bigvee_{\substack{y_0\\Q}} x_0 \overset{\overset{*}{R}_P}{\underset{P}{}} \overset{\cap}{\leq} \overset{S}{\underset{Q}{}} y_0 \quad (: P x_0 \dot{\rightarrow} {}_R P_{x_0} x \leftrightarrow P x \wedge \overset{*}{R}_P x x_0,$$

$$(o\,38),\ (o\,8),\ (o\,14) \text{ und } (o\,45),\ (o\,8),\ (o\,14))$$

$$\leftrightarrow \neg\neg \bigvee_{\substack{y_0\\Q}} x_0 \overset{R}{\underset{P}{}} \overset{\cap}{\leq} \overset{S}{\underset{Q}{}} y_0 \quad (: (o\,45),\ (o\,8),\ (o\,14)),$$

womit die Induktionsbehauptung erzielt ist.

**2.39** Ohne Beweis seien noch diejenigen Daten zusammengestellt, welche sich auf die Nachfolgerbildung und die Limes- (bzw. Supremum-)-bildung beziehen, auf jene Operationen also, die naiv zur Erzeugung der zweiten Zahlklasse dienen, falls die Limesbildung auf Ordinalzahlfolgen vom Wohlordnungstyp $\omega_0$ eingeschränkt wird, deren Glieder bereits erzeugt sind.

**Satz 9.** Vom 2-stelligen Prädikat F sei vorausgesetzt, daß

$$\vdash F x y \vee \neg F x y \quad \text{und} \quad \vdash \bigvee_{\substack{y\\Fxy}} \bigwedge_{\substack{z\\Fxz}} y = z, \quad \vdash F x_1 y \wedge F x_2 y \rightarrow x_1 = x_2.$$

Mit ${}^F P y \leftrightharpoons \bigvee_x P x \wedge F x y$ und ${}^F R y_1 y_2 \leftrightharpoons \bigvee_{x_1 x_2} R x_1 x_2 \wedge F x_1 y_1 \wedge F x_2 y_2$ gilt dann

(o 48) $\vdash I(P, R) \leftrightarrow I({}^F P, {}^F R)$,

(o 49) $\vdash I(P, R) \dot{\rightarrow} F x y \rightarrow x \overset{R}{\underset{P}{\leq}} \overset{{}^F R}{\underset{{}^F P}{}} y \wedge y \overset{{}^F R}{\underset{{}^F P}{\leq}} \overset{R}{\underset{P}{}} x$,

(o 50) $\vdash I(P, R) \rightarrow PR \equiv {}^F P\, {}^F R$.

Es sei

$${}^+ P x \leftrightharpoons \bigvee_{x_0} x = {}'x_0 \wedge P x_0 \mathbin{\dot{\vee}} x = 0$$

und

$${}^+ R(P)\, x y \leftrightharpoons \bigvee_{x_0 y_0} x = {}'x_0 \wedge y = {}'y_0 \wedge R x_0 y_0 \mathbin{\dot{\vee}}$$
$$\dot{\vee} \bigvee_x x = {}'x_0 \wedge P x_0 \wedge y = 0;$$

dann gilt

(o 51) $\vdash I(P, R) \rightarrow I({}^+ P, {}^+ R(P))$,

(o 52) $\vdash I(P, R) \wedge I(Q, S) \dot{\rightarrow} PR \leq QS \rightarrow {}^+ P {}^+ R(P) \leq {}^+ Q {}^+ S(Q)$,

(o 54) $\vdash I(P, R) \rightarrow {}^+ P {}^+ R(P) > PR$,

(o 55) $\vdash I(P, R) \wedge I(Q, S) \dot{\rightarrow} QS > PR \rightarrow {}^+ P {}^+ R(P) \leq QS$.

Mit $\underline{R} x \leftrightharpoons \bigvee_y R x y \vee R y x$ sei von den Prädikatformen P(z), R(z) vorausgesetzt, daß $\vdash z_1 = z_2 \leftarrow \bigvee_x P(z_1)\, x \wedge P(z_2)\, x \mathbin{\dot{\vee}} \underline{R(z_1)}\, x \wedge \underline{R(z_2)}\, x$; Q(z), S(z) seien ebenso disjunkt vorausgesetzt. Für ${}^{\vee} P x \leftrightharpoons \bigvee z\, P(z)\, x$ und ${}^{\vee} R x y \leftrightharpoons \bigvee z\, R(z)\, x y$ gilt dann

(o 57) $\vdash \bigwedge z\, I(P(z), R(z)) \rightarrow I({}^{\vee} P, {}^{\vee} R)$,

(o 58) $\vdash \bigwedge z\,(I(P(z), R(z)) \wedge I(Q(z), S(z))) \dot{\rightarrow}$
$\dot{\rightarrow} \bigwedge z_1 \bigvee z_2\, P(z_1)\, R(z_1) \leq Q(z_2)\, S(z_2) \rightarrow {}^{\vee}P\, {}^{\vee}R \leq {}^{\vee}Q\, {}^{\vee}S,$

(o 60) $\vdash \bigwedge z\, I(P(z), R(z)) \rightarrow P(z)\, R(z) \leq {}^{\vee}P\, {}^{\vee}R,$

(o 61) $\vdash \bigwedge z\, I(P(z), R(z)) \wedge I(Q, S) \dot{\rightarrow} \bigwedge z\, P(z)\, R(z) \leq QS \rightarrow {}^{\vee}P\, {}^{\vee}R \leq QS.$

(o 53), (o 56) und (o 59), (o 62) wie (o 52), (o 55) und (o 58), (o 61), bloß mit $\overset{\cap}{\leq}$, $\overset{\supset}{>}$ anstelle $\leq, >$.

Damit sind $^+$ und $^{\vee}$ als mit $\equiv$ wie mit $\overset{\cap}{\equiv}$ verträgliche Operationen nachgewiesen, die den kennzeichnenden Nachfolger- bzw. Supremum-Eigenschaften $\alpha < \alpha^+$, $\alpha < \beta \rightarrow \alpha^+ \leq \beta$ und $\alpha(z) \leq \sup_z \alpha(z)$, $\bigwedge z\, \alpha(z) \leq \beta \rightarrow \sup_z \alpha(z) \leq \beta$ entsprechen.

**2.4** Es kann sein, daß jede in $\mathfrak{S}_1$ erreichte Ordinalzahl majorisierbar ist durch eine in $\mathfrak{S}_1$ erreichte Ordinalzahl mit einem uneingeengten, primitiv rekursiv dargestellten Repräsentanten $\bigvee$ R. Der Limes der in $\mathfrak{S}_1$ erreichbaren primitiv rekursiv dargestellten Ordinalzahlen ist ohne weiteres nämlich erst in der reflektorischen Ergänzung $\mathfrak{S}_1^1$ aus **0.52** deduktiv zu erreichen: man gehe, analog [16] S. 105—108, von einer eingeschachtelten zweifachen Rekursion aus, die eine Aufzählung $R(n)$ der 2-stelligen primitiv rekursiven Prädikate leistet; anschließend übertrage man die R(n) „isomorph" (: (o 50)) auf disjunkte ${}^{F}R(n)$, wobei also

$$\vdash z_1 = z_2 \leftarrow \bigvee x(\underbrace{{}^{F}R(z_1)}\, x \wedge \underbrace{{}^{F}R(z_2)}\, x);$$

anhand $Tz \leftrightarrows \bigvee z_0\, (d\, d_{\mathfrak{S}_1^0}(z_0, cf_0 I(\bigvee, {}^{F}R({}^{(z)}0))))$ stellt dann QS mit $Qx \leftrightarrows \bigvee z_T\, \underbrace{{}^{F}R(z)}\, x$ und $Sxy \leftrightarrows \bigvee z_T\, {}^{F}R(z)\, xy$ den gewünschten Limes dar, der in $\mathfrak{S}_1^1$, wo per Reflexion $Tz \rightarrow I(\bigvee, {}^{F}R(z))$ folgt, erreicht wird (: (o 57) für $z_T$ statt z).

Die unter **2.3*** vorexerzierte konstruktive Ordinalzahltheorie würde von solchen primitiv rekursiven Majoranten $\bigvee$ R nicht wesentlich vereinfacht. Bei Einengungen PR ist ja weiterhin mit P zu rechnen, die in $\mathfrak{S}_1$ weder entscheidbar noch stabil sind; auch für primitiv rekursive $\bigvee$ R kann man im allgemeinen weder auf (${}^{\cap}$D 1, 2) noch auf (${}^{\supset}$D 3, 4) verzichten. Und wer durchweg das tertium-non-datur benutzen will, verwischt dabei den konstruktiven Wahrheitsgehalt von $\vdash I(\bigvee, R)$, der als Definitionsprämisse für einen Niveauvergleich $\underset{\bigvee}{R} \overset{[\cap]}{\leq} \underset{\bigvee}{S}$ unentbehrlich ist.

## 3. Anhang: Zur transfiniten und exorbitanten Vollständigkeit der Sphäre $\mathfrak{S}_1$

Um eine ungefähre Abschätzung des Vollständigkeitsgrads von $\mathfrak{S}_1$ in knapper Form anzugeben, rückt der Text bis zum Schluß weiter von der Präzisionsstufe ab, die **2.** gegenüber **1.** aufrechterhalten hat. Ohne Kenntnis der Zielsetzung unter **3.2*** bliebe übrigens der bisherige kalkulatorische Aufwand unmotiviert.

**3.1** Was $\mathfrak{S}_1$ mit alleiniger Hilfe von Übersetzungen — also mit eindeutigen Umdeutungen (ohne „echte" vieldeutige) — mindestens auszudrücken vermag, fassen wir zusammen im

*Hauptsatz 1.* (Transfinite Vollständigkeit.)

Es läßt sich ein finites Übersetzungsverfahren aufweisen, das jede Deduktion in der formalisierten klassischen Ordinalzahltheorie ohne Ersetzungs- und Kardinalzahl-Axiomschema (: z.B. [20], Axiomgruppen 1.1.—3. und Axiomschema 2.1.) auf eine Deduktion in $\mathfrak{S}_1$ abbildet. Die Übersetzung verknüpft total verschiedene Objekte —etwa 0 mit $\bigwedge\bigwedge$, $\omega$ mit $\bigvee <$ etc.—, aber die Aussagen ($\bigwedge$), ($\bigvee$) mit sich selbst.

*Beweisskizze.* „Primitive" (: vgl. **1.62**) Aussageformen $\underset{+}{A}$ über Ordinalzahlen werden halbstabilisiert in $\mathfrak{S}_1$ übersetzt, indem man zuerst ein[17] ($\bigvee_\alpha$-freies) $^{[\bullet]\circ}\underset{+}{A}$ bildet und dann ähnlich wie bei Elimination einer bestimmten Variablensorte vorgeht, nur daß transfinite Variablen $\alpha$ durch zwei (1- und 2-stellige) Marken $\pi\rho$ mit der Bedingung $I(\pi, \rho)$ vertreten werden und auch das transfinite Primprädikat $\leq$ durch $\stackrel{\cap}{\leq}$ zu ersetzen ist. Ist $\underset{+}{A}$ ein primitives Aussageschema, so wird darin $\underset{+}{\pi}\alpha_1 ..$ mit $\underset{+}{\pi}(\pi_1\rho_1, ..)$, d.h. mit passenden $I(\pi_1, \rho_1) \wedge .. \overset{\wedge}{\underset{\vee}{\rangle}}_{{}^0\Pi_m}\, {}^0\Pi_l$ (wo $l \neq m$) übersetzt, für die im Übersetzungsantezedens noch Verträglichkeitsbedingungen bezüglich $\stackrel{\cap}{\equiv}$ hinzutreten:

$$\pi_1\rho_1 \stackrel{\cap}{\equiv} \dot{\pi}_1\dot{\rho}_1 \dot{\rightarrow} \underset{+}{\pi}(\pi_1\rho_1, ..) \rightarrow \underset{+}{\pi}(\dot{\pi}_1\dot{\rho}_1, ..).$$

(Beim Induktions-Aussageschema in (o 44) war diese Verträglichkeitsbedingung für $\underset{+}{\pi}$ entbehrlich, da $\underset{+}{\pi}\alpha$ dort deduzierbar bisubjunktiv einheitlich „verläuft".) Die Ergänzung der Übersetzung bei nicht primitiven $\underset{+}{A}$, die mit Hilfe der in **2.38** erwähnten unbeschränkt transfinit rekursiv (aber mit beschränkt aufgebautem Definiens) definierten Prädikate $\underset{+}{P}$ zusammengesetzt sind, ergibt sich auf dem Weg zu dem dort vorgemerkten Rekursionssatz in Gestalt einer Übersetzung von verallgemeinerten Kennzeichnungen ${}^{C}_{P}\iota^{\sim}_{x}$; diese relativen bedingten abstrakten Deskriptionen stehen unter Hypothesen C und Bedingungen P, die von der Halbstabilisierung auszunehmen sind, und beziehen sich auf eine abstrakte Gleichheit $\sim$, d.h. auf ein deduzierbar reflexives komparatives Prädikatschema (wie beispielsweise $\stackrel{\cap}{\equiv}$). — Wie solche $\iota_*$-Ausdrücke zu übersetzen und zu eliminieren sind, führt [10] I (S. 422—457) für einen einfacheren Fall aus; [11] p. 407—415 behandelt diesen Fall bei sukzessive eingeführten Funktionszeichen.

Werden Repräsentanten für variable transfinite Terme $\underset{+}{t}$ eingeführt, so sind sie auf Verträglichkeit mit $\stackrel{\cap}{\equiv}$ zu prüfen. Termmarken $\underset{+}{\vartheta}$ kann man durch Prädikatmarken $\underset{+}{\pi}$ vertreten, für die Eindeutigkeits- und Existenzeigenschaften $\bigwedge_{\alpha_1..} \underset{\underset{+}{\pi}\alpha_1..\alpha}{\bigvee_\alpha} \underset{\underset{+}{\pi}\alpha_1..\beta}{\bigwedge_\beta} \alpha = \beta$ vorauszusetzen sind, bevor die so entstandenen $\underset{+}{A}$ via ${}^{[\bullet]\circ}\underset{+}{A}$ in ein A von $\mathfrak{S}_1$ übersetzt werden. Verschiedene transfinite Terme $\underset{+}{t}, \underset{+}{s}$, für die $\bigwedge_{\alpha_1..} \underset{+}{t}\alpha_1.. = \underset{+}{s}\,\alpha_1..$, stellen „dieselbe" Funktion dar: $\lambda_{\alpha_1..}\underset{+}{t} = \lambda_{\alpha_1..}\underset{+}{s}$. Gleichungen zwischen Funktionen sind also Abkürzung für Aussagen über Ordinalzahlwerte von Termen. Funktionsmarken $\underset{+}{\varphi}$ sind Abstraktionen $\lambda_{\alpha_1..}\underset{+}{\vartheta}$ aus Termmarken. Terme, die wiederum in der formalisierten transfiniten Arithmetik beschränkt gekennzeichnet sind (: $\iota_{\alpha}\underset{<\gamma^+}{A}(\alpha)$), lassen sich mit ${}^{\mathrm{C}}_{\mathrm{P}}\tilde{\iota}_{\mathrm{x}}$ in $\mathfrak{S}_1$ ausdrücken.

Vorbehaltlich des Beweises für den in **2.38** vorgemerkten Rekursionssatz, für den sich nach (o 1—47) aber ebenfalls genügend methodische Anleitung vorfindet, ist daraufhin klar, daß und wie die Axiome [20] 1.1.—3. und 2.1. zu übersetzen und zu beweisen sind, wenn Repräsentanten für $\alpha^+$ und j vorliegen, also für den Nachfolgerterm und GÖDELS P-Funktion [6] (p. 29, 7.9).[18]

$\alpha^+$ kann $\triangleq$-verträglich durch ${}^+\pi{}^+\rho(\pi)$ vertreten werden laut **2.39.** Da $P\alpha\beta = P\beta 0 + \beta + \alpha, = P\alpha 0 + \beta$, je nachdem $\alpha \leq \beta$ oder $\alpha > \beta$, da $P\alpha 0 = \sum_{\underset{<\alpha}{\gamma}}(\gamma \cdot 2 + 1)$ und da $\sum_{\underset{<\alpha^+}{\gamma}}t(\gamma)$ die additive und multiplikative Termbildung umfaßt (: $\underset{+}{t}(0) + \underset{+}{t}(1) = \sum_{\underset{<2}{\gamma}}\underset{+}{t}(\gamma)$, $\beta \cdot \alpha = \sum_{\underset{<\alpha}{\gamma}}\underset{+}{t}(\gamma)$ für $\underset{+}{t}(\gamma) = \beta$), genügt es, beschränkte transfinite Summen $\triangleq$-verträglich darzustellen:

PR repräsentiere $\alpha$; $\underset{+}{t}(\underset{<\alpha}{*})$ sei durch Prädikat*formen* Q(z) S(z) repräsentiert, die für z mit Pz definiert sind und die disjunkte Prädikate zusammenfassen, d.h. $\bigvee_x Q(z_1)x \wedge Q(z_2)x \,\dot{\vee}\, \underbrace{S(z_1)}x \wedge \underbrace{S(z_2)}x. \rightarrow z_1 = z_2$; dann ist $P^{\Sigma}x \leftrightharpoons \underset{P}{\bigvee}{}_z Q(z)x$, $R^{\Sigma}xy \leftrightharpoons \underset{P}{\bigvee}{}_z S(z)\overline{xy} \vee \underset{P}{\bigvee}{}_{z_1 z_2} Rz_1z_2 \wedge Q(z_1)x \wedge$ $\wedge\, Q(z_2)y$ eine Darstellung für $\sum \underset{+}{t}$. Die für + und · benötigte disjunkte Zusammenfassung zweier Ordinalzahlrepräsentanten bzw. eines Repräsentanten für einen beschränkten konstanten Term läßt sich mit geraden und ungeraden Grundzahlen bzw. mit der Cauchy-Diagonalfolge, in der zx bekanntlich die Nummer $\frac{1}{2}((z + x)^2 + z + x \cdot 3)$ erhält, leicht bewerkstelligen — und die Rechenterme + und · für Grundzahlen sind sehr einfach mit 3-stelligen Prädikaten rekursiv definierbar (: S. 191 f.).

Um zu zeigen, daß die Regeln des elementaren klassischen Prädikatenkalküls mit Gleichheit über transfinite Aussageschemata $\underset{+}{A}$ in $\mathfrak{S}_1$ eliminierbar sind, muß man mit ${}_{S}w$ statt ${}_{\angle}w_1w_2$ arbeiten, sonst wäre das Übersetzungsantezedens manchmal lückenhaft; $\alpha = \beta \,\dot{\rightarrow}\, \underset{+}{A}(\alpha) \rightarrow \underset{+}{A}(\beta)$

[18] $g^1$, $g^2$ lassen sich dann beschränkt kennzeichnen, Iq auch. Bei 0,1-Bewertung von ($\bigvee$), ($\bigwedge$) braucht man max nicht, um Konjunktionen zu bewerten: setze $\curlyvee\alpha\beta = \mathrm{Iq}\,0\,\mathrm{Iq}\,\alpha\,\mathrm{Iq}\,0\,\beta$ wonach $\mathrm{Cj}\,\alpha\beta = \curlyvee\ \curlyvee\alpha\beta\curlyvee\alpha\beta$, wie geläufig (indem $\curlyvee$ der Bewertung beim Sheffer-Strich entspricht).

folgt dem $\underset{+}{\mathrm{A}}$-Aufbau parallel aus $\alpha = \beta \dot{\rightarrow} \alpha = \gamma \rightarrow \beta = \gamma$, und dies mit (o 15) aus $\beta \leq \alpha \leq \gamma \leq \alpha \leq \beta$.

**3.2** Die Entwicklung der arithmetischen Theorie ist bis hierhin auf äußerste Allgemeinheit ausgerichtet. Dies hat den Vorteil, daß das logische Gefüge möglichst durchsichtig wird — die zwar leichter faßlichen, aber logisch irgendwie lästigen Eindeutigkeitseigenschaften, wie sie etwa der Isomorphiebegriff mit sich bringt, brauchten bei (o 1—47) ersichtlich nicht mitgeschleppt zu werden.

Nachträglich könnte man das Fundierungsmaterial sieben. Klassisch beurteilt hätte jede Ordinalzahl — extrem linear — eine wohlgeordnete Darstellung (: $\vdash \mathrm{I}(\underset{\smile}{\mathrm{R}}, \mathrm{R}^{<}) \wedge \bigwedge_{xyz} \mathrm{R}xy \wedge \mathrm{R}yz \rightarrow \mathrm{R}xz \,.\, \wedge \underset{\underset{\smile}{\mathrm{R}}}{\bigwedge}_{xy} \mathrm{R}xy \vee \mathrm{R}yx. \wedge \bigwedge_{xy} \mathrm{R}xy \wedge \mathrm{R}yx \rightarrow x = y$)[19] und — extrem verzweigt — eine doppelt fundierte konsekutive Darstellung (: $\vdash \mathrm{I}(\mathrm{P}, \mathrm{R}) \wedge \mathrm{I}(\mathrm{P}, \mathrm{R}^{-}) \wedge \mathrm{R} \subseteq \mathrm{R}_{\mathrm{P}}^{\lhd}$, wo $\mathrm{R}_{\mathrm{P}}^{\lhd} xy \leftrightarrows \overset{1}{\mathrm{R}}_{\mathrm{P}} xy \wedge \neg \underset{>1}{\bigvee}_{z} \overset{z}{\mathrm{R}}_{\mathrm{P}} xy$)[20].

Interessant wird eine Siebung des in $\mathfrak{S}_1$ ableitbaren Fundierungsmaterials aber erst, wenn es darum geht, außer der „niederen" Ordinalzahltheorie (wie in **3.1**) auch die „höhere" Ordinalzahltheorie mit Ersetzungs- und Kardinalzahl-Axiomschema (z.B. [20] 2.2. und 2.3.) und mit ihr die axiomatische Mengenlehre in den Griff zu bekommen.

**3.21** Um zu zeigen, daß die Ausdrucksfähigkeit von $\mathfrak{S}_1$ weit über die in Hauptsatz 1 behaupteten Übersetzungsmöglichkeiten hinausgreift, lege ich zunächst einen Repräsentanten für Prädikatmarken vor, die über unbeschränkten Ordinalzahlen einen beliebigen, transfinit hohen wie langen Typ einnehmen:

$$\underset{(\neg \bigvee_y \mathrm{P}y \wedge \mathrm{R}yx)}{\underset{\mathrm{P}}{\bigwedge}_x} \mathrm{I}(\pi(x), \rho(x)) \overset{\mathrm{R}}{\underset{\mathrm{P}}{\rangle}}_{\sigma x_0} (\underset{\mathrm{P}}{\bigwedge}_x \mathrm{R}xx_0 \dot{\rightarrow}$$
$$\dot{\rightarrow} \underset{\mathrm{P}}{\bigwedge}_y \mathrm{R}yx_0 \dot{\rightarrow} \mathrm{S}yx \rightarrow \sigma y \,.\rightarrow \sigma x) \overset{\wedge}{\underset{\vee}{\rangle}}_{{}^0\Pi_m} {}^0\Pi_l \,;$$

die Fundierung R auf P stellt den Markentyp dar, insoweit es die Vorgängerbeziehungen zwischen (den Nummern von) seinen Untertypen betrifft, und die Wohlordnung S auf P bringt diejenigen Typen, welche einem Typ mit der Nummer $x_0$ jeweils unmittelbar vorangehen, in die gewünschte wohlgeordnete Reihenfolge; vorgängerfreie x indizieren Ordinalzahlvariablen.

---

[19] Nähme man I bezüglich R anstelle $\mathrm{R}^{<}$, so fehlt ein Repräsentant für die Ordinalzahl 1. Eben darum in [21] S. 268 auch fd($R$) mit $\overset{R}{<}$.

[20] Ordinalzahlen $\geq \omega_1$ sind, klassisch interpretiert, freilich nur „halbfinit" darstellbar, denn die fächerartige Vorgängerverzweigung steigt dann noch über unabzählbar viele, jedoch allemal endliche Fäden ab und (invers) auf, oder die Fäden verteilen sich auf unabzählbar viele disjunkte Büschel.

Konkrete Prädikate höheren Typs über Ordinalzahlen ergeben sich mit iterierten Rekursoren, die auf ausdrucksreichere Definiensschemata angewendet werden — z. B. in der Form $* \overset{*}{\underset{*}{\rightarrow}}\!\!\rangle_{\rho y} * \overset{*}{\underset{*}{\rightarrow}}\!\!\rangle_{\pi x} A(x, \pi;\ y, \rho)$, wobei die innere Fundierung über x auch durch $\rho$ verlaufen kann, also durch einen Parameter der inneren Rekursifikation hindurch. Uneingeschränkte Quantifikationen vermag $\mathfrak{S}_1$ allerdings ausschließlich über transfinite Objekte vom niedrigsten Typ auszudrücken, nämlich über Ordinalzahlen.

Was $\mathfrak{S}_1$ trotzdem unter Heranziehung von Umdeutungen mindestens aufzulösen imstande ist, formulieren wir im

*Hauptsatz 2.* (Exorbitante Vollständigkeit.)

Es läßt sich ein finites Umdeutungsverfahren aufweisen, das jeder Deduktion $d_\in$ in der ZERMELO-FRAENKEL-SKOLEMschen Mengenlehre beziehungsweise jeder Deduktion $d_<$ in der nach TAKEUTI formalisierten klassischen Ordinalzahltheorie (mit Ersetzungs- und Kardinalzahl-Axiomschema) eine Deduktion aus $\mathfrak{S}_1$ zuordnet. Die Umdeutung stimmt ordinalzahltheoretisch bei quantorenfreien Aussageformen mit der Übersetzung im Hauptsatz 1 überein, aber schon dabei hängt der zur Lösung der Umdeutung benötigte Substitutionsbezirk für freie Variablen von der Ausgangsdeduktion $d_<$ ab.

**3.22** *Beweisvorbereitungen.* Die Grundrelationen $\in$ und $=$ zwischen GÖDELS konstruktiblen Mengen $F^c\alpha$ (: [6] 9.3, p. 37) fließen innerhalb einer eigenständigen Ordinalzahltheorie aus TAKEUTIS Prädikat fn (: [19] p. 208), das mit beschränkt aufgebautem Definiens transfinit rekursiv definiert ist. Transfinite Rekursionen werden also in dem unter **2.38** erwähnten Umfang benutzt: $(\vee)\overset{\leq}{\underset{\vee}{\rightarrow}}\!\!\rangle_{\pi\alpha} \underset{\alpha}{A}$, wo $\underset{\alpha}{A}$ andeutet, daß $_{<\alpha}$ jeden Operator beschränkt, der am Aufbau von A beteiligt ist — ebenso nachher bei $\underset{\alpha}{\tau}$.

fn sei nun im Anschluß an die stabilisierende Übersetzung $^{\bullet\circ}$ einer Umdeutung $^{\vee\!\wedge}$ unterworfen, welche die GÖDELsche [7] transfinit fortsetzt und auf transfinite Rekursoren der eben genannten Form überträgt. Im folgenden bringe ich eine Verbalfassung der erforderlichen Typenkonstruktion und daraufhin den Aufbau der Kerne der umdeutenden Formeln; der Zweck der normierten Typenrechnung beim Umdeuten wird aus **3.23** ersichtlich.

*Typen.* [$\circ$ sei vom Typ $_\circ$.] Ordinalzahlen — konstante, beschränkt variable, markenmäßige — seien vom Typ $_0$. Mit $\tau_1, \tau_2$ seien auch $\smile\tau_1\tau_2$ und $\frown\tau_1\tau_2$ Typen; $\tau = \smile\tau_1\tau_2$ ist der Typ für die „Verbindung“ $c^\tau = \smile c_1^{\tau_1} c_2^{\tau_2}$ zweier Funktionale, $\tau = \frown\tau_1\tau_2$ ist der Typ für ein Funktional $c^\tau$ mit $\tau_1$-typigem Argument und $\tau_2$-typigem Wert (: ${}^{c}c^\tau c_1^{\tau_1} = c_2^{\tau_2}$). Ferner seien eingeführt: 1. eine Liste einstelliger Typmarken $\overset{0}{\tau}, \overset{1}{\tau}, \ldots$, 2. Typformen $\tau(\alpha)$, zunächst mittels Marken $\dot{\tau}$, später auch mittels

Typtermen $\tau = \Xi_{\dot{\tau}\alpha} v$, 3. beschränkte Typfolgen $\underset{<\gamma}{\lambda_\alpha} \tau(\alpha)$, 4. unbeschränkte Typfunktionen $\lambda_\alpha \tau(\alpha)$, 5. Typschemata $v(\alpha, .., \dot{\tau}, ..)$, die sich aus den konstanten Typen $_{\text{o}}$, $_{0}$ sowie aus Typformen $\dot{\tau}(\alpha)$ mit Hilfe der Typfunktoren $\smile$, $\frown$ und der Typoperatoren $\lambda_\alpha$, $\underset{<\gamma}{\lambda_\alpha}$ einschließlich des noch folgenden Typrekursors $\Xi_{\dot{\tau}\alpha}$ aufbauen lassen, 6. Typterme $\Xi_{\dot{\tau}\alpha}\tau(\alpha, \dot{\tau})$, die für beschränkt aufgebaute Typschemata $\tau = \underset{\alpha}{\tau}$ eine Typform rekursiv definieren, und zwar gemäß $\Xi_{\dot{\tau}\alpha}\tau(\alpha, \dot{\tau})(\beta) = \tau(\beta, \Xi_{\dot{\tau}\alpha}\tau)$; zudem simultan definierte Typterme $\Xi_{\smile\dot{\tau}_1\dot{\tau}_2\alpha} \smile \tau_1(\alpha, \dot{\tau}_1, \dot{\tau}_2)\tau_2(\alpha, \dot{\tau}_1, \dot{\tau}_2)$ derselben Art.

Zu jedem so erzielten Typschema $\tau$ werden außer Funktional„konstanten" $c^\tau$ noch Marken $a^\tau$ einkalkuliert. Es bezeichne $\Delta_{\substack{1\\2}} \smile c_1 c_2 = c_{\substack{1\\2}}$ den ersten bzw. zweiten Bestandteil bei Zerlegung einer Verbindung, und es stehe $\Delta_{..mn}$ für $..\Delta_m \Delta_n$; entsprechend bei $\smile a_1 a_2$ sowie $\smile\tau_1\tau_2$.

*Kerne.* Die mit $\vee\!\wedge$ bezeichneten operatorenfreien Kerne der erweiterten Gödelschen Umdeutung von $\rightarrow$, $\wedge$, $\vee$, $\underset{[<\gamma]}{\bigvee_\alpha}$, $\underset{[<\gamma]}{\bigwedge_\alpha}$, $(\bigvee)\overset{\leq}{\underset{\bigvee}{\Rightarrow}}_{\pi\alpha}$ bauen sich dann folgendermaßen auf:

$$\vee\!\wedge(\pi\alpha_1 ..) \leftrightharpoons (\vee\!\wedge \pi\alpha_1 .. c^{\tau(\alpha_1,..)}(\alpha_1, ..) a^{v(\alpha_1,..)}(\alpha_1, ..)),$$

mit passenden Marken[21] $\overset{k}{\tau}$, $\overset{l}{\tau}$ als Konstantentyp $\tau$ und Markentyp $v$, kürzer $\leftrightharpoons \vee\!\wedge\pi(\alpha_1, ..; c, a)$; $\vee\!\wedge\leq \,\leftrightharpoons\, \leq$ mit $\tau = v =_{\text{o}}$; ebenso $\vee\!\wedge\bigvee \leftrightharpoons \bigvee$, $\vee\!\wedge\bigwedge \leftrightharpoons \bigwedge$; wenn bereits $\vee\!\wedge B_1 \leftrightharpoons D_1(c_1, a_1)$ und $\vee\!\wedge B_2 \leftrightharpoons D_2(c_2, a_2)$, dann sei

$$\vee\!\wedge(B_1 \rightarrow B_2) \leftrightharpoons D(c, a) \leftrightharpoons D_1(\Delta_1 a, {}^c\Delta_2 c\ a) \rightarrow D_2({}^c\Delta_1 c\, \Delta_1 a, \Delta_2 a)$$
$$\text{mit } \tau = \smile\frown\tau_1\tau_2\frown\smile\tau_1 v_2 v_1,\ v = \smile\tau_1 v_2,$$

speziell $\vee\!\wedge(\neg B_1) \leftrightharpoons \neg D_1(a, {}^c c\, a)$ mit $\tau = \frown\tau_1 v_1$, $v = \tau_1$, sofern

$$\frown\acute{\tau}_{\text{o}} = {}_{\text{o}},\ \smile_{\text{o}}\acute{\tau} = \smile\acute{\tau}_{\text{o}} = \acute{\tau}[= \frown_{\text{o}}\acute{\tau}],$$

weiter $\vee\!\wedge(\neg\neg B_1) \leftrightharpoons \neg\neg D_1({}^c c\, a, {}^c a\, {}^c c\, a)$ mit $\tau = \frown\frown\tau_1 v_1 \tau_1$, $v = \frown\tau_1 v_1$; ebenso sei dann

$$\vee\!\wedge(B_1 \wedge B_2) \leftrightharpoons D(c, a) \leftrightharpoons D_1(\Delta_1 c, \Delta_1 a) \wedge D_2(\Delta_2 c, \Delta_2 a)$$

mit $\tau = \smile\tau_1\tau_2$, $v = \smile v_1 v_2$,[22]

$$\vee\!\wedge(B_1 \vee B_2) \leftrightharpoons D(c, a) \leftrightharpoons \Delta_1 c = 0 \wedge D_1(\Delta_{12} c, \Delta_1 a)\ \dot{\vee}$$
$$\dot{\vee}\ \Delta_1 c = 1 \wedge D_2(\Delta_{22} c, \Delta_2 a) \quad \text{mit} \quad \tau = \smile_0\smile\tau_1\tau_2,\ v = \smile v_1 v_2;$$

---

[21] Einstellige $\pi$ genügen, wenn man sich auf fn konzentriert. Sonst sind auch Listen mehrstelliger Typmarken nötig.

[22] In Spectors Verifikation von [7] ist die Auflösung der Umdeutung für die Schlußregel A 7 nicht „similar to A 6" (: [18] p. 14, Zeile 4) zu konstatieren. In Fußnote 7 dazu (: [18] p. 10) meint Kreisel, A 7 könne im Sinne von [7] S. 286 durch Perm ersetzt werden. Zur Elimination von A 7 gehe ich aus von Sum für $\wedge$ anstelle $\vee$, und ich benötige außer Syll (d.i. Spectors A 4) die Dualisierung von Taut wie Perm. Sum für $\wedge$ ist in [7] nicht postuliert; es zu eliminieren, gelang mir nicht — aber dessen Umdeutung ist, im Unterschied zu der von A 7, leicht aufzulösen.

wenn bereits ${}^{\curlywedge}\mathrm{B}_0(\alpha) \leftrightharpoons \mathrm{D}_0(\alpha;\, \mathrm{c}_0, \mathrm{a}_0)$, dann sei

$$ {}^{\curlywedge}(\underset{[<\gamma]}{\vee_\alpha} \mathrm{B}_0) \leftrightharpoons \mathrm{D}(\mathrm{c}, \mathrm{a}) \leftrightharpoons [\Delta_1 \mathrm{c} < \gamma \wedge]\, \mathrm{D}_0(\Delta_1 \mathrm{c};\, \Delta_2 \mathrm{c}, \mathrm{a}) $$

$$ \text{mit} \quad \tau = \underset{[<\gamma]}{\smile_0 \lambda_\alpha} \tau_0,\; v = \underset{[<\gamma]}{\lambda_\alpha} v_0, $$

$$ {}^{\curlywedge}(\underset{[<\gamma]}{\wedge_\alpha} \mathrm{B}_0) \leftrightharpoons \mathrm{D}(\mathrm{c}, \mathrm{a}) \leftrightharpoons [\Delta_1 \mathrm{a} < \gamma \rightarrow]\, \mathrm{D}_0(\Delta_1 \mathrm{a};\, {}^{\mathrm{c}}\mathrm{c}\,\Delta_1 \mathrm{a}, \Delta_2 \mathrm{a}) $$

$$ \text{mit} \quad \tau = \underset{[<\gamma]}{\frown_0 \lambda_\alpha} \tau_0,\; v = \underset{[<\gamma]}{\smile_0 \lambda_\alpha} v_0 $$

(beachte, daß in ${}^{\mathrm{c}}\mathrm{c}\,\Delta_1 \mathrm{a}$ vom Typ $\underset{[<\gamma]}{\lambda_\alpha} \tau_0$ noch $\Delta_1 \mathrm{a}$ substituiert wird, so daß ${}^{\mathrm{cc}}\mathrm{c}\,\Delta_1 \mathrm{a}\,\Delta_1 \mathrm{a}$ vom Typ $\tau_0(\Delta_1 \mathrm{a})$ ist, ähnlich ${}^{\mathrm{c}}\Delta_2 \mathrm{a}\,\Delta_1 \mathrm{a}$ vom Typ $v_0(\Delta_1 \mathrm{a})$, und zuvor ${}^{\mathrm{c}}\Delta_2 \mathrm{c}\,\Delta_1 \mathrm{c}$ vom Typ $\tau_0(\Delta_1 \mathrm{c})$, ${}^{\mathrm{c}}\mathrm{a}\,\Delta_1 \mathrm{c}$ vom Typ $v_0(\Delta_1 \mathrm{c})$); wenn bereits ${}^{\curlywedge}\underset{\alpha}{\mathrm{B}_0}(\alpha, \pi) \leftrightharpoons \mathrm{D}_0(\alpha, \mathrm{D}_\pi, \mathrm{c}_\pi, \mathrm{a}_\pi;\, \mathrm{c}_0, \mathrm{a}_0)$ mit $\tau_0(\alpha, \overset{k}{\tau}, \overset{l}{\tau})$, $v_0(\alpha, \overset{k}{\tau}, \overset{l}{\tau})$, dann sei ${}^{\curlywedge}((\vee) \overset{\leq}{\underset{\vee}{\rightarrow}}_{\pi\alpha} \underset{\alpha}{\mathrm{B}_0} \bullet \beta) \leftrightharpoons \mathrm{D}(\beta;\, \mathrm{c}, \mathrm{a}) \leftrightarrow \mathrm{D}_0\big(\beta, \mathrm{D}, \overset{\beta}{\mathrm{c}}, \overset{\beta}{\mathrm{a}};\, \mathrm{c}_0, \mathrm{a}_0\big)$, wo $\overset{\beta}{\mathrm{c}} = \underset{<\beta}{\lambda_\alpha} \mathrm{c}$, $\overset{\beta}{\mathrm{a}} = \underset{<\beta}{\lambda_\alpha} \mathrm{a}$ und $\mathrm{c}(\alpha) = \mathrm{c}_0(\alpha, \overset{\alpha}{\mathrm{c}}, \overset{\alpha}{\mathrm{a}})$, $\mathrm{a}(\alpha) = \mathrm{a}_0(\alpha, \overset{\alpha}{\mathrm{c}}, \overset{\alpha}{\mathrm{a}})$, mit

$$ \smile \tau v = \equiv \smile_{\overset{k}{\tau}\overset{l}{\tau}\beta} \smile \tau_0(\beta, \underset{<\beta}{\lambda_\alpha} \overset{k}{\tau}, \underset{<\beta}{\lambda_\alpha} \overset{l}{\tau})\, v_0(\beta, \underset{<\beta}{\lambda_\alpha} \overset{k}{\tau}, \underset{<\beta}{\lambda_\alpha} \overset{l}{\tau}). $$

Die Typenrechnung für ${}^{\curlywedge\bullet\circ}\mathrm{fn}$ hängt ordinalzahlarithmetisch noch von den in die Definition von fn eingehenden transfiniten Funktionen j (und $\mathrm{g}^1$, $\mathrm{g}^2$), $\mathrm{g}_1$, $\mathrm{g}_2$ (und $\mathrm{g}_0$) ab (: [19] p. 202, 206 oder [20] p. 296, 301), welche GÖDELS Funktionen P, $\mathrm{K}_1$, $\mathrm{K}_2$ entsprechen (: [6] 7.9, 9.24; p. 29, 36). Für Argumente, die in CANTORscher Normalform gegeben sind, lassen sich diese Funktionen direkt auswerten, da man $\mathrm{P}\beta 0 = \sum_{\gamma < \beta} (\gamma \cdot 2 + 1)$ nach meiner vom Januar 1964 datierenden Notiz [24] explizit in $+$, $\cdot$, $\bigwedge$ (: **) ausdrücken kann. Die dortige Summationsformel mit 9 Fällen — explizit spaltet sich der 3. Fall in 4 Unterfälle, dem 2., 4., 5., 6. Fall entsprechend — ziehe ich inzwischen[23] auf 2 Fälle zusammen:

Es sei $k, l, m < \omega$, $0^0 = 1$, $g\alpha = \iota_\nu \omega^\nu \leq 1 + \alpha < \omega^{\nu+1}$ und $\eta < \omega^\xi$, $g(\omega \cdot \beta^*) < g(\omega^{\omega^\xi \cdot (1+k)} \cdot (1 + \omega^\eta \cdot l))$; dann wird $\mathrm{P}\beta 0 = l^2$,

$$ = \omega^{\omega^\xi \cdot (1+k)} \cdot (\omega^{\omega^\xi \cdot k} \cdot (1 + \omega^{\omega^\xi + \eta} \cdot l) + (1 + \omega^\eta \cdot l) \cdot (\omega \cdot \beta^* + 2m)) + $$
$$ + \omega \cdot \beta^* \cdot (1 - 0^m) + m, $$

je nachdem $\beta = l, = \omega^{\omega^\xi \cdot (1+k)} \cdot (1 + \omega^\eta \cdot l) + \omega \cdot \beta^* + m$; für jedes $\beta$ existiert genau ein Tupel $\xi, k, \eta, l, \beta^*, m$, falls $\eta > 0 \rightarrow l > 0$.

Beweisanleitung: die Induktionsaufgabe vereinfacht sich durch ein Lemma, wonach $\mathrm{P}\beta_0 0 = \mathrm{P}\beta_1 0 + \omega^{g\beta_1} \cdot \beta_2$, wenn

---

[23] Anläßlich einer internationalen Tagung über „Universelle Algebra“ im Mathematischen Forschungsinstitut Oberwolfach habe ich am 29. Juli 1966 einen Vortrag „Über eine formal-universelle Arithmetik“ gehalten, bei dem die neue Version ohne Beweis bekanntgegeben wurde.

$$\beta_0 = \beta_1 + \beta_2 \wedge (g\,\beta_2 < g\,\beta_1 \wedge \beta_2 = \omega \cdot \beta_2^* \dot{\vee}$$
$$\dot{\vee}\, 0 < g\,\beta_1 \wedge 0 < l' < \omega \wedge 0 \leq \mu < \omega^{g g \beta_1} \wedge \beta_2 = \omega^{g \beta_1 + \mu} \cdot l').$$

Zur Lösung der ⋁∧•°-Umdeutung von $d_<$, die (gemäß [20] p. 316) aus $d_\in$ hervorgehen, werden in erster Linie primitiv rekursive Funktionale jeden durch ⋁∧•°fn geforderten Typs benutzt. Abweichend von [20] wird die beschränkte transfinite Minimumbildung $\mu_{\underset{<\alpha}{\xi}}$ nicht zu den primitiv rekursiven transfiniten Operationen gezählt, weil konstruktiv im allgemeinen unentschieden bleibt, ob $\bigvee_{\underset{<\alpha}{\xi}} \mathrm{f} = 0$ oder $\bigwedge_{\underset{<\alpha}{\xi}} \mathrm{f} \neq 0$. (Eben deshalb nehmen wir auch das Prädikat fn in [19] p. 208 zum Ausgangspunkt und nicht die Funktion fn in [20] p. 302.)

Um innerhalb $\mathfrak{S}_1$ primitiv rekursive Funktionale für jeden erforderlichen Typ zu erhalten, chiffriere man zunächst, wenn $\gamma$ durch PR repräsentiert wird, auf P bezüglich R die „bis" $\gamma$ beschränkt aufgebauten Typschemata. Rekursive Definitionen von Funktionalrepräsentanten stützen sich dann auf eine teils zur Fundierung abgestellte, aber gleichzeitig als Typnumerierung $\ulcorner\tau\urcorner$ fungierende Variable; die einfache Verschachtelung $^{\mathrm{c}}\mathrm{c}^{\tau}\mathrm{c}_1^{\tau_1} = \mathrm{c}_2^{\tau_2}$ wird entflochten zu einer Aussage

$$\mathrm{B}(\mathrm{Q}(\ulcorner\tau\urcorner, ..), \mathrm{Q}_1(\ulcorner\tau_1\urcorner, ..), \mathrm{Q}_2(\ulcorner\tau_2\urcorner, ..)),$$

welche $=$ auf $\stackrel{\frown}{\equiv}$ zwischen Repräsentanten vom Typ $_0$ zurückspult; die zusammenfassende Überlagerung begrenzter transfiniter Schachtelung mittels $\underset{<\beta(\leq\gamma)}{\lambda_\alpha}$, im allgemeinen via $\equiv_{\smile\dot{\tau}_1\dot{\tau}_2\alpha}$ selbst transfinit geschachtelt, ist mit iterierten Rekursoren zu bewerkstelligen. Bei Einführung von Repräsentanten für Funktionalmarken der „bis" $\gamma$ numerierten Typschemata dient die $=$-entflechtende Übersetzung der Eindeutigkeitsforderung als Hypothese. Nachdem für festgehaltenes $\gamma$ alle gewünschten Termarten dargestellt sind, kann man PR noch schematisch durch $\pi\rho$ unter $\mathrm{I}(\pi, \rho)$ ersetzen und bei verschiedenen Darstellungen „desselben" Abschnitts gewisse Isomorphieeigenschaften beweisen, die den Übergang von $\underset{<\gamma}{\lambda_\alpha}$ zu $\lambda_\alpha$ rechtfertigen.

**3.23** *Abriß der Beweisausführung (Obstruktionstheorie).* Der weitere Gang der Überlegung sei im vorliegenden Rahmen wenigstens noch umrissen. Zur besseren Orientierung über die Vielfalt der (letztlich wieder kanonisch) zu regelnden Denkvorgänge diene eine zeichnerische Wiedergabe ihres kybernetischen Schemas.

Der äußere Kreis, bei dem Prädikate $P$, Aussagen $A$ und Sätze $S$ miteinander verflochten sind, ist in **1.** ausgiebig behandelt; die Zebrastreifen auf den Wegen $PA$ und $AS$ deuten die Regeln 9.2.$*$ bzw. 9.3.$*$ an, die Streifendichte steht in grobem Verhältnis zur Regelzahl, der Pfeil $SP$ entspricht der Regel 9.1.5. (Für 9.1.1—4 fehlt ein zeichnerisches Gegenstück ebenso wie für 1. bis 7. und 8.1.—3..) Der Übergang $SO$ zu transfiniten Ordinalzahlen $O$ ist aus **2.** geläufig.

Den Schritt von $O$ und $P$ zu Chiffrierungen $C$, sowie von da zurück nach $S$, jedesmal per Reflexion, kennen wir (wenn auch nicht kanonisch) aus **0.52.** Dieser koerzible Perfektionsprozeß läßt sich — analog **0.51** — ebenfalls durch Einfädelung von Ordinalzahlmarken via *PO* aufstocken. Soweit es sich um das Zermelo-Fraenkel-Skolemsche System handelt, brauchen wir *C* aber nicht und können von *SO* allein starten (ohne *OP—C*, *CS* und erst recht ohne *PO*).

Die Punktspiralen sollen die aus den Regeln produzierten Formeln wiedergeben; die im Verschlüsseln steckende Widerspiegelung klingt im Wechsel des Drehsinns der Spiralen an. Die unterschiedliche Breite der Pfeile soll die Häufigkeit der Regelanwendung zeigen — z.B. wird der dünne Weg *PO* am seltensten herangezogen.

Von der restlichen Linienführung in der Zeichnung können wir uns vorerst bloß eine bildhafte Verdeutlichung machen, ehe wir daraus die im Vorwort erwähnten Bezirke $\mathrm{T}(d_<)$ von Ordinalzahlen *O*, Typen *T* und Funktionen *F* konkretisieren.

Die Sperre auf dem Weg *PO* rührt vom Kontrollzentrum *K* her, das seine obstruktive, dem freien Produzieren entgegenwirkende Tendenz — von der Verzögerung über die Hemmung bis zur Drosselung — aber erst einmal auf die kleineren Schnecken *T*, *F*, *N* (Numerierung) und auf *L* (Limitation) ausübt. Während *K* die von der Quelle *O* gespeiste Ausfuhr steuert, hat das Zählregister *Z* die wieder in *O* mündende Einfuhr, die teils vom Wohlordnungswerk *W* abfließt, zu sichten und die Neueingänge vor der Abfertigung ihrer Herkunft nach zu etikettieren. Der von *K* ausgehende Riegel ist wegen *OC* unterbrochen und kehrt an der Bruchstelle seine Durchlaßrichtung um — was auch die Strichelung zeigen soll. Die von *Z* ausgehende Einlaßschranke hat demgegenüber keine obstruktive Wirkung, sondern eine über-

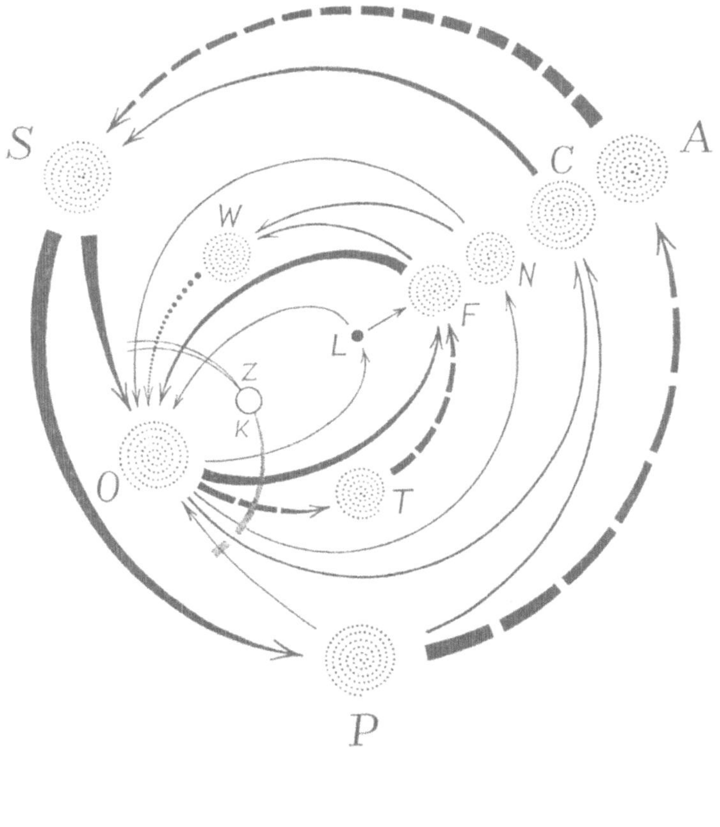

wachende Aufgabe. $Z$ und $K$ sind zu einem Leitkreis aufeinander abgestimmt: $K$ richtet sich bei der Steuerung nach der Etikettierung und darf seine Anweisungen an die Adressen von $O$, $F$, $T$, $N$, $W$, $L$ nur geben, solange $Z$ noch registrierte Einsatznummern zum Abruf frei vorrätig hat.

Zur Konstruktion von $\mathrm{T}(d_<)$ gehen wir aus von den Typen, die in den $\mathsf{\Lambda}^{\bullet\circ}$-Kernen der Formeln von ganz $d_<$ vorkommen. (Eine nachträgliche Umformung oder Verlängerung von $d_<$ wird nicht erlaubt, weil dann die schließliche Lösung der Anfangszahlfunktion $\omega_\alpha$ i. a. wechselt.) Das Ziel ist nun, eine aus möglichst wenig Ordinalzahlen, Typen und Funktionen bestehende Konstanten-Hülle zu konstruieren, innerhalb der $\mathsf{\Lambda}^{\bullet\circ}d_<$ lösbar ist.

Wegen des Kardinalzahl-Axiomschemas muß das aufzuzählende System $\mathrm{T}(d_<)$ so ausgeglichen sein, daß man darin in nuce eine Spielart von GÖDELS $\in$-isomorpher „Kompression" [6] 12.3—6 (p. 54—61) durchführen kann, nur daß hier, obstruktiv, die für Marken zur Substitution zuzulassenden Konstanten im Aufzählbaren selbst einzuschränken sind und das Komprimat isomorph zu sein hat gegenüber einer Reihe von Operationen mit Ordinalzahlen, Typen und Funktionen. Im Unterschied zu GÖDEL [6] p. 37, wo die Operationen $\mathfrak{F}_{1-8}$ und eine Zusammenfassung vorgängiger Erzeugnisse als 9. Operation zyklisch zur Anwendung kommen, arbeiten wir mit etwas mehr rhythmisch akzentuierten Einsätzen, um die Produktion neu anzukurbeln, nachdem sie sich anderwärts in Teilabgeschlossenheiten erschöpft hat.

Die besagten Operationen sind bei den Ordinalzahlen: die Rechenoperationen; die Bildung des Wohlordnungstyps von Funktionswert-Nummern zu konstanten Argumenten, deren Nummern einem Ordinalzahlabschnitt angehören, dessen Grenze $L$ bereits ausgegeben hat; die Limitation selbst in Anwendung auf das noch zu besprechende Schaltschema zwischen $O—N—O$ und $O—F—O$. Bei den Typen sind es: $\smile$, $\frown$, $\underset{<\gamma}{\lambda_{\alpha}}$; ein Typformschema, das derjenigen transfiniten Schachtelung entspricht, die zur Lösung der umgedeuteten Schlußregel der transfiniten Induktion[24] erforderlich ist — dieses Schema setzt die finite Schachtelung mit konstantem Werttyp in SPECTORS Lemma zur Verifikation von [7] (: [18] 9. D, p. 14/15) transfinit fort, indem es die jeweils vorangegangenen Typen zu einer Folge zusammenfaßt, zumindest an den Limesstellen. Sonstige Typfunktionen, -schemata und -terme aus $\mathsf{\Lambda}^{\bullet\circ}d_<$ dienen lediglich als zusätzliche Anfänge. Bei den Funktionen sind es: die primitiv rekursiven Funktionalschemata (ohne unberechenbare, auf transfinit viele Auswertungen reflektierende Modifikation); die beschränkte Ab-

---

[24] Beim Umdeuten der transfiniten Induktion ist es bequemer, Prämisseneigenschaften heranzuziehen und demgemäß nicht mit einem Axiomschema zu arbeiten, sondern mit einem Schlußschema, dessen Prämisse die $<$-Progressivität ist.

straktion $\lambda_{\alpha \atop <\gamma}$; eine Funktion, welche die Limitationsausgaben (als Werte) ihren Eingabenummern (als Argumenten) zuordnet.

Der obstruktive Arbeitsvorgang sei grob wie folgt charakterisiert.

Wenn $K$ (nach Abruf in $Z$) bei $O$ neue Ordinalzahlen anfordert, so wendet $O$ die Rechenoperationen auf obstruktiv schon registrierte Ordinalzahlen an; die restlichen Ordinalzahlbildungen nehmen $W$ und $L$ vor. Bei Anforderung von Typen wendet $T$ die Typoperationen und -anfänge auf dort erzielte Typen an; diese Typen sind zugleich für die Typmarken einzusetzen, und als Substitute für transfinite Variablen werden registrierte Ordinalzahlen herangezogen ($OT$). $T$ gibt neben geschlossenen Typen auch Typformen an $F$ weiter ($TF$).

Bevor $K$ jedoch Typen anfordert, erteilt $K$ entsprechende Aufträge an $F$ zwecks Funktionsbildungen; dabei kann $F$ auch jenen Abschnitt der vorhin letztgenannten Funktion anwenden, den $L$ bereits sukzessive angewiesen hat. $F$ bildet selbsttätig alle Funktionswerte für Argumente, die dort schon erzielt worden sind oder die, sofern $_0$ ihr Typ ist, den Weg $OF$ dank Registriervermerk seitens $Z$ passiert haben; die entstandenen Werte vom Typ $_0$ läßt $F$ wieder durch $Z$ überprüfen ($FO$). Registriert $Z$ via $O-F-O$ keine Neueingänge mehr, fragt $K$ bei $Z$ an, ob $T$ noch einsatzfähig ist. Im bejahenden Fall wird der Kreis $O-F-O$ neu angetrieben durch Typen, die $O-T-F$ einschleust.

Ist auch die Bahn $O-F-O\dot{-}T-F$ abgeschlossen, dann versucht $K$ den Abruf von $N$. Im Erfolgsfall ($ON$) liefert $N$ eine transfinite Numerierung aller bis dahin obstruktiv angesammelten Ordinalzahlen, Typen und Funktionen; diese „Auszählung", die mit Rücksicht auf spätere Kompressionsakte rechnerisch besonders ausgefeilt ist, läßt sich mit einem Prädikat aus $\mathfrak{S}_1$ vollziehen. Die als Nummern benutzten Ordinalzahlen werden wieder registriert ($NO$). In Zusammenarbeit mit $F$ bildet $N$ selbsttätig solche Funktionen vom Typ $^\frown{}_{00}$, die Nummernbilder der bei $F$ gespeicherten Funktionen sind, und schickt sie nach $W$ ($NF-W$). $W$ bildet daraus die Wertwohlordnungen für Argumentabschnitte, die $L$ bereits begrenzt hat; diese Wohlordnungen stellen Ordinalzahlen dar, die in $\mathfrak{S}_1$ deduktiv zu erreichen sind — und zwar mit (o 2) in **2.321**, sobald die Grenze des Nummernbezirks erreicht ist. Alle Ordinalzahlen unterhalb solcher Wohlordnungstypen überweist $W$ an $Z$, wonach lückenlos besetzte, „inkompressible" Ordinalzahlabschnitte zur Verarbeitung kommen ($WO$). So wird $O-F-O\dot{-}T-F$ neu belebt durch $O-N-O\dot{-}NF-W-O$.

Ist das Wechselspiel zwischen diesen beiden Teilschaltungen wiederum zur Stagnation gekommen, so fordert $K$ bei $Z$ die nächste registrierte Ordinalzahl an, die noch nicht als Eingabenummer für $L$ abgerufen wurde; $Z$ gibt diese selbst aber nur dann frei, wenn sie mitsamt allen vorgängigen Eingabenummern einem bereits von $W$ ausgegebenen Ordinalzahlabschnitt entstammt — sonst behält sich $Z$ vor, aufgrund späterer Ausstöße von $W$ sämtliche bis dahin aus Ersatznummern entsprungenen Produkte umzudatieren. Im Freigabefall ($OL$) liefert $L$ den Limes aller von $W$ schon obstruktiv erreichten Wohlordnungen; dieser Limes ist wieder in $\mathfrak{S}_1$ deduktiv zu erreichen, weil die benutzten Abschließungen letzten Endes über eine Iteration mit Grundzahlindex durch gewöhnliche vollständige Induktion beherrscht werden. $L$ gibt $F$ den neuen Limeswert zur Ergänzung der limitierenden Funktion bekannt ($LF$) und läßt ihn registrieren ($LO$).

Wenn $Z$ schließlich — ohne die äußere Hilfe weiterer Schaltelemente — über keine Ordinalzahl mehr verfügt, die $L$ zum Einsatz bringt, dann sind wir fertig.

Mit dem so erhaltenen System $\mathrm{T}(d_<)$ ist $\bigwedge^{\bullet\circ} d_<$ lösbar. Um dies für die in $d_<$ auftretenden Anwendungen des Kardinalzahl-Axiomschemas einzusehen, muß nachgewiesen werden, daß jede in $F$ vorrätige Funktion vom Typ $\frown_{00}$ nach Einengung auf einen Argumentabschnitt, dessen Grenze ein von $L$ ausgegebener Limes ist, Wertlücken aufweist, die dem nächsten von $L$ ausgegebenen Limes vorangehen. Und das geschieht durch ein Kompressionsverfahren innerhalb $\mathrm{T}(d_<)$.

Zur Konstruktion dieser Hierarchie der $\mathrm{T}(d_<)$ könnte man die Hilfsformen und -schemata nachträglich abstreifen, um sich einerseits auf den jeweils entstandenen Konstantenbezirk zu beschränken und um andererseits innerhalb $\mathfrak{S}_1$ ohne quantifikativ gebundene Marken sowie ohne relative Definitionen vorzugehen. Damit kommen wir einer Inspiration von WEYL entgegen, der sich, wie bekannt (: etwa [26], insbesondere 5., 6.), dagegen gesträubt hat, die erste Schicht einer allein mit Quantifikation über Grundzahlen errichteten Typenkonstruktion zu überschreiten und sich für einen möglichst weiten, auf einen Schlag eingesetzten Vorrat an Grundoperationen ausgesprochen hat.

Will man mengentheoretische Systeme, die über das ZERMELO-FRAENKEL-SKOLEMsche System hinausreichen, aber den elementaren klassischen Prädikatenkalkül mit Gleichheit durch extralogische Axiome oder Axiomschemata verstärken, nach entsprechender Umdeutung auflösen, so hat man in einfacheren Fällen — etwa beim BERNAYS-GÖDEL-System — zusätzliche Schaltstellen für Limitationen einzuführen oder in verwickelteren Fällen solche Schaltstellen mit Hilfe eines Schaltelements für Indizierungen zu staffeln.

**3.24** *Digression.* Eine nähere Inspektion des Wf.beweises führt zu einer Ordinalzahl, die „relativ exorbitant" ist, indem sie die Deduktionsstärke des ZERMELO-FRAENKEL-SKOLEMschen Systems (: ZF) mißt oder majorisiert. Es ist denkbar (oder vielleicht zu vermuten), daß diese Grenzzahl verhältnismäßig niedrig ist — so klein, daß sie auch innerhalb ZF oder gar in einem Teilsystem der klassischen Analysis deduktiv zu erreichen ist. Da die Lösung der Umdeutung für eine solche Erreichbarkeits-Deduktion keine ebensolche ist, ergäbe dies keinen unmittelbaren Widerspruch zu GÖDELs Wf.unableitbarkeit [5].

Die in [5] benutzten Beweismittel sind zwar höchst konstruktiv (: [14] S. 129 bis 132), aber nicht finit, jedenfalls nicht direkt anschaulich[25]. Finit steckt in "$r$ ist ableitbar" der Aufweis einer konkreten Ableitung, die mit $r$ endet; und solche Aussagen $\vdash r$ sind finit nicht negationsfähig, sofern nicht zufällig ein direktes Entscheidungsverfahren (vgl. **0.12**) vorliegt. Es steht jedoch frei, $\nvdash f$ finit als Redeweise für "jede Ableitung endet mit einer Figur $r$, die semiotisch ungleich $f$" zu verstehen. Diese semiotische Ungleichheit kommt nicht unter den Primfiguren vor, mit denen der Ableitbarkeitsbegriff arbeitet, und ist leicht direkt zu entscheiden. Finite Unableitbarkeitsbeweise erfolgen dann durch direkte Ableitungsinduktion für die vereinbarte Redeweise; der Induktionsschritt muß dabei von Annahmen ausgehen, die jeweils schon anschaulich zu verwirklichen sind. Die Unableitbarkeitsbeweise in [5] erfolgen demgegenüber indirekt: die hypothetische Annahme einer Ableitung wird widerlegt.

---

[25] Diese Ansicht ist nicht unverträglich mit Äußerungen von BERNAYS und GÖDEL. BERNAYS bemerkte 1957 in Amsterdam während einer öffentlichen Diskussion über Intuitionismus, die hypothetische Annahme einer Konstruktion sei keine anschauliche Annahme. (Vgl. auch seine Ausführungen in *L'Enseignement mathématique* 34 (1935), p. 62f.) Zu solchen Konstruktionen zählt aber bereits der Aufweis einer Ableitung nach gegebenen Regeln: vgl. dazu GÖDELs Fußnote 2 in [7] S. 281.

Der finit-direkte Kern des Gödelschen Beweisgangs ist der Aufweis eines rein arithmetischen Terms $t(x)$, für den im Peano-Formalismus folgende Formel ableitbar ist: B sei ein primitiv rekursives Prädikat, für das, ähnlich der Symbolik in **0.52** und **2.4**, $B(k, \mathrm{cf}\, A(\xi))$ konstantenweise $\mathrm{dd}(k, \mathrm{cf}\, A(^{(\mathrm{cf}\, A(\xi))}0))$ ausdrückt, ferner sei $g \leftrightarrows \mathrm{cf} \bigwedge x \neg B(x, \xi)$ und $w \leftrightarrows \mathrm{cf} \bigwedge$, dann gilt $\vdash B(x, {}^{(g)}0) \rightarrow B(t(x), {}^{(w)}0)$. (Vgl. dazu [10] II, speziell § 5.2.)

Will man der indirekt erschlossenen Behauptung $\nvdash \neg \bigvee x\, B(x, {}^{(w)}0)$ eine finit-direkte Wendung geben, so ist $\nvdash$ offenbar entweder selbst zu formalisieren (unter Einbettung des Peano-Formalismus in ein formales Metasystem) oder erneut zu verschlüsseln. Die bei wiederholter Verschlüsselung entstehende Situation sei flüchtig skizziert am Beispiel der ZF-Ableitbarkeit, um uns so zugleich die denkbaren Konsequenzen des in **3.23** umrissenen Wf.beweises zu vergegenwärtigen.

In ZF sei $^\urcorner\nvdash_{ZF} \wedge^\ulcorner$ auf die Form $w = \bigcap$ gebracht, d.h. eine Menge $w$ ($\subseteq \omega$) wird leer, deren Elemente Gödelnummern von ZF-Deduktionen sind, die mit $\wedge$ enden. Unter erneuter Verschlüsselung von $\nvdash_{ZF}$ können die metamathematisch für $\nvdash_{ZF} w = \bigcap$ benutzten konstruktiven Schlüsse erst recht in ZF deduziert werden: $\vdash_{ZF} {}^{◹}\nvdash_{ZF} w = \bigcap^{◸}$. Bei ZF-Erreichbarkeit einer relativ exorbitanten Ordinalzahl wäre unter anderer Verschlüsselung, welche außer $\vdash_{ZF}$ noch die Umdeutung der ersten Verschlüsselung $w = \bigcap$ umgreift, eine „verkappte" Wf.-ableitbarkeit in ZF deduzierbar: $\vdash_{ZF} {}^{◥}\vdash_{ZF} {}^{(\mathsf{W})}w = \bigcap^{◤}$.

Die Entzifferung der widersprüchlichen direkten Wf.ableitbarkeit $\vdash_{ZF} w = \bigcap$ scheitert, auch wenn $\vdash_{ZF} {}^{(\mathsf{W})}w = \bigcap \leftrightarrow w = \bigcap$, von ZF aus bereits daran, daß das $=$ in ${}^{(\mathsf{W})}w = \bigcap$ nicht aus der in ${}^{◥}\vdash_{ZF} {}^{(\mathsf{W})}w = \bigcap^{◤}$ eingehenden Nummer von $=$ in das $=$ des ZF-Klartextes zurückzuverwandeln ist; denn $=$ ist ja keine Menge, kann also nicht Gegenstand von ZF-Definitionen sein. (Nach nochmaliger Verschlüsselung wäre die „verbotene" finit-direkte Metaentschlüsselung auch wieder in ZF zugänglich.)

${}^{◹}\nvdash_{ZF} w = \bigcap^{◸}$ und ${}^{◥}\vdash_{ZF} {}^{(\mathsf{W})}w = \bigcap^{◤}$ seien in ZF auf die Form $u = \bigcap$ bzw. $v \neq \bigcap$ gebracht. Der Wf.beweis sollte dann eigentlich ausschließen, daß die Implikation $W \leftrightarrows v \neq \bigcap \rightarrow u \neq \bigcap$ von ZF aus deduzierbar ist. Verschlüsselt mag $\vdash_{ZF} {}^{◺}\nvdash_{ZF} W^{◺}$ und $\vdash_{ZF} {}^{◢}\vdash_{ZF} W^{◣}$ deduzierbar sein, dies würde die formale ZF-Widerspruchsfreiheit ZF-intern nicht stören.

Vom finiten Standpunkt aus ist nun $\mathfrak{S}_1$ wieder ein Wortspiel (und auch $\mathfrak{K}_1$ in **1.8** würde nichts weiter sein, obwohl $\mathfrak{K}_1$ finite Inhalte kodifiziert), infolgedessen weiß niemand, inwieweit unsere Intentionen mit ihrer Formalisierung tatsächlich konform gehen. Von vornherein ist deswegen nicht auszuschließen, daß finit-direktes Deduzieren am Ende doch $\vdash_{ZF} W$ ergibt — etwa, weil sich die Verschlüsselungen ${}^{◹}.{}^{◸}$ und ${}^{◥}.{}^{◤}$ hinreichend aufeinander abstimmen lassen.

Um völlige Klarheit über diesen möglichen negativen Aspekt der Wf.theorie zu gewinnen, bei dem es auf genaue Zahl und Steuerung jeder benutzten bzw. einfach oder doppelt verschlüsselt benutzten Primfigur ankommt, muß zuerst der umrissene positive Aspekt entsprechend detailliert ausgearbeitet vorliegen — ein Grund mehr, von Anfang an mit größtmöglicher Präzision vorzugehen.

**3.3** *Ausklang.* Eingedenk solcher Präliminarien erscheint die klassische Mathematik eher als eine sekundäre Wahrheit; ganz abgesehen davon, wie sich die in **3.24** skizzierte Situation entwickelt. Wenn aber die Konsistenz der Mengenlehre so zweischneidig ist wie es bei semantischer Explikation der Formeln scheint, indem ein (arithmetischer) Ausschnitt der Mengenlehre gewissermaßen zum Beweis ihrer Widerspruchsfreiheit hinreichen soll, wie ist dann die in Geometrie und Physik (und Zahlentheorie) erprobte Kraft der klassischen Analysis zu verstehen? Und wie wäre deren scheinbare Wahrheit mit der handfesten Funktionstüchtigkeit unserer technischen Apparatur zu vereinbaren? Wo kann, da vom finiten Standpunkt noch nicht ohne weiteres eine direkte Gefahr interner Widersprüche besteht, ein externer Gesichtspunkt aufgezeigt werden, von dem aus die gewohnten analytischen Beschreibungen — unabhängig von der konstruktiven Deutung, die wir $\mathfrak{S}_1$ gegeben haben, und von der Umdeutung, die die Mengenlehre in $\mathfrak{S}_1$ erfährt — auch unmittelbar als sekundär wahr erscheinen.

Da das Symposion unter ein Thema gestellt ist, das Beziehungen zur Naturwissenschaft und Philosophie einschließt, mag es gestattet sein, einen am Schluß des Amsterdamer Vortrags [21] gestreiften Gedanken aufzugreifen und einige Äußerungen zu „dem, was gemeinsam zwischen Geometrie und Physik zu suchen sein könnte", anzuknüpfen.[26]

Die intuitionistische und die konstruktive Kritik an der klassischen Analysis gleiten nach meinem Dafürhalten beide an der Tatsache vorbei, daß der Funktionsbegriff unvermittelt in ein (mutwillig) cartesisch gleichmäßig gedachtes Kontinuum hineingezwängt wird: nur uninteressante Funktionskeime münden nach WEIERSTRASS und RIEMANN nicht in singuläre Punkte, und diese zerstören die Homogenität der Teile eines Kontinuums sowie seine Unzerlegbarkeit im ganzen, auf die es bei BROUWER ankommt.

EINSTEIN fordert zwar Singularitätenfreiheit für die Lösungen von (allgemeinen) Feldgleichungen und beklagt, daß die Mathematik keine Näherungsverfahren bereit hält, die eine Singularitätenfreiheit der entstehenden exakten Lösung garantieren, aber er hat versäumt, zunächst die fundamentale Singularität für $v \geq c$ aufzulösen und von dort den Weg für ein singularitätenfreies Kontinuum zu öffnen.

Versuchen wir, solche funktionalen Verhältnisse zu erfassen, die aus einem geometrisch „elastischen", asymptotisch eingespannten Kontinuum, während es noch im Entstehen begriffen ist, gleichsam organisch hervorgehen, so werden wir anscheinend zur Auffindung physisch bedeutsamer geometrischer Konstruktionsmöglichkeiten angeleitet. [25] schildert in gedrängter Fassung und mehr durch Verweis auf bildliche Darstellungen, wie die beiden Grundphänomene der messenden Physik, Relativität und Komplementarität, im einfachsten Fall als vordergründige Erscheinungen einer rein geometrischen Begriffswelt denkbar wären, die erst dann physikalische Züge annimmt, wenn man die Beschreibung der Teile eines „vollkommenen" Kontinuums von einem einebnenden (statt einbettenden) Meßfeld her aufzäumt. Die dort veranschaulichte Anlage der Definition eines solchen Kontinuums durch Induktion, bei der approximative Prozesse mit geometrisch-kombinatorischen Prozessen zu (nach innen und außen) rückwirkenden Bedingungsformen verwoben sind, muß ständig in der Schwebe gehalten werden, damit sie sich geometrisch ohne Riß und Bruch durchhalten läßt.

Soviel dürfte wenigstens zu erkennen sein: ohne Wert und Notwendigkeit der wissenschaftlichen Arbeit mit intern widerspruchsfreien, aber unanschaulichen Kompromissen anzutasten oder zu schmälern — in der Physik will dies schwerwiegender anmuten als in der Mathematik —, so ist doch zu bezweifeln, daß eine derartige Haltung auf die Dauer den Anspruch erheben kann, aus prinzipiellen Gründen die allein mögliche zu sein; vielmehr wird begriffliche Tieferlegung, die das Alte unter Aufhebung von Totalitätsansprüchen zu bewahren trachtet, unseren Blick von selbst auf neue Erkenntnisse lenken, die sonst unzugänglich geblieben wären.

Es wird freilich Geduld kosten, das Leitmotiv bei der Erarbeitung des Sphärenkalküls $\mathfrak{S}_1$ nochmals zu befolgen und jeden „Teil" im Ansatz beständig auf die

---

[26] Die in [21] angeschnittenen mengentheoretischen Unabhängigkeitsprobleme haben sich dank der Ergebnisse von COHEN 1963 als — vom konstruktiven Standpunkt aus beurteilt — nicht so tief liegend erwiesen wie ich damals dachte.

„Ganzheit“ der Denkmöglichkeiten einer Wissenschaft auszubalancieren, soweit es die Tendenz der Systematik angeht, die nach ihrem gegenwärtigen Selbstverständnis das Fazit ihrer bisherigen historischen Entwicklung zieht. Ich bin jedoch optimistisch genug, zu glauben, daß es menschenmöglich ist, nicht nur über unsere eigenen Denkmittel endgültig Rechenschaft abzulegen, sondern auch den theoretischen Extrakt der Daten unserer eigenen Meßwerkzeuge zu durchschauen — bei der Dimensionszahl von Raum und Zeit sowie der Signatur der Metrik im Meßfeld angefangen —, statt sie als Fakten anzuerkennen, die sich einem Zugriff der reinen Mathematik auf immer entziehen.

Sollte der formale Extrakt einiger mathematischer Disziplinen — die hier von mir vertretene konstruktive Arithmetik keineswegs ausgenommen — wirklich von einer „paradoxen“ Widerspruchsfreiheit sein, die je nach formalem Beweisverlauf und formaler Definitionsverwicklung zu einem Widerspruch umschlagen kann, so erblicke ich darin kein Unglück. Im Gegenteil: wenn das statische Bewußtsein vom Wahren in der Mathematik nach so geraumer Zeit noch einmal in Bewegung geriete, dann wäre dies ein Zeichen für eine überwältigende Spannkraft und Regenerationsfähigkeit der exakten Wissenschaft vom Unendlichen; und durch Widerspiegelung des 'Unendlichen' im Endlichen würden dann unerhörte Harmonien der natürlichen Zahl in Erscheinung treten, deren Einfluß auf eine finit-arithmetische (logik- und analysisfreie) Geometrie heute kaum jemand im voraus zu erahnen weiß.

Zumindest erwächst uns gegenwärtig die Aufgabe, eine kritische Selbstbesinnung auf die exakte Methode anzustreben — mit dem Ziel einer finiten Selbstverwirklichung der Mathematik.

## Nachtrag

Während zweier Tagungen, die im Mathematischen Forschungsinstitut Oberwolfach stattfanden, ergab sich in der Woche vom 30. III. bis 7. IV. 1967 Gelegenheit, mit Professor Bernays einige Ausschnitte dieser Arbeit zu besprechen. Daraufhin sei um größerer Klarheit willen folgendes nachgetragen.

*1.* Die Stellenzahl der Prädikate $\vee$ und $\wedge$ läßt sich mit notieren. Dazu ersetze man die Regeln 9.1.2, 3 durch die Regeln ${}_k\mathrm{w} \rightarrow {}_P\mathrm{w}{}^{\cup}\mathrm{w}\vee$, ${}_P\mathrm{w}{}^{\cup}\mathrm{w}\wedge$. Bei dieser Fassung werde anstelle ${}^{\cup}\mathrm{k}\vee$, ${}^{\cup}\mathrm{k}\wedge$ inoffiziell ${}^{k}\vee$, ${}^{k}\wedge$ geschrieben. Dann hat man ${}^{\cup}\wedge\circ$ in 9.3.11, 18, 22 und ${}^{\cup}\vee\circ$ in 9.3.12, 17, 19 durch ${}^{\cup\cup}0\wedge\circ$ bzw. ${}^{\cup\cup}0\vee\circ$ zu ersetzen.

Die 4 Regeln 9.3.11, 18 und 9.3.12, 17 lassen sich zu 2 Regeln zusammenziehen: ${}_{\cup *k}\mathrm{wv}$, ${}_A\mathrm{u} \rightarrow {}_{\angle}{}^{\cup\cup}\mathrm{w}\wedge\mathrm{vu}$, ${}_{\angle}\mathrm{u}{}^{\cup\cup}\mathrm{w}\vee\mathrm{v}$.

Es ist nicht möglich, den Relator ${}_2$ auf den Relator ${}_+$ zu reduzieren. Versucht man, $\circ$ ($\varkappa\varepsilon\nu\acute{o}\nu$) wegzulassen und mit natürlichen Zahlen zu arbeiten, die bei 1 statt bei 0 anfangen, so müßte man 0-stellige Prädikatschemata (wie etwa die in (D2), (${}^{\cap}$D2) unter **2.311** relativ explizit definierten) durch besondere, von 9.1.5 und 9.3.24, 25 getrennte Regeln als Aussageschemata einführen und den implikativen Umgang mit ihnen gesondert postulieren — sie sind ja im relativen Fall nicht auf dem Wege von Abkürzungskonventionen zu beseitigen.

*2.* Ohne Durchführung der Elimination von $\iota$-Termen bietet $\mathfrak{S}_1$ keinen Ausdruck für Terme wie $\mathrm{x} + \mathrm{y}$. Die Aussageform $+\mathrm{xyz}$ läßt sich mit folgendem Ausdruck für das 3-stellige (absolut unbedingt definierte) Prädikat $+$ gewinnen:

$$+ \leftrightharpoons ((\bigvee)\,{}^{3}\bigvee \mathrm{R}_{+} \longrightarrow \pi\,\mathrm{x\,y\,z}\,(\mathrm{y}=0 \wedge \mathrm{x}=\mathrm{z} \;\dot{\vee}\; \underset{\mathrm{y}='\mathrm{y}_0}{\bigvee_{\mathrm{y}_0}} \;\; \underset{\mathrm{z}='\mathrm{z}_0}{\bigvee_{\mathrm{z}_0}} \pi\,\mathrm{x\,y_0 z_0})),$$

worin

$$\mathrm{R}_{+} \leftrightharpoons ((\bigvee)\,{}^{6}\bigvee\,{}^{12}\bigwedge \longrightarrow \rho\,\mathrm{x_1 y_1 z_1 x_2 y_2 z_2}\,(\mathrm{x_1}=\mathrm{x_2} \wedge '\mathrm{y_1}=\mathrm{y_2} \wedge '\mathrm{z_1}=\mathrm{z_2})).$$

Zur Legitimation dieses Ausdrucks von der Form

$$+ \leftrightharpoons ((\bigvee)\,{}^{3}\bigvee \mathrm{R}_{+} \longrightarrow \pi\,\mathrm{x\,y\,z}\,\mathrm{A}_{+}(\mathrm{x,y,z};\pi)),$$

worin $\mathrm{R}_{+}$ von der Form $((\bigvee)\,{}^{6}\bigvee\,{}^{12}\bigwedge \longrightarrow \rho\,\overset{6}{\mathrm{x}}\,\mathrm{B}(\overset{6}{\mathrm{x}}))$ ist, muß vorher $\vdash \mathrm{I}({}^{6}\bigvee, {}^{12}\bigwedge)$ und $\vdash \mathrm{I}({}^{3}\bigvee, \mathrm{R}_{+})$ deduziert werden, ferner

$$\vdash \bigwedge \overset{6}{\mathrm{y}}\,({}^{12}\bigwedge \overset{6}{\mathrm{y}}\,\overset{6}{\mathrm{x}} \dot{\rightarrow} \pi_1 \overset{6}{\mathrm{y}} \leftrightarrow \pi_2 \overset{6}{\mathrm{y}}) \dot{\rightarrow} \mathrm{B}(\overset{6}{\mathrm{x}}) \leftrightarrow \mathrm{B}(\overset{6}{\mathrm{x}})$$

und

$$\vdash \bigwedge \mathrm{x_1 y_1 z_1}(\mathrm{R}_{+}\mathrm{x_1 y_1 z_1 x y z} \dot{\rightarrow} \pi_1 \mathrm{x_1 y_1 z_1} \leftrightarrow \pi_2 \mathrm{x_1 y_1 z_1}) \dot{\rightarrow}$$
$$\dot{\rightarrow} \mathrm{A}_{+}(\mathrm{x,y,z};\pi_1) \leftrightarrow \mathrm{A}_{+}(\mathrm{x,y,z};\pi_2).$$

Die dritte Deduktion ist trivial. Die vierte Deduktion ergibt sich unter Berücksichtigung von $\mathrm{y}=0 \vee \bigvee \mathrm{y_0}\,\mathrm{y}='\mathrm{y_0} \;\dot{\wedge}\; \mathrm{z}=0 \vee \bigvee \mathrm{z_0}\,\mathrm{z}='\mathrm{z_0}$; dabei ist ein Rückgriff auf das $\dot{\rightarrow}$-Antezedens nur im Fall $\bigvee \mathrm{y_0}\,\mathrm{y}='\mathrm{y_0} \wedge \wedge \bigvee \mathrm{z_0}\,\mathrm{z}='\mathrm{z_0}$ nötig. Die erste Deduktion ist einfach: es gilt $\bigwedge \overset{6}{\mathrm{x}}\,({}^{12}\bigwedge \overset{6}{\mathrm{x}}\,\overset{6}{\mathrm{x}}_0 \rightarrow \rightarrow \pi \overset{6}{\mathrm{x}})$, so daß $\bigwedge \overset{6}{\mathrm{x}}_0(\bigvee \rightarrow \pi \overset{6}{\mathrm{x}}_0) \rightarrow \pi \overset{6}{\mathrm{x}}$ zu zeigen bleibt; 9.3.15 liefert $\bigwedge \overset{6}{\mathrm{x}}_0(\bigvee \rightarrow \pi \overset{6}{\mathrm{x}}_0) \dot{\rightarrow} \bigvee \rightarrow \pi\,\overset{6}{\mathrm{x}}$, und mit 9.3.10 reicht dann $\mathrm{D} \wedge \bigvee \leftrightarrow \mathrm{D}$ hin. Die zweite Deduktion ergibt sich nach Reduktion auf

$$(*) \qquad \vdash \bigwedge \mathrm{y_0}(\bigwedge \mathrm{y}('\mathrm{y}=\mathrm{y_0} \rightarrow \rho\mathrm{y}) \rightarrow \rho\mathrm{y_0}) \rightarrow \rho\mathrm{y}$$

(wie bei (o 1) unter **2.32**) durch formale Induktion nach $\mathrm{y}$. In der Tat gilt $'\mathrm{y}=\mathrm{y_0} \rightarrow \bigwedge \mathrm{x} \bigwedge \mathrm{z}\,\pi\,\mathrm{xyz} \dot{\rightarrow} \mathrm{x}=\mathrm{x_0} \wedge '\mathrm{y}=\mathrm{y_0} \wedge '\mathrm{z}=\mathrm{z_0} \rightarrow \pi\,\mathrm{xyz}$, also auch $\bigwedge_{\mathrm{x_0 y_0 z_0}} \bigwedge_{\mathrm{xyz}} \mathrm{x}=\mathrm{x_0} \wedge '\mathrm{y}=\mathrm{y_0} \wedge '\mathrm{z}=\mathrm{z_0} \rightarrow \pi\,\mathrm{xyz}. \rightarrow \pi\,\mathrm{x_0 y_0 z_0} \dot{\rightarrow} \dot{\rightarrow} \bigwedge_{\mathrm{y}} '\mathrm{y}=\mathrm{y_0} \rightarrow \bigwedge \mathrm{x} \bigwedge \mathrm{z}\,\pi\,\mathrm{xyz}. \rightarrow \bigwedge \mathrm{x} \bigwedge \mathrm{z}\,\pi\,\mathrm{x y_0 z}$ und folglich $\mathrm{I}({}^{3}\bigvee, \mathrm{R}_{+})$ aus (*). Induktion für (*) nach $\mathrm{y}$: der Induktionsanfang ergibt sich wegen $\bigwedge \mathrm{y}('\mathrm{y}=0 \rightarrow \rho\mathrm{y})$ aus $\bigwedge \mathrm{y_0}(\bigwedge \mathrm{y}('\mathrm{y}=\mathrm{y_0} \rightarrow \rho\mathrm{y}) \rightarrow \rho\mathrm{y_0}) \dot{\rightarrow} \bigvee \rightarrow \rho\,0$; der Induktionsschritt $\bigwedge \mathrm{y_0} * \rightarrow \rho\mathrm{y_1} \dot{\rightarrow} \bigwedge \mathrm{y_0} * \rightarrow \rho'\mathrm{y_1}$ beginnt mit $\bigwedge \mathrm{y_0} * \dot{\rightarrow} \bigwedge \mathrm{y}('\mathrm{y}='\mathrm{y_1} \rightarrow \rho\mathrm{y}) \rightarrow \rho'\mathrm{y_1}$, wonach aus $\bigwedge \mathrm{y_0}*$ und $\rho\mathrm{y_1}$ (wegen $'\mathrm{y}='\mathrm{y_1} \leftrightarrow \mathrm{y}=\mathrm{y_1}$) zunächst $\rho'\mathrm{y_1}$ folgt — dies ist dann auf $\mathrm{D_1} \dot{\rightarrow} \mathrm{D_2} \rightarrow \mathrm{D_3} \ddot{\rightarrow} \mathrm{D_1} \rightarrow \mathrm{D_2} \dot{\rightarrow} \mathrm{D_1} \rightarrow \mathrm{D_3}$ für $\mathrm{D_1} \leftrightharpoons \bigwedge \mathrm{y_0}*$, $\mathrm{D_2} \leftrightharpoons \rho\mathrm{y_1}$, $\mathrm{D_3} \leftrightharpoons \rho'\mathrm{y_1}$ anzuwenden.

*3.* Aus der Notation $(\mathrm{CPR} \longrightarrow \pi \overset{k}{\mathrm{x}} \mathrm{A})$ geht nicht hervor, für welche Variablentupel $\overset{k}{\mathrm{z}}\overset{k}{\mathrm{y}}$ die Deduktionen von (1), (2) laut **1.632** geführt worden sind und für welche Marken $\rho, \pi_1, \pi_2$, obwohl diese Angaben in die Prämissen von 9.1.5 eingehen. Im Extremfall würde sich anbieten, die Deduktionen $\vdash_{\mathfrak{S}_1}(1)$, $\vdash_{\mathfrak{S}_1}(2)$ selbst als aufweispflichtige Argumente eines Rekursors einzuführen. Aus [14] S. 101—104 ist zu entnehmen, daß und wie man Ableitungen nach einem Kalkül in einem neuen, mit einem etwas umfassenderen Alphabet arbeitenden Kalkül herstellen kann. Für $\mathfrak{S}_1$ empfiehlt sich, die beim Aufbau von Ableitungs-Stammbäumen benutzten Regeln und Einsetzungen für Wortmitteilungen der Eindeutigkeit halber mit einzukalkulieren; ein Rekursor $(d_{(1)} d_{(2)} \twoheadrightarrow \pi \overset{k}{x} A)$ würde natürlich den Kalkül für die $\mathfrak{S}_1$-Ableitungen mit $\mathfrak{S}_1$ selber verflechten.

Weil die in $d_{(1)}, d_{(2)}$ frei vorkommenden Variablen und Marken nur insoweit für $(d_{(1)} d_{(2)} \twoheadrightarrow \pi \overset{k}{x} A)$ relevant sind, als sie in C, P, R und A aus (1), (2) frei vorkommen, ist unter die Hilfsbegriffe dabei auch das Tupel der in einem Prädikat- bzw. Aussageschema frei vorkommenden Variablen wie Marken aufzunehmen. $\twoheadrightarrow$ bindet $\pi$ und x wieder ausschließlich in A, und die Regeln zur Umbenennung von $\pi$, x auf kongruente Rekursifikationen fließen aus (3) bzw. 9.3.24, 25. Im Fall $\twoheadrightarrow$ ist nicht nur die bisubjunktive Unabhängigkeit des rekursiv definierten Prädikats von der Fundierung zu zeigen (wie bei Theorem 1, 2.11), sondern auch die von $d_{(1)}$ und $d_{(2)}$. Derlei Unabhängigkeiten, die sich hinterher im formalen Zusammenhang herausstellen, machen aber diejenigen (figürlich-wortmäßigen) Abhängigkeiten nicht rückgängig, welche mit der ursprünglich eingeführten Setzung einmal verbunden und notiert sind.

Bemerkenswerterweise wird der Prädikatbegriff in $\mathfrak{S}_1$ beim Überwechseln von $\rightarrow$ zu $\twoheadrightarrow$ „direkt" entscheidbar (in jenem Sinn, der unter 1., 2. im letzten Satz von **1.2** erwähnt ist).

*4.* Zur dialogischen Deutung ziehe man außer [14] noch den Anhang der 3. Auflage von LORENZENs Göschenband „Formale Logik" (Berlin 1967, S. 160—176) heran und die unabhängig lesbare Auslegung von W. STEGMÜLLER in Notre Dame Journal of Formal Logic V (1964), p. 81—112.

*Zusatz.* Im Zusammenhang mit meinem Vortrag „Widersprüche in der axiomatischen Mengenlehre aufgrund eines systeminternen Widerspruchsfreiheitsbeweises für eine formalisierte ordinalzahltheoretische Analysis" am 3. April 1968 im Mathematischen Forschungsinstitut Oberwolfach wurde ich privat auf imprädikative Definitionsmöglichkeiten im System $\mathfrak{S}_1$ hingewiesen. Diese sind jedoch keineswegs gravierend, denn die Rechnungen und Überlegungen unter **2.***** und **3.**** funktionieren ebensogut (: vgl. **1.52**) bei folgender Verschärfung in den Prämissen der Hauptregel 9.1.5:

ersetze die 6. Prämisse durch $_{-A}u$.

Für den 5. einstelligen Relator $_{-A}$ sind 6 Regeln hinzuzufügen:

9.4.1 $\quad {}_{\cup}{*}_{k}w\,v, {}_{P}w\,u \rightarrow {}_{-A}{}^{\cup}u\,v;$

9.4.2, 3, 4 $\quad {}_{-A}u, {}_{-A}v \rightarrow {}_{-A}{}^{\rightarrow}u\,v, {}_{-A}{}^{\wedge}u\,v, {}_{-A}{}^{\vee}u\,v;$

9.4.5, 6 $\quad {}_{x}v, {}_{-A}u \rightarrow {}_{-A}{}^{\bigvee}v\,u, {}_{-A}{}^{\bigwedge}v\,u.$

Intendierte Deutung von ${}_{-A}\mathrm{w}$: $w$ ist Aussageschema über Grundzahlen ohne Prädikatquantifikation.

Quantoren mit Prädikatmarken dienen hiernach innerhalb des Wirkungsstücks eines Rekursors nicht zum logischen Aufbau, eventuell indes zum definitorischen Aufbau (und zwar zum logischen Aufbau der C in den CPR vor Rekursoren). Diese Maßnahme schaltet „generalized inductive definitions" aus.

Als Regelzahl des Sphärenkalküls $\mathfrak{S}_1$ ergibt sich nun 97 (und 129 für die in **1.2** erwähnte Neufassung): laut *1.* sind 2 Regeln unter 9.3. eingespart, und bei Hinzufügung von

9.2.0 $${}_{-A}\mathrm{u} \rightarrow {}_{A}\mathrm{u}$$

entfallen 9.2.1 — 6 dank 9.4.1 — 6.

## Literatur


1. BERNAYS, P.: A system of axiomatic set theory. JSL **2**, 65—77 (1937); **6**, 1—17 (1941); **7**, 65—89, 133—145 (1942); 8, 89—106 (1943); **13**, 65—79 (1948); **19**, 81—96 (1954).
2. — and A. A. FRAENKEL: Axiomatic set theory. Amsterdam 1958/63.
3. DEDEKIND, R.: Was sind und was sollen die Zahlen? Braunschweig 1888.
4. GENTZEN, G.: Die Widerspruchsfreiheit der reinen Zahlentheorie. Math. Annalen **112**, 493—565 (1936).
5. GÖDEL, K.: Über formal unentscheidbare Sätze der Principia Mathematica und verwandter Systeme I. Monatshefte Math. Phys. **38**, 173—198 (1931).
6. — The consistency of the axiom of choice and of the generalized continuum-hypothesis with the axioms of set theory. Ann. Math. Studies 3, Princeton 1940/51/53/58/61.
7. — Über eine bisher noch nicht benützte Erweiterung des finiten Standpunktes. Dialectica **12**, 280—287 (1958).
8. HEYTING, A.: Some remarks on intuitionism. Constructivity in mathematics, Proceedings of the Colloquium held at Amsterdam 1957, pp. 69—71. Amsterdam 1959.
9. HILBERT, D.: Über das Unendliche. Math. Annalen **95**, 161—190 (1926).
10. — u. P. BERNAYS: Grundlagen der Mathematik. Berlin 1934 (Bd. I), 1939 (Bd. II).
11. KLEENE, S. C.: Introduction to metamathematics. Amsterdam, Groningen 1952/59.
12. — and R. E. VESLEY: The foundations of intuitionistic mathematics. Amsterdam 1965.
13. LORENZEN, P.: Einführung in die operative Logik und Mathematik. Berlin Göttingen-Heidelberg 1955.

13a. Referat zu [13] in Zbl. Math. **66**, 248—254 (1957).

14. LORENZEN, P.: Metamathematik. Mannheim 1962.
15. — Differential und Integral. Frankfurt a.M. 1965.
16. PÉTER, RÓZSA: Rekursive Funktionen. Budapest 1951/57.
17. SCHÜTTE, K.: Beweistheorie. Berlin-Göttingen-Heidelberg 1960.

18. SPECTOR, C.: Provably recursive functionals of analysis: a consistency proof of analysis by an extension of principles formulated in current intuitionistic mathematics. Recursive function theory. Proc. Symposia Pure Math. 5, Amer. Math. Soc., pp. 1–27. Providence R. I. 1962.
19. TAKEUTI, G.: Construction of the set theory from the theory of ordinal numbers. J. Math. Soc. Japan **6**, 196–220 (1954).
20. — A formalization of the theory of ordinal numbers. JSL **30**, 295–317 (1965).
21. WETTE, E.: Von operativen Modellen der axiomatischen Mengenlehre. Constructivity in mathematics, Proceedings of the Colloquium held at Amsterdam 1957 (hrsg. von A. HEYTING), pp. 266–277. Amsterdam 1959. Druckfehler: S. 267, Zeile 36: stadium; S. 273, Zeile 22: Beweismittel; S. 275, Zeile 4: pflügten; S. 276, Zeile 10: $\geq$ anstelle des unteren $\leq$; Zeile 24: Zahlen.
22. — Intuitionistisch-rekursiver Konsistenzbeweis für die axiomatische Mengenlehre. International Congress for logic, methodology and philosophy of science. Abstracts of contributed papers, pp. 31–33. Stanford 1960. Druckfehler: S. 31, Zeile 16 heißt 'letztere angewendet auf logisch-rekursive'; Zeile 18 heißt 'Logik, Vorgängerinduktion für $<$, Totalordnungseigenschaften von $\leq$ und Rekursionsaxiomschema für Relatoren $\rightarrow_{\pi(\overset{m}{\xi})} {}^*\mathfrak{A}$'; S. 32, Zeile 3: $W_\zeta$ anstelle $W_\xi$.
23. — Intuitionistic-recursive consistency proof for the axiomatic set theory. (Unveröffentlichter Text des Vortrags gemäß Proceedings of the 1960 International Congress, Stanford 1962 (hrsg. von E. NAGEL, P. SUPPES u. A. TARSKI), p. 658.)
24. — Eine Anmerkung zu GÖDELS Komprehensionstheorem. Archiv für mathematische Logik und Grundlagenforschung **9**, 59–60 (1967).
25. — On the mathematical and the physical aspects of the continuum problem. (Abriß: XI Congrès international d'histoire des sciences, Warszawa-Kraków 1965, Sommaires p. 167. Druckfehler: Zeile 23: we anstelle w; Zeile 25: $\rightarrow$ zwischen $2 = 2$ und $1 = 1$.) Actes du XI$^{e}$ Congrès International d'Histoire des Sciences, vol. III, pp. 283–293 (incl. 10 fig.). Ossolineum, Wrocław 1968. Errata: p. 284 below: exchange $^1$ and $^2$; note $^1$: spheres, $\sum_\eta$; note $^4$: $[c] \sim 1 = s^0 \sim [h]$, $\sim s^{-2} \sim$ Gaussian curvature K; note $^{10}$: expanding/straining.
26. WEYL, H.: Mathematics and logic. A brief survey serving as a preface to a review of "The Philosophy of Bertrand Russell". Amer. math. monthly **53**, 2–13 (1946).
27. Selecta HERMANN WEYL. Basel und Stuttgart 1956.